Lecture Notes in Bioinformatics 5267

Subseries of Lecture Notes in Computer Science

T0238833

Craig E. Nelson   Stéphane Vialette (Eds.)

Comparative Genomics

International Workshop, RECOMB-CG 2008
Paris, France, October 13-15, 2008
Proceedings

 Springer

Series Editors

Sorin Istrail, Brown University, Providence, RI, USA
Pavel Pevzner, University of California, San Diego, CA, USA
Michael Waterman, University of Southern California, Los Angeles, CA, USA

Volume Editors

Craig E. Nelson
University of Connecticut
Division of Genetics and Genomics, Molecular and Cell Biology
Beach Hall, 354 Mansfiled Rd, Storrs, CT 06269, USA
E-mail: craig.nelson@uconn.edu

Stéphane Vialette
Université Paris-Est, IGM-LabInfo
5 Bd Descartes, Champs sur Marne, 77454 Marne la Vallée, France
E-mail: vialette@univ-mlv.fr

Library of Congress Control Number: Applied for

CR Subject Classification (1998): F.2, G.3, E.1, H.2.8, J.3

LNCS Sublibrary: SL 8 – Bioinformatics

ISSN   0302-9743
ISBN-10   3-540-87988-9 Springer Berlin Heidelberg New York
ISBN-13   978-3-540-87988-6 Springer Berlin Heidelberg New York

Springer is a part of Springer Science+Business Media

springer.com

© Springer-Verlag Berlin Heidelberg 2008
Printed in Germany

Typesetting: Camera-ready by author, data conversion by Scientific Publishing Services, Chennai, India
Printed on acid-free paper   SPIN: 12533763   06/3180   5 4 3 2 1 0

Preface

Over the last decade great investments have been made in the acquisition of enormous amounts of gene sequence data from a diverse collection of organisms. Realizing the full potential of these investments will require the continued development of computational tools for comparative genomics and the intelligent application of these tools to address biologically relevant questions. The RECOMB Workshop on Comparative Genomics (RECOMB-CG) is devoted to bringing together scientists working on all aspects of comparative genomics, from the development of new computational approaches to genome sequence analysis and comparison, to the genome-wide application of computational tools to study the evolutionary dynamics of prokaryotic and eukaryotic genomes.

This volume contains the 19 papers presented at the 6th Annual RECOMB-CG workshop held during October 13–15 at the École Normale Supérieure, in Paris, France. The papers selected for presentation and published in these proceedings were selected from 48 submissions from scientists around the world. Each paper was reviewed by at least three members of the Program Committee in a stringent and thoughtful peer-review process.

The conference itself was enlivened by invited keynote presentations from Laurent Duret (Université Claude Bernard), Aviv Regev (Broad Institute), Chris Ponting (University of Oxford), Olga Troyanskaya (Princeton University), and Patricia Wittkopp (University of Michigan). These talks were supplemented by both presentation of the papers in this volume and a series of "late-breaking talks" selected from a wide-ranging and provocative poster session. Together, these talks and papers highlighted the acceleration of comparative genomics tools and applications. From the inference of evolution in genetic regulatory networks, to the divergent fates of gene and genome duplication events, to the importance of new computational approaches to unraveling the structural evolution of genomes, these presentations illustrate the crucial role of comparative genomics in understanding genome function.

RECOMB-CG 2008 would not have been possible without the participation of the many scientists who contributed their time and effort to making the conference a success. We thank the scientists who submitted their work for presentation at the conference, those members of the Program Committee who made every effort to ensure fair and balanced review of the many papers submitted for consideration at this year's workshop, the members of the local Organizing Committee for arranging all the myriad details of the organizational aspects of the event, and the continued efforts of the Steering Committee for their ongoing dedication and guidance. RECOMB-CG 2008 is also deeply indebted to its sponsors including the Centre National de la Recherche Scientifique (CNRS), the GdR BiM, the Université Paris-Est Marne-la-Vallée, the Université de Nantes and the Institut National de Recherche en Informatique et Automatique (INRIA

Rhône-Alpes), and to the École Normale Supérieure Paris for hosting the conference.

It is the continued support and dedication of this community that allows RECOMB-CG to bring together comparative genomics researchers from across the globe to exchange ideas and information and focus the force of comparative genomics on improving our understanding genome evolution and function.

July 2008   Craig E. Nelson
                                                                         Stéphane Vialette

Organization

Program Chairs

Craig E. Nelson   University of Connecticut, USA
Stéphane Vialette   Université Paris-Est, France

Program Committee

Lars Arvestad   KTH, Sweden
Véronique Barriel   Muséum National d'Histoire Naturelle, France
Anne Bergeron   Université du Québec Montréal, Canada
Guillaume Blin   Université Paris-Est, France
Guillaume Bourque   Genome Institute of Singapore, Singapore
Jeremy Buhler   Washington University in St. Louis, USA
Pierre Capy   Université Paris-Sud, France
Cédric Chauve   Simon Fraser University, Canada
Avril Coghlan   Sanger Institute, UK
Miklós Csuros   Université de Montréal, Canada
Aaron Darling   University of Queensland, Australia
Sankoff David   University of Ottawa, Canada
Bernard Dujon   Institut Pasteur, France
Dannie Durand   Carnegie Mellon University, USA
Nadia El-Mabrouk   Université de Montréal, Canada
Niklas Eriksen   Göteborg University, Sweden
Guillaume Fertin   Université de Nantes, France
Olivier Gascuel   Université de Montpellier, France
Henri Grosjean   Université Paris-Sud, France
Matthew Hahn   Indiana University, USA
Tao Jiang   University of California - Riverside, USA
Jens Lagergren   Stockhom Bioinformatics Centre and KTH, Sweden
Emmanuelle Lerat   Université Claude Bernard, France
Aoife McLysaght   University of Dublin, Ireland
Bernard Moret   École Polytechnique Fédérale de Lausanne,
                          Switzerland
Michal Ozery-Flato   University of Tel-Aviv, Israel
Pierre Pontarotti   Université de Provence, France
Eduardo Rocha   Université Paris 6 et Institut Pasteur, France
Hugues Roest Crollius   École Normale Supérieure, France
Antonis Rokas   Vanderbilt University, USA
Marie-France Sagot   INRIA Rhône-Alpes, France
Cathal Seoighe   University of Cape Town, South Africa

Jens Stoye   Bielefeld University, Germany
Chuan-Yi Tang   National Tsing Hua University, Taiwan
Eric Tannier   INRIA Rhône-Alpes, France
Glenn Tesler   University of California - San Diego, USA
Louxin Zhang   National University of Singapore, Singapore

External Reviewers

Max Alekseyev   Charles Hébert   Irena Rusu
Sébastien Angibaud   Jian Ma   Maureen Stolzer
David Enard   Matthieu Muffato   Sara Vieira-Silva
Claire Herrbach   Aida Ouangraoua

Steering Committee

Jens Lagergren   Stockhom Bioinformatics Centre and KTH, Sweden,
Aoife McLysaght   Trinity College Dublin, Ireland
David Sankoff   University of Ottawa, Canada

Sponsoring Institutions

CNRS   http://www.cnrs.fr/index.html
Université Paris-Est   http://www.univ-paris-est.fr
Université de Nantes   http://www.univ-nantes.fr
GDR Bioinformatique Moléculaire http://www.gdr-bim.u-psud.fr
INRIA Rhône-Alpes   http://www.inrialpes.fr

Local Organization

Sèverine Bérard   Université de Montpellier, France
Guillaume Blin   Université Paris-Est France
Maxime Crochemore   Université Paris-Est France
Guillaume Fertin   Université de Nantes, France
Hugues Roest Crollius   École Normale Supérieure, France
Éric Tannier   INRIA Rhône-Alpes, France
Jean-Stéphane Varré   INRIA Futurs - Université de Lille, France
Stéphane Vialette   Université Paris-Est France

RECOMB CG 2008 website   http://igm.univ-mlv.fr/RCG08/

Table of Contents

Algorithms for Exploring the Space of Gene Tree/Species Tree Reconciliations

Jean-Philippe Doyon[2], Cedric Chauve[1], and Sylvie Hamel[2]

[1] Department of Mathematics, Simon Fraser University, 8888 University Drive,
V5A 1S6, Burnaby (BC), Canada
cedric.chauve@sfu.ca
[2] DIRO, Université de Montréal, CP6128, succ. Centre-Ville,
H3C 3J7, Montréal (QC), Canada
{hamelsyl,doyonjea}@iro.umontreal.ca

Abstract. We describe algorithms to explore the space of all possible reconciliations between a gene tree and a species tree. We propose an algorithm for generating a random reconciliation, and combinatorial operators and algorithms to explore the space of all possible reconciliations between a gene tree and a species tree in optimal time. We apply these algorithms to simulated data.

1 Introduction

Genomes of contemporary species, especially eukaryotes, are the result of an evolutionary history, that started with a common ancestor from which new species evolved through evolutionary events called speciations. One of the main objectives of molecular biology is the reconstruction of this evolutionary history, that can be depicted with a rooted binary tree, called a *species tree*, where the root represents the common ancestor, the internal nodes the ancestral species and speciation events, and the leaves the extant species. Other events than speciation can happen, that do not result immediately in the creation of new species but are essential in eukaryotic genes evolution, such as gene duplication and loss [12]. Duplication is the genomic process where one or more genes of a single genome are copied, resulting in two copies of each duplicated gene. Gene duplication allows one copy to possibly develop a new biological function through point mutation, while the other copy preserves its original role. A gene is said to be lost when it has no function or is fully deleted from the genome. (See [12] for example). Other genomic events such as lateral gene transfer, that occurs mostly in bacterial genomes, will not be considered here. Genes of contemporary species that evolved from a common ancestor, through speciations and duplications, are said to be homologs [9] and are grouped into a gene family. The evolution of a gene family can be depicted with a rooted binary tree, called a *gene tree*, where the leaves represent the homologous contemporary genes, the root their common ancestral gene and the internal nodes represent ancestral genes that have evolved through speciations and duplications.

C.E. Nelson and S. Vialette (Eds.): RECOMB-CG 2008, LNBI 5267, pp. 1–13, 2008.
© Springer-Verlag Berlin Heidelberg 2008

Given a gene tree G and the species tree of the corresponding genomes S, an important question is to locate in S the evolutionary events of speciations and duplications. A *reconciliation* between G and S is a mapping of the genes (extant and ancestral) of G onto the nodes of S that induces an evolutionary scenario, in terms of speciations, duplications and losses, for the gene family described by G. In this perspective, the notion of reconciliation was first introduced in the pioneering work of [10] and a first formal definition was given in [17] to explain the discrepancies between genes and species trees. The LCA-mapping, that maps a gene u of G onto the most recent species of S that is ancestor of all genomes that contain a gene descendant of u, is the most widely used mapping, as it depicts a parsimonious evolutionary process according to the number of duplications or duplications and losses it induces. It is widely accepted that parsimony is a pertinent criterion in evolutionary biology, but that it does not always reflects the true evolutionary history. This lead to the definition of more general notions of reconciliations between a gene tree and a species tree [2,11,1] and the natural problem of exploring non-optimal (for a given criterion) reconciliations, and then alternative evolutionary scenarios for gene families.

The main concern of our work is the development of algorithms for exploring the *space of the reconciliations* between a gene tree and a species tree. After introducing a very general notion of reconciliation (Section 2), we describe in Section 3 an algorithm that generates a random reconciliation under the uniform distribution, in Section 4.1 combinatorial operators that are sufficient to explore the complete space of reconciliations between a gene tree and a species tree, and in Section 4.2 an algorithm that explores exhaustively this space and computes in optimal time the distribution of reconciliation scores in the duplication, loss, and mutation (duplication + loss) cost models. (All proofs will be given in a future technical report [6]). There are several applications of our algorithms in functional and evolutionary genomics, such as inferring orthologs and paralogs [8,14], the gene content of an ancestral genome [16], or in the context of Markov Chain Monte Carlo analysis of gene families [1]. We illustrate our algorithms with experiments on simulated gene families in Section 5 computed using duplication and loss rates taken from [13]. Our experiments suggest that, at least for some real datasets, the use of a parsimony model may be justified.

2   Preliminaries

Let T be a binary tree with vertices $V(T)$ and edges $E(T)$, and such that only its leaves are labeled. Let $r(T)$, $L(T)$, and $\Lambda(T)$ respectively denote its root, the set of its leaves, and the set of the labels of its leaves. We will adopt the convention that the root is at the top of the tree and the leaves at the bottom. A *species tree* S is a binary tree such that each element of $\Lambda(S)$ represents an extant species and labels exactly one leaf of S (there is a bijection between $L(S)$ and $\Lambda(S)$). A *gene tree* G is a binary tree. From now on, we consider a species tree S, with $|V(S)| = n$ and a gene tree G such that $\Lambda(G) \subseteq \Lambda(S)$ and $|V(G)| = m$. Let $\sigma : L(G) \to L(S)$ be the function that maps each leaf of G to the unique leaf of S with the same label.

For a vertex u of T, we denote by u_1 and u_2 its children and by T_u the subtree of T rooted at u. For a vertex $u \in V(T) \setminus \{r(T)\}$, we denote by $p(u)$ its parent. A *cell* of a tree T is either a vertex of T or an edge of T. Given two cells c and c' of T, $c' \leq_T c$ (resp. $c' <_T c$) if and only if c is on the unique path from c' to $r(T)$ (resp. and $c \neq c'$); in such a case, c' is said to be a *descendant* of c. The *LCA-mapping* $M : V(G) \to V(S)$ maps each vertex u of G to the unique vertex $M(u)$ of S such that $\Lambda(S_{M(u)})$ is the smallest cluster of S containing $\Lambda(G_u)$.

Definition 1. A reconciliation between a gene tree G and a species tree S is a mapping $\alpha : V(G) \to V(S) \cup E(S)$ such that

1. (Base constraint) $\forall u \in L(G), \alpha(u) = M(u) = \sigma(u)$.
2. (Tree Mapping Constraint) For any vertex $u \in V(G) \setminus L(G)$,
   (a) if $\alpha(u) \in V(S)$, then $\alpha(u) = M(u)$.
   (b) If $\alpha(u) \in E(S)$, then $M(u) <_S \alpha(u)$.
3. (Ancestor Consistency Constraint) For any two vertices $u, v \in V(G)$, such that $v <_G u$,
   (a) if $\alpha(u), \alpha(v) \in E(S)$, then $\alpha(v) \leq_S \alpha(u)$,
   (b) otherwise, $\alpha(v) <_S \alpha(u)$.

Remark 1. This definition of reconciliation differs slightly from the classical ones as vertices of G can be mapped onto edges of S, in order to represent duplication events (see explanations below). However, it is equivalent to the definitions given in [1,11], that are the most complete ones known so far, and it is more general than the Inclusion-Preserving mapping of [2].

The whole set of reconciliations between a gene tree G and a species tree S is denoted $\Psi(G, S)$. A reconciliation α of $\Psi(G, S)$ implies an evolutionary scenario for the genes of G in terms of gene duplications, gene losses, and speciations. A vertex u of G that is mapped onto an edge (x, y) of S (where $x = p(y)$) represents a gene of the ancestral species $p(y)$ that has been duplicated in y. If u is mapped onto an internal vertex x of S, then this represents a gene that will be present in a single copy in the two genomes x_1 and x_2 following a speciation event that happened to x. It is important to point out that the number of reconciliations is finite. Briefly, a reconciliation α between G and S represents any birth-and-death scenario along S such that the resulting gene tree is consistent with G and each duplication event that implies an internal node u of G is consistent with the mapping $\alpha(u)$. (See Figure 1).

We denote by $dup(\alpha)$ and $los(\alpha)$ respectively the number of duplications and losses induced by a reconciliation α. $dup(\alpha)$ is the number of vertices of G that are mapped onto an edge of S (see below[1]). Given two cells $c, c' \in V(S) \cup E(S)$, where $c' <_S c$, $D(c, c')$ is the number of vertices $x \in V(S)$ such that $c' <_S x <_S c$. Also, if $c = c'$, then $D(c, c') = 0$. The number of losses associated to a vertex

---

[1] To consider duplication that precedes the first speciation event represented by $r(S)$, we can insert in S an "artificial" cell c such that $r(S) <_S c$. For space reason, we assume here that no duplication occurs in the most ancestral species. The details to account for such early duplications will be described in the full version of this paper.

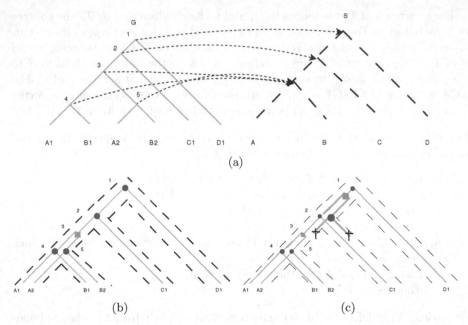

Fig. 1. (a) Left: gene tree G. Right: species tree S. The arrows represent the LCA-mapping between G and S. (b) A reconciliation between G and S. The red circles represent speciation events, and the green squares, duplications. (c) A birth-and-death scenario that is consistent with the reconciliation. A cross represents a gene loss. The right lineage of the first duplication has no extant gene that descents from it, as opposite to its left lineage. We then say that this duplication is hypothetical, because it is not a useful information for the evolutionary scenario of the extant genes of G along S. Hence, such duplication is not depicted by the reconciliation.

$u \in V(G) \setminus L(G)$ is noted l_u and equal to $D\big(\alpha(u), \alpha(u_1)\big) + D\big(\alpha(u), \alpha(u_2)\big)$ (see [15] for example). $los(\alpha)$ is then the sum of l_u over all internal vertices u. The third constraint of Definition 1 leads to the notion of *forced duplication*, that corresponds to vertices of G that can only be mapped onto an edge of S: an internal vertex $u \in V(G) \setminus L(G)$ is said to be a forced duplication if and only if $M(u) = M(u_1)$ or $M(u) = M(u_2)$.

For a vertex $u \in V(G)$, a cell of S *covers* it if u can be mapped onto this cell according to Definition 1. The set of cells that can cover it is denoted by $A(u)$ and is defined below.

$$A(u) = \begin{cases} \{M(u)\} & \text{if } u \in L(G) \text{ or } u = r(G) \\ \{c \in E(S) : M(u) <_S c\} & \text{if } u \text{ is a forced duplication} \\ \{c \in E(S) : M(u) <_S c\} \cup \{M(u)\} & \text{otherwise} \end{cases}$$

It is important to point out that there is three mappings that are considered here: $M(u)$, $\alpha(u)$, and $A(u)$. From now on, except when indicated, the term mapping will refer to the reconciliation mapping $\alpha(u)$ of Definition 1.

Finally, combinatorial and probabilistic criteria can be used to compare the different possible reconciliations and pick one that is supposed to reflect the most the true evolution of G according to S. Three parsimonious cost models, that aim to minimize the number of genomic events, have been proposed so far: duplication [15], loss [4], and mutation (duplication+loss) [15]. Arvestad *et al.* also introduced a notion of likelihood of a reconciliation in the framework of birth-and-death processes [1].

3   Counting and Uniform Random Generation

In this section, we describe an efficient algorithm that computes a random reconciliation between G and S following the uniform distribution. This problem is important in the context of MCMC analysis for gene families, as a major issue is to analyze if the Markov chain converges to the true posterior probabilities. One of the most popular and simple tests of convergence is to run several Markov chains, each starting at a different state in the space, which motivates our random generation algorithm.

As usual in uniform random generation, it is based on a preprocessing that computes the cardinality of $\Psi(G, S)$ [5]. We first address this problem, then describe the random generation algorithm.

For every node $u \in V(G)$ and cell $c \in A(u)$, we denote by $Nb(u, c)$ the number of reconciliations of G_u and S_c for which u is mapped on c. It follows immediately that $|\Psi(G, S)| = Nb(r(G), r(S))$.

Lemma 1. *Let $u \in V(G)$ and $c \in A(u)$ be a cell that covers u. Then $Nb(u, c) = 1$ if $u \in L(G)$, and otherwise*

$$Nb(u, c) = \sum_{c_1 \in A(u_1),\ c_1 \leq_S c} Nb(u_1, c_1) \sum_{c_2 \in A(u_2),\ c_2 \leq_S c} Nb(u_2, c_2). \tag{1}$$

Proposition 1. *$|\Psi(G, S)|$ can be computed in $O(mn)$ time and space.*

It follows from the work [4] that there is a single optimal reconciliation for the loss and mutation costs, but that there can be several ones for the duplication cost. Building on Lemma 1, we also get the following result, that is of interest with respect to this point.

Proposition 2. *The number of reconciliations of $\Psi(G, S)$ that minimizes the duplication cost can be computed in $O(mn)$ time and space.*

The algorithm 1.1 below computes a random reconciliation between G and S. For a node $u \in V(G)$ and a cell $c \in A(u)$, let $f(c)$ $(d(c))$ be the ancestor (resp. descendant) cell of c in $A(u)$, that is the cell of $A(u)$ that is the closest one to c and above (resp. below) it. The lowest cell of $A(u)$ is the one that has no descendant cell in $A(u)$.

Algorithm 1.1. Uniform random generation in $\Psi(G, S)$.

1: Let α be an empty reconciliation.
2: Perform a prefix traversal of G, and let $u \in V(G)$ be the current node.
3:     **if** $u = r(G)$ or $u \in L(G)$ **then** $\alpha(u) \leftarrow M(u)$
4:     **else**
5:         Let $\hat{c} \leftarrow \alpha(p(u))$.
6:         {Choose randomly a cell $c \in A(u)$ such that $c \leq_S \hat{c}$}
7:         Let $k \leftarrow \sum_{c \in A(u), c \leq_S \hat{c}} Nb(u, c)$
8:         Generate randomly and uniformly an integer $n \in \{1, \ldots, k\}$.
9:         $c \leftarrow$ lowest cell in $A(u)$ {If u is a forced duplication, then $M(u) \notin A(u)$}
10:        $l \leftarrow Nb(u, c)$
11:        **while** $l < n$ **do** $c \leftarrow f(c)$, $l \leftarrow l + Nb(u, c)$
12:        $\alpha(u) \leftarrow c$
13: **return** α

Theorem 1. *Given a reconciliation $\alpha \in \Psi(G, S)$, Algorithm 1.1 returns α with probability $\frac{1}{|\Psi(G,S)|}$. Given the table Nb and the sets $A(u)$ for every node u of G, it can be implemented to run in $O(mn)$ space and $\Theta(mn)$ time in the worst case and $\Theta(m)$ time in the best case.*

Hence, the preprocessing time of our algorithm (computing the table Nb and the sets $A(u)$) requires $O(mn)$ time and space. However, it needs to be done once and can be used for generating several random reconciliations.

Our algorithm is useful for sampling the space of reconciliations, but not for exhaustive enumeration of that space. Therefore, in the next section, an algorithm for enumeration is introduced.

4   Exhaustive Exploration of the Whole Space $\Psi(G, S)$

We first define combinatorial operators used to explore the space of all possible reconciliations, and then give an algorithm, based on these operators, that explores exhaustively this space.

4.1   Space Exploration Operators

We present in this section a type of operator, called *Nearest Mapping Change* (NMC), acting on a reconciliation between a gene tree G and a species tree S. This movement is similar to the ones described in [11]. We show that this operator is sufficient to explore the space of all possible reconciliations.

Definition 2. Let $\alpha : V(G) \rightarrow V(S) \cup E(S)$ be a given reconciliation between G and S, and u a vertex of $V(G) \setminus L(G)$ such that $u \neq r(G)$. Let $\hat{c}, c, c_1,$ and c_2 respectively denote $\alpha(p(u))$, $\alpha(u)$, $\alpha(u_1)$, and $\alpha(u_2)$.

1. An *upward* NMC (uNMC) can be applied to u if $c <_S \hat{c}$, and if $\hat{c} \in V(S)$ and $c \in E(S)$, then $D(\hat{c}, c) > 0$. It changes $\alpha(u)$ into its ancestor cell $f(\alpha(u))$ of $A(u)$.

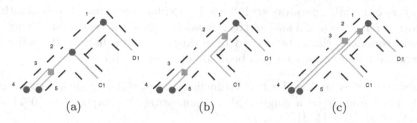

(a)   (b)   (c)

Fig. 2. (a) A section of the reconciliation depicted in figure 1. Here, the mapping of node 2 forbids to move up node 3. (b) The node 2 changes from a speciation to a duplication by moving it up. (c) Then, node 3 can be moved up and still is a duplication.

2. A *downward* NMC (dNMC) can be applied to u if $c_1 <_S c$, $M(u) <_S c$, and if $c_1 \in V(S)$ and $c \in E(S)$, then $D(c, c_1) > 0$ (idem for c_2). It changes $\alpha(u)$ into its descendant cell $d(\alpha(u))$ of $A(u)$.

It follows immediately from the definition of NMC operators that, given $\alpha \in \Psi(G, S)$, applying an NMC operator to a vertex u of G results in a reconciliation α' between G and S. More precisely, it can induce the following changes in the evolutionary scenario for the gene family (see Figure 2).

- Changing a speciation by a duplication (uNMC, $\alpha(u) = M(u)$).
- Changing a duplication by a speciation (dNMC, $\alpha'(u) = M(u)$).
- Moving a duplication upward (uNMC, $\alpha(u) \neq M(u)$).
- Moving a duplication downward (dNMC, $\alpha'(u) \neq M(u)$).

For $u \in V(G)$, and $c, c' \in A(u)$, $d_u(c, c')$ is the number of cells of $A(u)$ between c and c', where $d_u(c, c') = 0$ if and only if $c = c'$. For two reconciliations α and α', $D_{NMC}(\alpha, \alpha') = \sum_{u \in V(G)} d_u(\alpha(u), \alpha'(u))$. We call $D_{NMC}(\alpha, \alpha')$ the *NMC distance* between α and α'. A valid (according to Definition 2) NMC application to α can be encoded by a pair (u, c), where $u \in V(G)$ is the node being moved and $c \in V(S) \cup E(S)$ is its new mapping. We denote by $NMC(\alpha)$ the set of such pairs for a given reconciliation α.

Theorem 2. *Let α and α' two reconciliations of $\Psi(G, S)$. There exists a sequence of $D_{NMC}(\alpha, \alpha')$ NMC that transforms α into α'. No shorter sequence of NMC can transform α into α'.*

We denote by $\mathcal{G}_{NMC}(G, S)$ the graph with vertex set $\Psi(G, S)$ and where two reconciliations are linked by an edge if they differ by a single NMC. Let α_{min} be the unique reconciliation where, for each vertex u of G, $\alpha_{min}(u)$ is the unique cell of $A(u)$ that has no descendant in $A(u)$, and α_{max} be the unique reconciliation where, for each vertex u of G, $\alpha_{max}(u)$ is the unique cell of $A(u)$ that has no ancestor in $A(u)$. The following results shows that although $\Psi(G, S)$ can have an exponential size, NMC operators are sufficient to define a structure on this space of polynomial diameter.

Corollary 1. *The diameter of $\mathcal{G}_{NMC}(G, S)$ is equal to $D_{NMC}(\alpha_{min}, \alpha_{max})$ and is in $O(nm)$.*

Finally, as our NMC operators are intended to explore the space of reconciliations between a gene tree and a species tree, we address now the issue of updating the classical combinatorial criteria used to evaluate a reconciliation: the following observation implies that they can be easily updated in constant time.

Property 1. Let α and α' be two reconciliations of $\Psi(G, S)$ such that α' can be obtained from α by a single NMC. Then, $|dup(\alpha) - dup(\alpha')| \in \{0, 1\}$ and $|los(\alpha) - los(\alpha')| \in \{1, 2\}$.

4.2   Algorithm for the Exhaustive Exploration

We present in this section a simple algorithm, based on the NMC operator, that computes the set of all possible reconciliations between a gene tree G (with $|V(G)| = m$) and a species tree S (with $|V(S)| = n$) in time $\Theta(|\Psi(G, S)|)$ (see Theorem 3), which gives a CAT (Constant Amortized Time) algorithm to generate $\Psi(G, S)$.

For a node $u \in V(G)$, let $id(u)$ be the number of nodes that precede u according to the prefix traversal of G, where the left child u_1 of a node $u \in V(G) \setminus L(G)$ is visited before the right child u_2. Let $\mathcal{T}_{NMC}(G, S)$ be the tree defined as follows (see Figure 3):

- The root is the reconciliation α_{min} and its children are the reconciliations that can be obtained from α_{min} by applying a single uNMC from $NMC(\alpha_{min})$,
- Given a reconciliation α, that differs from its parent by an uNMC (u_i, c), its children are the reconciliations that can be obtained from α by applying a single uNMC (u_j, c') from $NMC(\alpha)$ such that $id(u_j) \geq id(u_i)$.

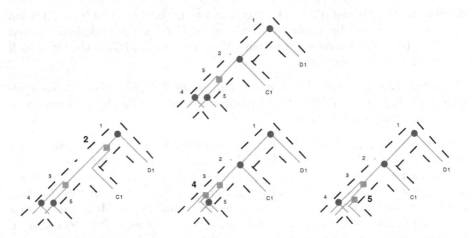

Fig. 3. The subtree of $\mathcal{T}_{NMC}(G, S)$ rooted at α_{min} for the trees G and S depicted in Figure 1. α_{min} and its children respectively are at the top and bottom of the figure. For each child, the node that has been moved upward is in boldface.

Proposition 3. $\mathcal{T}_{NMC}(G, S)$ *is a spanning tree of* $\mathcal{G}_{NMC}(G, S)$.

The exhaustive exploration algorithm of the whole space $\Psi(G, S)$ is based on the tree $\mathcal{T}_{NMC}(G, S)$. It follows immediately from the definition of $\mathcal{T}_{NMC}(G, S)$ that the main tasks for a given reconciliation α is 1) to know the list of allowed uNMC operators that can be applied to obtain the children of α, and 2) to keep in order its nodes according to the increasing value of their indexes id. We denote by $P(\alpha)$ this ordered list. The key to achieve this efficiently is the Property 2 below, that follows easily from the definitions of NMC operators and of $\mathcal{T}_{NMC}(G, S)$.

Property 2. Let α and α' be two reconciliations of $\Psi(G, S)$ such that α' is a child of α in $\mathcal{T}_{NMC}(G, S)$, and differs from α by an uNMC (u, c). Then $P(\alpha)$ and $P(\alpha')$ differ by at most three uNMC, that involve u, u_1 and u_2.

Based on this property, we describe below an algorithm that performs a prefix traversal of $\mathcal{T}_{NMC}(G, S)$, where the children of a reconciliation α are visited according to the ordered list $P(\alpha)$, in such a way that each time an edge from α to a reconciliation α' is followed, $P(\alpha)$ is updated into $P(\alpha')$. To perform this update in constant time, we encode P using two disjoint lists P_ℓ and P_r and two cursors u_ℓ and u_r on these lists, in such a way that a node u is in P if and only if u is in the sublist of P_ℓ (or P_r) that starts at u_ℓ (resp. u_r).

Algorithm 1.2 below describes the general recursive function, where the main tasks for the current reconciliaton α are i) select the next node $u \in P(\alpha)$ with the smallest id from the sublists of P_ℓ or P_r (lines 4,5); ii) define the child reconciliation α' by moving u upward (line 6); iii) define $P(\alpha')$ from $P(\alpha)$ by updating P_ℓ, P_r, u_ℓ, and u_r (lines 7-12). The function is first called with $\alpha = \alpha_{min}$, $P_\ell = NMC(\alpha)$, $P_r = \emptyset$, $u_\ell = first(P_\ell)$, and $u_r = end(P_r)$, that are

---

Algorithm 1.2. Exhaustive exploration algorithm of the space $\Psi(G, S)$

```
 1:  RecurExplore (α, uₗ, uᵣ)
 2:      Let u'ₗ ← uₗ and u'ᵣ ← uᵣ
 3:      while u'ₗ ≠ end(Pₗ) or u'ᵣ ≠ end(Pᵣ) do
 4:          if u'ₗ = end(Pₗ) then u ← u'ᵣ
 5:          else if u'ᵣ = end(Pᵣ) or id(u'ₗ) < id(u'ᵣ) then u ← u'ₗ, else u ← u'ᵣ
 6:          α'(u) ← f(α(u))
 7:          if u₁ ∉ Pₗ and u = u'ₗ then insert u₁ in Pₗ after u'ₗ
 8:          else if u₁ ∉ Pₗ and u = u'ᵣ then insert u₁ in Pₗ before u'ₗ, u'ₗ ← u₁
 9:          if u₂ ∉ Pₗ ∪ Pᵣ and u = u'ᵣ then insert u₂ in Pᵣ after u'ᵣ
10:          else if u₂ ∉ Pₗ ∪ Pᵣ and u = u'ₗ then insert u₂ in Pᵣ before u'ᵣ, u'ᵣ ← u₂
11:          if (u, f(α'(u))) ∉ NMC(α') and u = u'ₗ then u'ₗ ← succ(u'ₗ, Pₗ)
12:          if (u, f(α'(u))) ∉ NMC(α') and u = u'ᵣ then u'ᵣ ← succ(u'ᵣ, Pᵣ)
13:          RecurExplore (α', u'ₗ, u'ᵣ)
14:          Retrieve old values of Pₗ, Pᵣ, u'ₗ, u'ᵣ by performing the inverse operations of
             lines 7 to 12.
15:          α(u) ← d(α'(u)) {Backtrack}
16:          if u = u'ₗ then u'ₗ ← succ(u'ₗ, Pₗ) else u'ᵣ ← succ(u'ᵣ, Pᵣ)
```

computed during a preprocessing phase. Here, $first()$ and $end()$ respectively represents the first cursor of the considered list and a null one located at the end of the list. For a node $u \in V(G)$ and a cell $c \in A(u)$, recall that $f(c)$ and $d(c)$ respectively are its ancestor and descendant cells in $A(u)$.

Theorem 3. *Algorithm 1.2 visits all reconciliations of $\Psi(G, S)$. Given α_{min}, and $P_\ell = NMC(\alpha_{min})$, it can be implemented to run in time $\Theta(|\Psi(G, S)|)$ and space $O(nm)$.*

Together with Property 1, that implies that updating the number of duplications and/or losses after a single NMC can be done in constant time, this algorithm allows to compute efficiently the exact distribution of the duplication, loss and mutation costs in optimal time $\Theta(|\Psi(G, S)|)$ (see Section 5).

5   Experimental Results

We considered the phylogenetic tree of 12 *Drosophila* species and the branch lengths, and gene gain/loss rates that are given in [13, Figure 1]. We generated 1000 synthetic gene trees according to the birth-and-death process (with a single ancestral gene) along this species tree, and removed multiple copies of each gene tree. This resulted in 249 unique gene trees having from 6 to 22 leaves (Figure 4). Figure 5 describes the cardinality and diameter of $\Psi(G, S)$ for these 249 gene trees.

For each of the 249 unique gene trees, we used the algorithm 1.2 to explore the whole space $\Psi(G, S)$ focusing on the duplication cost (for the loss and mutation criteria, the results are similar). For the duplication criterion, 237 gene trees have a unique global minimum, and 12 have two. In each of these 12 cases, the NMC distance between the two global minimums is one. Over all the 249 gene trees, the LCA reconciliation α_{min}, that is a global minimum, is either identical or, in the worst case, at a distance of a single NMC to the true evolutionary scenario induced by the birth-and-death and noted α_{real}. However, it is important to point out that this is probably due to the low duplication and loss rates given in [13].

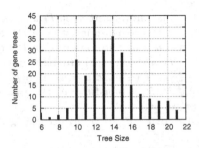

Fig. 4. Distribution of the 249 gene trees according to their number of leaves

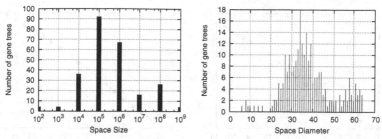

Fig. 5. Left. Distribution of the number of gene trees (y axis) according to the reconciliation space size (x axis). A gene tree G is counted in the bar 10^i iff $10^{i-1} \leq |\Psi(G, S)| < 10^i$. Right. Distribution of the 249 gene trees according to the diameter of $\Psi(G, S)$.

For a reconciliation $\alpha \in \Psi(G, S)$, let $d_{\text{cost}}(\alpha) = dup(\alpha) - dup(\alpha^*)$, where α^* is a global minimum, according to the duplication cost, that minimizes $D_{NMC}(\alpha, \alpha^*)$. We denote by $N(k)$ the number of reconciliations $\alpha \in \Psi(G, S)$ such that $d_{\text{cost}}(\alpha) = k$, for a given $k \in \mathbb{N}$. Figure 6 shows that, on average over all gene trees, $N(k)$ is proportional to k from $k = 0$ to $k = 13$ and inversely proportional from $k = 13$ to $k = 18$. This can be explained by the following facts: the maximum value of d_{cost} is equal to the number of internal nodes u of G that can be mapped on $M(u)$, and the average number of such nodes is 13. All this suggests that, for a given gene tree, $N(k)$ is maximized at this maximum value of $d_{\text{cost}} = k$.

We analyzed the relationship between the NMC and cost distances using the average value of $D_{NMC}(\alpha, \alpha^*)$ over all gene trees G and all reconciliations $\alpha \in \Psi(G, S)$ such that $d_{\text{cost}}(\alpha) = k$, for a given $k \in \mathbb{N}$. We also computed the number of nodes $u \in V(G)$ such that $\alpha(u) \neq \alpha_{\text{real}}(u)$. According to Figure 7, we observe that the cost distance of a reconciliation α is proportional both to

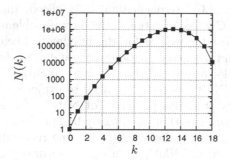

Fig. 6. Over all 249 gene trees, average distribution of the number $N(k)$ (y axis) of reconciliations α such that $d_{\text{cost}}(\alpha) = dup(\alpha) - dup(\alpha^*) = k$, for $k \in \mathbb{N}$ (x axis). α^* is a global minimum that minimizes the NMC distance $D_{NMC}(\alpha, \alpha^*)$.

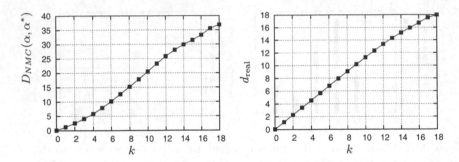

Fig. 7. Left: over all gene trees G, average value of $D_{NMC}(\alpha, \alpha^*)$ (y axis) for all reconciliations $\alpha \in \Psi(G, S)$ such that $d_{\text{cost}}(\alpha) = k$, for a given $k \in \mathbb{N}$ (x axis). Right: the same distribution with the real distance d_{real}, that is the number of nodes u of G such that $\alpha(u) \neq \alpha_{\text{real}}(u)$.

the NMC distance with the closest optimal reconciliation α^* and to how much α differs from the real reconciliation α_{real}.

6   Conclusion

We described in this work several algorithms related to exploring the space of all reconciliations between a gene tree and a species tree. From an algorithmic point of view, our exhaustive exploration algorithm is optimal as it requires an (amortized) constant time between successive reconciliations. Our experiments on a realistic simulated dataset with low duplication/loss rates (we will consider simulated datasets with higher duplication/loss rates in the full version of this paper) show that even in this case the number of reconciliations can be very large, but that for all three combinatorial criterion considered there are relatively few optimal or near-optimal reconciliations, always located close (in terms of NMC distance) to the LCA reconciliation. It is known that for the loss and mutation costs, this LCA reconciliation is the only possible minimum. However, for the duplication cost (as well as for the maximum likelihood cost), it can happen that several optimal reconciliations exist and our exhaustive exploration algorithm was able to locate them. This motivates our current work to modify our algorithm to handle dNMC operators in order to explore efficiently alternative but close evolutionary scenarios (in terms of NMC) of a given reconciliation (work in progress). Our algorithm can already be applied to this task when the starting reconciliation is α_{min} by visiting only the reconciliations that are at a fixed distance (in terms of NMC) from α_{min}. Natural generalizations of the algorithms we described in the present work include handling either non-binary gene or species trees [3,18] (or both) and attacking the more difficult problem of multiple gene duplications [7]. Moreover, we are now developing our exhaustive exploration algorithm for the maximum likelihood cost.

Acknowledgements. Cédric Chauve akcnowledges the support of NSERC through an Individual Discovery Grant and of Simon Fraser University through a Startup Grant.

References

1. Arvestad, L., Berglund, A.-C., Lagergren, J., Sennblad, B.: Gene tree reconstruction and orthology analysis based on an integrated model for duplications and sequence evolution. In: RECOMB 2004, pp. 326–335 (2004)
2. Bonizzoni, P., Della Vedova, G., Dondi, R.: Reconciling a gene tree to a species tree under the duplication cost model. Theoret. Comput. Sci. 347, 36–53 (2005)
3. Chang, W.-C., Eulenstein, O.: Reconciling gene trees with apparent polytomies. In: Chen, D.Z., Lee, D.T. (eds.) COCOON 2006. LNCS, vol. 4112, pp. 235–244. Springer, Heidelberg (2006)
4. Chauve, C., Doyon, J.-P., El-Mabrouk, N.: Gene family evolution by duplication, speciation and loss. J. Comput. Biol. (to appear, 2008)
5. Denise, A., Zimmermann, P.: Uniform random generation of decomposable structures using floating-point arithmetic. Theoret. Comput. Sci. 218, 233–248 (1999)
6. Doyon, J.-P., Chauve, C., Hamel, S.: Algorithms for exploring the space of gene tree/species tree reconciliations. IRO Technical Report # 1323 (2008)
7. Fellows, M.R., Hallett, M.T., Stege, U.: On the multiple gene duplication problem. In: Chwa, K.-Y., H. Ibarra, O. (eds.) ISAAC 1998. LNCS, vol. 1533, pp. 347–356. Springer, Heidelberg (1998)
8. Fitch, W.M.: Distinguishing homologous from analogous proteins. Syst. Zool. 19, 99–113 (1970)
9. Fitch, W.M.: Homology - a personal view on some of the problems. Trends Genet. 16, 227–231 (2000)
10. Goodman, M., Czelusniak, J., Moore, G.W., Herrera, R.A., Matsuda, G.: Fitting the gene lineage into its species lineage, a parsimony strategy illustrated by cladograms constructed from globin sequences. Syst. Zool. 28, 132–163 (1979)
11. Górecki, P., Tiuryn, J.: DLS-trees: a model of evolutionary scenarios. Theoret. Comput. Sci. 359, 378–399 (2006)
12. Graur, D., Li, W.-H.: Fundamentals of Molecular Evolution, 2nd edn. Sinauer Associates, Sunderland (1999)
13. Hahn, M.W., Han, M.V., Han, S.-G.: Gene family evolution across 12 Drosophilia genomes. PLoS Genet. 3, e197 (2007)
14. Jensen, R.: Orthologs and paralogs - we need to get it right. In: Genome Biology, 2:interactions1002.1–interactions1002.3 (2001)
15. Ma, B., Li, M., Zhang, L.: From gene trees to species trees. SIAM J. Comput. 30, 729–752 (2001)
16. Ma, J., Ratan, A., Zhang, L., Miller, W., Haussler, D.: A heuristic algorithm for reconstructing ancestral gene orders with duplications. In: Tesler, G., Durand, D. (eds.) RECMOB-CG 2007. LNCS (LNBI), vol. 4751, pp. 122–135. Springer, Heidelberg (2007)
17. Page, R.D.: Maps between trees and cladistic analysis of historical associations among genes, organisms, and areas. Syst. Biol. 43, 58–77 (1994)
18. Vernot, B., Stolzer, M., Goldman, A., Durand, D.: Reconciliation with non-binary species tree. In: CSB 2007 pp. 441–452 (2007)

Limitations of Pseudogenes in Identifying Gene Losses

James C. Costello[1,2], Mira V. Han[1,2], and Matthew W. Hahn[1,2]

[1] School of Informatics, Indiana University, Bloomington, IN, 47408, USA
[2] Department of Biology, Indiana University, Bloomington, IN, 47405, USA

Abstract. The loss of previously established genes has been proposed as a major force in evolutionary change. While the sequencing of many new species offers the opportunity to identify cases of gene loss, the best method to do this with is unclear. A number of methods to identify gene losses rely on the presence of a pseudogene for each loss. If genes are completely or largely removed from the genome, however, such methods will fail to identify these cases. As the fate of gene losses is still unclear, we attempt to identify losses using nine Drosophila genomes and determine whether these lost genes leave behind pseudogenes in the lineage leading to *D. melanogaster*. We were able to find 109 cases of unambiguous gene loss. Of these, a maximum of 18 have identifiable pseudogenes, while the other 91 do not. We were also able to identify a large number of previously unannotated genes in the *D. melanogaster* genome, most of which also had evidence for transcription. Though our results suggest that pseudogene-based methods for finding gene losses will miss a large proportion of these events, we discuss the dependence of these conclusions on the divergence times among the species considered.

1 Introduction

Comparative genomic approaches to find evolutionarily important genes have traditionally involved comparisons between orthologous protein-coding sequences. Such comparisons can identify rapidly evolving genes whose high rate of evolution may indicate adaptive natural selection (*e.g.* ref. [1]). Recent extensions to this approach have further considered non-coding sequences and have uncovered several regions involved in human adaptation [2,3]. The availability of high-quality genome sequences has also allowed researchers to discover genes lost during evolution, where sequences are not necessarily shared between species. These changes may also have played important roles in adaptive evolution.

Gene loss is a ubiquitous phenomenon across all sequenced genomes, both eukaryotic and prokaryotic [4,5,6]. Gene loss generally refers to the loss of a functional gene present in a genome, rather than simply the creation of new pseudogenes by gene duplication. In humans, gene loss has been proposed to be an especially important source of adaptive change under the "less is more" hypothesis [7,8]. A number of well-studied examples of human-specific losses are known, including CMAH [9], ELN [10], Siglec-13 [11], and MYH16 [12]. In addition to these individual cases, several groups have conducted computational

C.E. Nelson and S. Vialette (Eds.): RECOMB-CG 2008, LNBI 5267, pp. 14–25, 2008.

searches to identify human- or primate-specific gene losses via comparative genomics [13,14,15]. These searches have collectively discovered over a hundred new gene losses in humans. Though the methods introduced in these papers differ in their details, they have one important thing in common: they all initialize their search for gene losses using sequences currently present in the focal (*i.e.* human) genome. This means that they use either previously annotated pseudogenes [14], annotate their own pseudogenes [15], or require there to be an EST for the pseudogene [13]. In each case, a pseudogene is defined as a genomic feature in the focal genome with homology to a functional gene in other species, but that has lost its ability to code for a protein. Any gene loss resulting from a complete or near-complete deletion of a gene, or any sequence that has been deleted since becoming a pseudogene is therefore missed.

It is currently unknown how many gene losses have gone undiscovered because of the limitations of these algorithms. There is a bias towards deletions in the human genome [16], which may result in the loss of many sequences no longer maintained by selection. Deletion bias is even stronger in Drosophila [17], which may cause methods requiring pseudogene sequences to have extremely high false negative rates when searching for gene losses. However, the publication of 12 Drosophila genomes [18,19] provides a novel comparative genomic dataset that offers the opportunity to identify recent gene losses with unprecedented resolution. Therefore, to determine the extent to which algorithms dependent on pseudogenes may miss gene losses, we conducted an extensive analysis of apparent losses among the genomes within the Sophophora sub-genus of Drosophila (which includes the model organism, *D. melanogaster*). We were able to identify a large number of gene losses along the lineage leading to *D. melanogaster*, only a small fraction of which are present as pseudogenes. Additionally, we examined two *D. melanogaster* genome assemblies and annotations in order to highlight the effect of genome annotation on identifying gene losses. Our results suggest that alternative algorithms may be needed to uncover the full extent of gene loss across species.

2   Data

2.1   Drosophila Genomes

The sequences of 12 Drosophila genomes were recently used to compare the complement of protein-coding genes among species [18,19]. In 11 of the 12 species (all except *D. melanogaster*) *de novo* gene prediction was conducted to establish the set of genes in each genome, including in the previously sequenced *D. pseudoobscura* [20]. We used both the reconciled set of predicted genes from the newly sequenced species in the Sophophora sub-genus and the assembly and annotations from *D. melanogaster* v4.3 to initially identify gene losses; these are the same set of genes used for these genomes in the main analyses of ref. [18] and ref. [19]. The genomes in the Sophophora sub-genus are: *D. melanogaster* (*Dmel*), *D. simulans* (*Dsim*), *D. sechellia* (*Dsec*), *D. yakuba* (*Dyak*), *D. erecta* (*Dere*), *D. ananassae* (*Dana*), *D. pseudoobscura* (*Dpse*), *D. persimilis* (*Dper*), and *D. willistoni (Dwil)*. The additional 3 sequenced Drosophila genomes are *D. grimshawi (Dgri)*, *D.virilis (Dvir)*, and *D. mojavensis (Dmoj)*.

2.2   Defining Gene Families

Gene families were defined using the Fuzzy Reciprocal BLAST (FRB) method introduced in ref. [19]. FRB compares all proteins in a reciprocal manner between

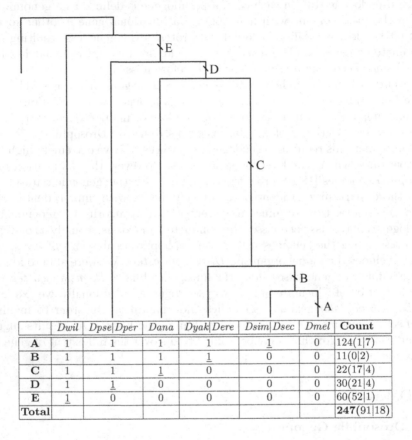

	Dwil	*Dpse|Dper*	*Dana*	*Dyak|Dere*	*Dsim|Dsec*	*Dmel*	**Count**
A	1	1	1	1	<u>1</u>	0	124(1|7)
B	1	1	1	<u>1</u>	0	0	11(0|2)
C	1	1	<u>1</u>	0	0	0	22(17|4)
D	1	<u>1</u>	0	0	0	0	30(21|4)
E	<u>1</u>	0	0	0	0	0	60(52|1)
Total							**247(91|18)**

Fig. 1. At the top of the figure is the phylogeny for the sub-genus Sophophora. The letters on the phylogeny represent the timing of candidate gene losses. The table below the phylogeny shows the breakdown of all the 247 candidate gene losses considered. A "1" indicates that at least one gene is present in a given gene family and a "0" indicates the absence of a gene defined for a given gene family. The underlined values represent the species from which one gene per gene family was used as a query to the *D. melanogaster* genome. In the case where two species are sister to *D. melanogaster*, genes from the better assembled genomes (*Dpse, Dyak, Dsim*) were taken if possible. The left-most column corresponds to the letters on the phylogeny. The right-most column shows the number of candidates in each category of gene loss, as well as the number of complete losses and pseudogenes in parentheses; displayed as (complete losses|pseudogenes). In total, there are 109 identified gene losses (91 + 18).

all pairs of genomes using BLASTp. Instead of using only the reciprocal best hit, FRB uses a rank-based method to identify potential homologs of each protein. The genes are then clustered based on their reciprocal similarity scores so that the resulting families are maximally connected and disjoint from one another. The method results in families that include both orthologs and paralogs, but has a propensity to break down families into 1:1:1...1 matches across species. This aspect of FRB allows us to easily identify homologs of candidate gene losses. Among the sequenced Drosophila genomes, FRB identified 11,434 families present in the most recent common ancestor of all 12 species, comprising a total of 148,326 genes. By comparing the number of genes within a family across species we were able to identify genes that appear to have been lost along each lineage as shown in Figure 1. See ref. [21] for further details.

2.3 Drosophila Sequences

To verify gene losses in *D. melanogaster* we searched against both the assembly and annotation of this genome used in the initial definition of gene families as well as an updated version. Both v4.3 and v5.3 *D. melanogaster* sequences were downloaded from the FlyBase ftp website.[1] Coordinates for *D. melanogaster* sequences (coding sequences and pseudogenes) were extracted from the fasta headers.

    D. melanogaster EST sequences were downloaded from the Berkeley Drosophila Genome Project (BDGP) website.[2] Gene models of the eight non-*melanogaster* Sophophora species were defined by the GLEANR consensus set of the Drosophila Genome Sequencing and Analysis Consortium [19].

3 Results

3.1 Gene Losses

We initially identified potential gene losses along the lineage leading to *D. melanogaster* since the split with *D. willistoni* (Figure 1) by using fuzzy reciprocal BLAST [21]. Because annotated *D. melanogaster* pseudogenes were not used as input to FRB, this method calls genes as absent whether or not a pseudogene can be found. Here we consider only those cases of potential gene losses where a single loss has occurred. This means that the gene family containing the lost gene is required to have at least one intact homolog present in all of the sister branches to the lineage of interest, including at least one homolog in *Dmoj*, *Dvir*, or *Dgri*. For example, for a gene to be considered lost in the *melanogaster* group (*Dmel, Dsec, Dsim, Dyak, Dere, Dana*), there must be at least one gene from the same family present in the obscura group (*Dpse, Dper*), one in the *willistoni* group (*Dwil*), and one among the Drosophila sub-genus species *Dmoj*, *Dvir*, or *Dgri* (case "D" in Figure 1). All cases involving the parallel loss of genes were therefore not considered. However, because of the low sequence coverage

---

[1] ftp://www.flybase.net/
[2] http://www.fruitfly.org/sequence/dlcDNA.shtml

of several of the Drosophila genomes, we used the annotations of closely related sister species to eliminate apparent parallel losses that were due to missed predictions in genomes with low sequence coverage. Therefore the following species were treated as individual lineages: *Dsim|Dsec*, *Dyak|Dere*, and *Dpse|Dper*. Figure 1 shows the counts of candidate gene losses in relation to *D. melanogaster*. In total, 247 gene families from the FRB results met the criteria listed above.

For each of the 247 gene families, one gene was selected as a query sequence and used for further analysis. The gene sequence selected was taken from the most closely related species that contained an intact protein-coding gene homologous to the lost gene. Figure 1 identifies the species from which query sequences were taken for each case of gene loss. Since gene families are defined only for protein-coding genes, the coding sequence for a given query gene was used in all subsequent analyses.

As a first step in confirming gene losses along the *D. melanogaster* lineage, the coding sequences of the 247 query genes were searched against the *D. melanogaster* genome using BLASTn. The results from this search constitute the first major division within the candidate gene losses. Of the starting 247 coding sequences, 133 have hits to the v4.3 *D. melanogaster* genome meeting our BLAST criteria (e-value $< 10^{-6}$, percent identity $> 80\%$, and hit length > 40), while 114 do not have a significant hit (Figure 2).

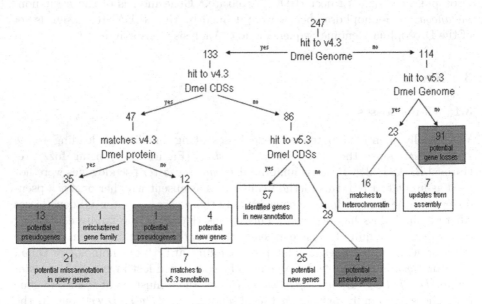

Fig. 2. Results of the gene loss analysis. The boxes shaded with horizontal stripes represent potential gene losses through either pseudogenization or complete removal of a gene from the genome. The white boxes represent genes that were not annotated or improperly annotated in v4.3 of the *D. melanogaster* genome assembly and annotation, but that are called as potential new genes in our analyses. Many of the genes missed in the v4.3 genome are in fact called gene models in the updated v5.3.

Query genes that do not hit the *D. melanogaster* **genome.** The 114 query genes not hitting the v4.3 *D. melanogaster* genome were first checked against the v5.3 genome assembly and annotation to determine if any of these potential gene losses are simply due to gaps in the v4.3 assembly. The 114 query sequences were searched against the v5.3 *D. melanogaster* genome using BLASTn with the same criteria as before. Interestingly, 23 query genes hit very strongly to genes predicted in the v5.3 genome. Of these 23, 16 mapped to heterochromatin and 7 mapped to euchromatin. The 7 hits to euchromatin are clear examples of gaps in the assembly that have been closed from v4.3 to v5.3 of the *D. melanogaster* genome. The 16 new genes found in heterochromatin are due in large part to recent efforts towards sequencing heterochromatic regions of the *D. melanogaster* genome [22,23].

As an additional verification that these 23 query genes do map to the *D. melanogaster* genome and are not gene losses, these sequences were searched against the *D. melanogaster* EST library using BLASTn with an e-value cutoff of 10^{-6}. Of these, 22 of the 23 query sequences have matches to ESTs, suggesting that they are true genes missed in previous assemblies. The one query sequence that did not map to an EST is dpse_GLEANR_9567, which hits a predicted gene located on an unmapped contig of the *D. melanogaster* genome.

The 91 query genes that do not have a hit to the *D. melanogaster* genome (both v4.3 and v5.3) meeting our requirements are likely losses of genes that were completely removed from the *D. melanogaster* genome. An alternative explanation for not finding these 91 genes is that any remaining remnants of the pseudogenes have been degraded beyond the detectable limits of the given BLAST parameters. To demonstrate that this is not the case, we ignored the percent identity and sequence length cutoffs and also lowered the BLASTn e-value cutoff from 10^{-6} to 10^{-3}. We did not recover a single additional hit to the v4.3 *D. melanogaster* genome using these criteria. A third potential reason for not being able to find these 91 proteins is that the coding regions lie in heterochromatic DNA that was not assembled into either the v4.3 or v5.3 *D. melanogaster* genome. Although this is unlikely given the progress that has been made on recent versions of the *D. melanogaster* genome where great efforts have been taken to fully sequence the heterochromatic regions [22]. We wanted to verify that this was not the case. As mentioned above, the 91 query genes (being a subset of the total 114) were searched against the v5.3 *D. melanogaster* genome with no hits to the heterochromatin; however, potential heterochromatic regions may still exist. Because genes located in heterochromatic regions are assumed to be transcribed, we reasoned that evidence for transcribed sequences could be used to find unassembled genes. In other words, a match to an unmapped EST would suggest a transcription unit that is not assembled into the current *D. melanogaster* genome release. We carried out this check by searching the set of 91 query genes against the *D. melanogaster* EST library with a BLASTn e-value cutoff of 10^{-6}, but no reliable hits were found. These 91 genes therefore represent good cases of gene loss with no identifiable pseudogenes.

Query genes that hit the *D. melanogaster* **genome.** The 133 sequences that hit the *D. melanogaster* genome were analyzed to determine whether they matched an annotated coding sequence. The physical chromosome coordinates from the v4.3 *D. melanogaster* BLAST results were checked against the physical coordinates of all *D. melanogaster* coding sequences, which resulted in 47 query sequences overlapping at least one *D. melanogaster* coding sequence and 86 not overlapping a *D. melanogaster* coding sequence. These two sets are further explored in the next two sections.

Query genes that do not overlap a *D. melanogaster* **coding sequence**
The set of 86 non-*melanogaster* query genes that hit part of the *D. melanogaster* genome but do not overlap with any *D. melanogaster* coding sequences were first tested against the v5.3 *D. melanogaster* genome to identify missed genes due to poor genome annotation. Coding sequences in v5.3 *D. melanogaster* were searched with the 86 query genes using BLASTn (e-value $< 10^{-6}$, percent identity $> 80\%$, and hit length > 40). Of the 86 query sequences, 57 unambiguously mapped to a putative coding sequence in *D. melanogaster* (v5.3). All of these genes represent annotations that were added from v4.3 to v5.3. To find evidence of gene expression for the 57 query genes we searched the EST sequence database using a BLASTn e-value cutoff of 10^{-6}, resulting in 50 sequences that had EST evidence and only 7 sequences that did not.

Improved annotations can explain 57 of the 86 query sequences, but does not explain the remaining 29 hits to the genome. These 29 cases are suggestive of either pseudogenes or missed annotations (*i.e.* new genes not included in v5.3). To test whether the regions hit by these 29 genes have evidence for transcription, we queried the non-*melanogaster* coding sequences against the *D. melanogaster* EST library. We found good matches to ESTs for 21 genes and no matches for the remaining 8 genes.

To evaluate the potential gene structure of these 29 regions in the *D. melanogaster* genome, we performed gene predictions using GeneWise [24]. We used translated query peptide sequences compared with two different lengths of *D. melanogaster* genomic regions ($\pm$ 2,000 bases or $\pm$ 5,000 bases from the BLAST hit) as input to GeneWise and used the output peptide of longest length as our gene model. Out of the 29 cases, 25 regions were identified as novel genes that are missing even in the v5.3 annotation of *D. melanogaster*. These 25 regions of the *D. melanogaster* genome have valid exon structures that align across the whole query gene without any nonsense or frameshift mutations. Eighteen of these newly predicted genes have independent supporting evidence, such as a perfect match to a third party annotated *D. melanogaster* protein in the non-redundant database of NCBI. Of these, 17 also overlap with new annotations predicted in ref. [18] and have EST evidence.

Finally, the four remaining regions have predicted exons that align only partially to the query gene or have nonsense/frameshift mutations, and are identified as pseudogenes. None of these four pseudogenes have any EST evidence.

Query genes that hit a *D. melanogaster* **coding sequence.** The set of 47 genes overlapping at least one *D. melanogaster* coding sequence suggests either misannotation or misclustering of the input genes, or requires some other explanation for their high similarity to genes present in *D. melanogaster*. To determine which of these scenarios may have occurred, we conducted further analyses.

In order to verify whether the *D. melanogaster* gene matching the query sequence is indeed a protein homolog, we again used GeneWise to predict exons in the genomic region using the query protein. We then used BLASTp to query the predicted peptide against the v4.3 proteins. We found 12 cases where the peptide matches the genomic nucleotide sequence but does not match an annotated protein in v4.3. Of these, 4 cases appear to be novel genes that overlap already annotated proteins. Because they are overlapping genes present in the current annotation we found significant nucleotide similarity, but no protein similarity. EST evidence was found for all four novel predictions. Another 7 cases match the nucleotide sequence of predicted genes in v4.3 that have since been updated with new predictions in v5.3. In all of these cases, the v5.3 predicted protein is in a different reading frame than the previously annotated gene, and this new protein has significant similarity to the peptide predicted by GeneWise. Our predicted peptides did not have significant protein similarity to the v4.3 annotations. The one remaining predicted peptide does not have a hit to v5.3 and only partially aligns to fragments of the query gene, and therefore is identified as a pseudogene.

The remaining 35 cases do have a matching *D. melanogaster* protein in v4.3, but still fail to cluster together in the same family. We found that 21 of the 35 peptides only partially match the *D. melanogaster* protein in the far 5' or 3' ends of the gene. For all of these cases the query gene is much shorter than the *D. melanogaster* gene it is aligned to. For a few cases the query gene matches a short first exon of the *D.melanogaster* protein that resides more than 10,000 bases upstream of the second exon. We suspect that these are misannotations in the other Drosophila species, where the *de novo* gene prediction program has predicted short exons at either end of long genes as separate genes. It is possible that a gene fusion event has occurred along the *D. melanogaster* lineage [25], though these generally do not occur between initially adjacent genes.

In 1 of the 35 cases, the gene family of the matching *D. melanogaster* gene appears to have one extra member, meaning that the matching *D.melanogaster* gene should have been placed with the query gene in order to explain the gene loss. This is the only case that appears to represent an apparent loss explained by the misclustering of gene families by FRB. For the remaining 13 cases (of the 35) there are one or more genes in the non-*melanogaster* species that are already clustered with the matching *D. melanogaster* gene, and the alignment among those genes is better than the alignment between the query gene and the *D. melanogaster* gene. These cases represent ancient duplications predating the base of the Drosophila tree, for which a gene is lost in one of the paralogous lineages and the query sequence is hitting the other paralog. These represent gene losses, though the high similarity to intact paralogs make it hard to unambiguously say whether a pseudogene is present in the *D. melanogaster* genome.

4 Discussion

Identifying cases where previously functional genes maintained by natural selection are lost is one of the novel and important challenges posed by comparative genomics. Though a large number of pseudogenes have been identified in many genomes (*e.g.* ref. [26]), the vast majority of pseudogenes identified are duplicated genes that were never maintained by selection. A number of new methods have been used to find true gene losses, but they require the remnants of lost genes to be identified in the target genome (*e.g.* refs. [13,14,15]). Alternatively, true gene losses can be found by identifying annotated genes in other species that do not have significant similarity to genes in the target genome [27,21]. Though this method does not require the presence of pseudogenes, it may misidentify gene losses when genes present in the target genome are not clustered with their homologous genes or when there are gaps in the genome sequence.

Here we have used this latter method to determine the utility of algorithms that require the presence of pseudogenes to identify gene losses. While we have not run any of these algorithms on the Drosophila dataset used here, by finding gene losses that do not have pseudogenes we are able to estimate the maximum number of genes that could be identified by such methods. By closely examining a number of cases, we are also able to extend previous results to judge the accuracy of methods based only on the lack of significantly similar genes (*i.e.* ref. [21]).

We initially identified 247 candidate gene losses along the lineage leading to *D. melanogaster*. Note that because we ignored parallel gene losses, these do not represent the full set of losses that have occurred along this lineage since the split with *D. willistoni*. It does mean, however, that we are unambiguously able to assign losses to a specific branch of the tree (Figure 1).

Of the 247 genes we initially identified as candidate gene losses, 109 appear to be unambiguous losses along the lineage leading to *D. melanogaster*. The vast majority of candidates that do not appear to be losses are instead genes that were not annotated in earlier versions of the *D. melanogaster* genome. Some of these were not annotated because of gaps in the genome assembly ($n = 7$), unsequenced heterochromatic regions ($n = 16$), or were simply not found by previous gene-finding algorithms ($n = 86$). The large majority of the annotation updates account for the 124 gene loss candidates between the *Dsim|Dsec* and *Dmel* lineages (Figure 1, row A), thus artificially inflating potential gene losses between sister species. We also found a large number of losses on branches D and E relative to C (Figure 1), a result consistent with previous estimates of loss rates along these lineages [21]. The v4.3 *D. melanogaster* genome, though out of date, still represents one of the most high quality assemblies and annotations available, particularly in a metazoan genome. These annotation updates illustrate the large influence that genome assembly and annotation can have on identifying gene losses. Additionally, that this "finished" genome can be missing so many gene annotations attests to the difficulties in identifying eukaryotic protein-coding genes in large genomes. In fact, 29 of the newly predicted proteins from this study are still not included in the v5.3 *D. melanogaster* annotation.

We were only able to identify 5 pseudogenes out of the 109 unambiguous gene losses, though for 13 cases this has not been determined definitively. This result implies that methods depending on the presence of pseudogenes to identify gene losses will find a maximum of 18 losses (5+13) along this lineage. Missing 83% of all gene losses would appear to be a major disadvantage of these methods.

However, the apparent failure of these methods in identifying gene losses masks a more complicated result. In the recent paper by Zhu et al. [15] the authors state that: "gene loss normally leaves behind a pseudogene." Motivated to determine the accuracy of this statement, we have examined the pattern of gene loss using nine Drosophila species with respect to the D. melanogaster lineage. Despite the 91 cases of total gene loss without the presence of a pseudogene, our results appear to at least partly support the Zhu et al. [15] supposition: only one of these 91 cases corresponds to the complete removal of a recently lost gene (Figure 1, row A). In other words, most of these losses may indeed have left behind a pseudogene, and only over time have these pseudogenes been degraded beyond recognition. Because there are only a few recent (< 10 million years) losses in D. melanogaster among the set considered here, it is hard to determine exactly what proportion initially leave behind a pseudogene as opposed to being completely deleted.

This result also raises the issue of the timeframe over which pseudogene-based methods can be used. For example, the Zhu et al. [15] study used the mouse genome to predict gene models of human pseudogenes. Though the divergence time between human and mouse is much greater than even the most distantly related Drosophila, the level of nucleotide divergence is equivalent to approximately the Dmel-Dyak split; comparing D. melanogaster and D. willistoni is equivalent to comparing the human genome to a lizard genome [18]. It is obvious that pseudogene-based methods cannot be used beyond the limits of our ability to identify the homologs of pseudogenes, and it may simply be that they are inappropriate or less useful in rapidly evolving lineages. It should be reiterated, however, that these problems do not result in any false positives, only false negatives.

In contrast to pseudogene-based methods, the clustering method used here identified a large number of gene losses across all time-scales of comparison. While we have not determined how many gene losses potentially identified by pseudogene-based methods were not identified by our clustering method, we expect this number to be small. If a pseudogene were present in the D. melanogaster genome, our method should also identify the loss of a homologous gene in the relevant gene family. The clustering method did result in a single false positive due to misclustering of genes into families, but this case was easily identified through follow-up analyses. Finally, the clustering method has the added property of finding a large number of previously unannotated genes initially identified by the lack of homologous proteins in D. melanogaster [21,18]; it also found a number of cases of misannotation in the other Drosophila species that can be fixed. These fortuitous results should be of benefit regardless of the divergence times among the genomes considered.

5 Funding

Computing resources provided by the Center for Genomics and Bioinformatics were supported in part by the METACyt Initiative of Indiana University, funded by a major grant from the Lilly Endowment. This research was supported by grants from the National Science Foundation (DBI-0543586) and National Institutes of Health (R01-GM076643A) to MWH.

References

1. Nielsen, R., Bustamante, C., Clark, A., Glanowski, S., Sackton, T., et al.: A scan for positively selected genes in the genomes of humans and chimpanzees. PLoS Biol. 3, e170 (2005)
2. Dermitzakis, E., Reymond, A., Lyle, R., Scamuffa, N., Ucla, C., et al.: Numerous potentially functional but non-genic conserved sequences on human chromosome 21. Nature 420, 578–582 (2002)
3. Pollard, K., Salama, S., Lambert, N., Lambot, M., Coppens, S., et al.: An RNA gene expressed during cortical development evolved rapidly in humans. Nature 443, 167–712 (2006)
4. Aravind, L., Watanabe, H., Lipman, D., Koonin, E.: Lineage-specific loss and divergence of functionally linked genes in eukaryotes. PNAS USA 97, 11319–11324 (2000)
5. Hughes, A., Friedman, R.: Recent mammalian gene duplications: robust search for functionally divergent gene pairs. J. Mol. Evo. 59, 114–120 (2004)
6. Roelofs, J., Van Haastert, P.: Genes lost during evolution. Nature 411, 1013–1014 (2001)
7. Olson, M.: When less is more: gene loss as an engine of evolutionary change, American journal of human genetics. Am. J. Human Genet. 64, 18–23 (1999)
8. Olson, M., Varki, A.: Sequencing the chimpanzee genome: insights into human evolution and disease. Nature Rev. 4, 20–28 (2003)
9. Chou, H., Takematsu, H., Diaz, S., Iber, J., Nickerson, E., et al.: A mutation in human CMP-sialic acid hydroxylase occurred after the Homo-Pan divergence. PNAS USA 95, 11751–11756 (1998)
10. Szabo, Z., Levi-Minzi, S., Christiano, A., Struminger, C., Stoneking, M., et al.: Sequential loss of two neighboring exons of the tropoelastin gene during primate evolution. J. Mol. Evo. 49, 664–671 (1999)
11. Angata, T., Margulies, E., Green, E., Varki, A.: Large-scale sequencing of the CD33-related Siglec gene cluster in five mammalian species reveals rapid evolution by multiple mechanisms. PNAS USA 101, 13251–13256 (2004)
12. Stedman, H., Kozyak, B., Nelson, A., Thesier, D., Su, L., et al.: Myosin gene mutation correlates with anatomical changes in the human lineage. Nature 428, 415–418 (2004)
13. Hahn, Y., Lee, B.: Identification of nine human-specific frameshift mutations by comparative analysis of the human and the chimpanzee genome sequences. Bioinformatics 21(suppl.1), 186–194 (2005)
14. Wang, X., Grus, W., Zhang, J.: Gene losses during human origins. PLoS Biol. 4, 52 (2006)
15. Zhu, J., Sanborn, J., Diekhans, M., Lowe, C., Pringle, T., Haussler, D.: Comparative Genomics Search for Losses of Long-Established Genes on the Human Lineage. PLoS Comput. Biol. 3, 247 (2007)

16. Kvikstad, E., Tyekucheva, S., Chiaromonte, F., Makova, K.: A macaque's-eye view of human insertions and deletions: differences in mechanisms. PLoS Comput. Biol. 3, 1772–1782 (2007)
17. Petrov, D., Hartl, D.: Patterns of nucleotide substitution in Drosophila and mammalian genomes. PNAS USA 96, 1475–1479 (1999)
18. Stark, A., Lin, M., Kheradpour, P., Pedersen, J., Parts, L., et al.: Discovery of functional elements in 12 Drosophila genomes using evolutionary signatures. nature 450, 219–232 (2007)
19. Clark, A., Eisen, M., Smith, D., Bergman, C., Oliver, B., et al.: Evolution of genes and genomes on the Drosophila phylogeny. Nature 450, 203–208 (2007)
20. Richards, S., Liu, Y., Bettencourt, B., Hradecky, P., Letovsky, S., et al.: Comparative genome sequencing of Drosophila pseudoobscura: Chromosomal, gene, and cis-element evolution. Genome Res. 15, 1–18 (2005)
21. Hahn, M., Han, M., Han, S.G.: Gene Family Evolution across 12 Drosophila Genomes. PLoS Genet. 3, e197 (2007)
22. Hoskins, R., Carlson, J., Kennedy, C., Acevedo, D., Evans-Holm, M., et al.: Sequence finishing and mapping of Drosophila melanogaster heterochromatin. Science 316, 1625–1628 (2007)
23. Smith, C., Shu, S., Mungall, C., Karpen, G.: The Release 5.1 annotation of Drosophila melanogaster heterochromatin. Science 316, 1586–1591 (2007)
24. Birney, E., Clamp, M., Durbin, R.: GeneWise and Genomewise. Genome Res. 14, 988–995 (2004)
25. Long, M.: A new function evolved from gene fusion. Genome Res. 10, 1655–1657 (2000)
26. Zhang, Z., Gerstein, M.: Large-scale analysis of pseudogenes in the human genome. Curr. Opin. Genet. Dev. 14, 328–335 (2004)
27. Demuth, J., De Bie, T., Stajich, J., Cristianini, N., Hahn, M.: The evolution of mammalian gene families. PLoS One 1, e85 (2006)

Duplication Mechanism and Disruptions in Flanking Regions Influence the Fate of Mammalian Gene Duplicates

Paul Ryvkin[1], Jin Jun[2], Edward Hemphill[3], and Craig Nelson[3]

[1] Genomics and Computational Biology Graduate Group, University of Pennsylvania,
Philadelphia, PA 19104, USA
pry@mail.med.upenn.edu
[2] Department of Computer Science and Engineering, University of Connecticut,
Storrs, CT 06269, USA
jinjun@engr.uconn.edu
[3] Department of Molecular and Cell Biology, University of Connecticut,
Storrs, CT, 06269, USA
{edward.hemphill_iii,craig.nelson}@uconn.edu

Abstract. Here we identify duplicated genes in five mammalian genomes and classify these duplicates based on the mechanisms by which they were generated. Retrotransposition accounts for at least half of all predicted duplicate genes in these genomes, with tandem and interspersed duplicates comprising the other half. Estimation of the evolutionary rates in each class revealed greater rate asymmetry between retrotransposed and interspersed segmental duplicate pairs than between tandem duplicates, suggesting that retrotransposed and interspersed segmental duplicates are diverging more quickly. In an attempt to understand the basis of this asymmetry we identified disruption of flanking DNA as an indicator of new duplicate fate. Loss of synteny accelerates the asymmetry of divergence of DNA-mediated duplicates duplicates. These findings suggest that the differential evolution of duplicate genes may be significantly influenced by changes in local genome architecture and synteny.

Keywords: duplication, retrotransposition, segmental, tandem, asymmetry.

1 Introduction

Gene duplication has long been recognized as an important contributor to the evolution of organismal complexity [1]. Large gene families, which comprise a major portion of mammalian genomes and carry out some of its most important functions, are the result of the gene duplication process. Hox genes, olfactory receptors, and many vertebrate growth factors are well-known examples of such families. Recent data reveals that lineage-specific gene family expansions and contractions comprise a large portion of the differences between closely related mammalian genomes. For example, while it is often stated that there is a 2-5% nucleotide difference between the human and chimpanzee genomes [2][3], this comparison ignores variation in gene copy number and the divergence of duplicated genes. [4][5][6].

C.E. Nelson and S. Vialette (Eds.): RECOMB-CG 2008, LNBI 5267, pp. 26–39, 2008.
© Springer-Verlag Berlin Heidelberg 2008

1.1 There Are Several Mechanisms of Gene Duplication

Several types of duplication have been observed in genomes to date, including whole genome duplication, segmental duplication, and retrotransposition. The last whole genome duplication in the mammalian lineage is thought to have occurred before the emergence of modern mammals [7][8]. In contrast, segmental and retrotransposed duplication are ongoing processes. Segmental duplications, in which sections of DNA are duplicated either in tandem or across chromosomes, are DNA-mediated processes that preserve varying amounts of the original gene's intron-exon structure and varying amounts of flanking intergenic DNA. Contrast this with retrotransposition, a process in which a spliced mRNA transcript is reverse transcribed into DNA and spontaneously integrated into a random genomic location [9]. In this case an intronless gene copy is created whose flanking DNA is unrelated to that of the parent gene. Since the ability of a gene to be transcribed depends on the existence of promoters and other cis-regulatory elements in proximity, such retrogenes are generally believed to be non-functional unless they are deposited within the range of influence of regulatory regions of other genes. However, recent studies have shown that many of these genes can, in fact, be functional [10][11][12][13].

1.2 Factors Influencing the Fate of Duplicated Genes Remain Unresolved

Despite our knowledge of the mechanisms by which duplicates are generated, the relative importance of various factors in determining the retention and evolution of duplicate genes remains a major unresolved question. One of the first models proposed for the evolution of gene duplicates was that of non- vs. neo-functionalization. This model assumes that the two genes are functionally redundant and that one copy either experiences disabling mutations or acquires a new function through positive selection. The more recently developed model of subfunctionalization allows for another possibility: the genes lose complementary functions or expression domains until both are retained because the sum of their individual functions maintains the original [14]. The duplication-degeneration-complementation (DDC) version of this model specifically predicts complementation of expression domains [15]. In addition to assuming redundancy, the DDC model's degeneration is often envisioned as a gradual process. However, since many mutations consist of transpositions, insertions, and deletions of vast stretches of DNA, changes in regulatory information may often be very abrupt.

While several paralogs with complementary expression domains have been observed [16][17], it is not clear that the DDC model is the best general predictor for the fate of most gene duplication events. Both the DDC and the neo-functionalization models assume that duplicate genes are equal at the time of duplication. However, this is clearly not true for retrotransposed gene copies. Furthermore, while the DDC model attempts to explain the retention of duplicate genes by changes in their regulation, it does not explain differences in the evolutionary rates of their protein coding regions. Studies have shown little to no coupling between regulatory and protein evolution in duplicates [18][19][20], despite others pointing out widespread asymmetry in evolutionary rates between the protein coding regions of duplicate genes [21][22]. One study has shown that duplication mechanism and genomic locality are important factors affecting rate asymmetry in rodent duplicates [23]. Another suggests that the variability in recombination rates across the genome cause distal yeast duplicates to evolve at different rates [24].

Here we show the relative contribution of each duplication mechanism to mammalian gene families and examine the distinct consequences of each duplication mechanism for duplicate fate. Our data suggest that retrotransposition produces nearly half of all predicted gene duplicates in the human, chimpanzee, mouse, rat, and dog genomes. We show that relocated and retrotransposed duplications yield genes whose coding regions evolve more asymmetrically than tandem duplications, and that retrogenes show greater rates of non-synonymous substitution and relaxed selective constraint compared to their parent genes. Finally we show that among distant segmental duplicates, disruptions in flanking regions correlate with a relaxation of selective constraint on these duplicates, providing evidence that abrupt changes in cis-regulatory regions can have profound effects on protein coding evolution in duplicate gene copies.

2 Methods

2.1 Sequence Retrieval and Protein Family Identification

Protein and DNA sequences for the five species analyzed (human, chimpanzee, dog, mouse, and rat) were obtained from Ensembl release 37 [25]. For genes with multiple alternatively spliced transcripts, we developed a collapsed gene model which merges all potential exons from the gene's transcripts. Protein families were established via Ensembl's family annotation [26]. The protein sequences of each family were aligned with CLUSTALW, Kimura distances were computed, and trees were produced via neighbor-joining [27][28]. Due to CLUSTALW's excessive computation time when aligning more than 50 sequences at once, protein families with more than 50 members were excluded from further analysis. Singleton families and families containing genes on unassembled contigs were also excluded.

2.2 Clustering of Orthologs

We used BranchClust [29] to determine orthologous groups with its one free parameter (number of species required for a full cluster) set to 3. This value was optimal given our species tree; it allowed for loss of either the primate or rodent lineages within clusters. Higher values would have disallowed such events and lower values would have counted primate/rodent duplications as separate clusters, which is not desirable. In short, BranchClust works by traversing the gene tree from leaf to root, building clusters such that a maximal number of species is represented in each cluster. It selects initial leaves which minimize the number of extraneous members (inparalogs) within each cluster. The method proved to be robust in the face of small differences between the gene phylogeny and species phylogeny because it is insensitive to the precise gene tree topology within any given cluster.

2.3 Inference of Duplication Events

For each orthologous group, the species tree was traversed and the number of genes from each species was counted. Orthologous groups (OG's) that contained greater than one gene from a species indicated duplications in that lineage. Using an assumption of simple parsimony, gains in related lineages were moved to appropriate parental nodes in the species tree. For example, duplicates in both human and chimp in one OG, are considered one duplication event on the branch leading from the primate-rodent split to the human-chimpanzee split. This procedure was applied recursively: For each species node, if all children have greater than one representative gene in the cluster, subtract from their counts the minimum count among them (minus one) and add this value to the parent node's count.

The Ensembl dataset contains some genes that are annotated as two or more separate gene fragments. These fragments appear on the phylogenetic trees as paralogs in 1:x relationships with other species, where x > 1. We consider genes to be fragments if they are apparent inparalogs on the phylogenetic tree yet they have a global-alignment identity of less than 25% with an ortholog in the closest species. As a consequence, any partial duplications (those that act on less than 25% of a gene) were removed from our analysis. Out of 57,733 genes, 1,123 (2.3%) were classified as fragments.

2.4 Distinguishing between Gene Duplication Mechanisms

Since a retrogene typically does not duplicate the parental gene's flanking region or introns, classification as DNA-mediated duplication or retrotransposition (RT) was based on two criteria: local gene synteny and exon-intron structure conservation. Local synteny between two genes is defined by the sequence similarity and collinearity of their neighboring genes.

We established local synteny between two genes by comparing their five neighboring genes on each side. For each side we counted the number of collinear neighbors for each gene which were orthologous. Orthology between these neighbors was defined by reciprocal BLASTP scores, requiring a minimum score of 50 and sequence identity of 80%. In order to be locally syntenic, two genes must have at least two collinear orthologous neighbors in total or one orthologous neighbor on each side.

Intron conservation was taken as the number of orthologous intron positions between two genes as defined by Rogozin et al.; orthologous introns are introns in two genes whose positions in the protein alignment are within a certain number of residues [30]. We generated protein alignments using MUSCLE [31], noting the locations of introns in the protein alignments, and intron positions were matched using a "slide" threshold of 4 residues.

We considered a duplication event to be DNA-mediated if either of the following two conditions held: 1) the genes are locally syntenic or 2) they have at least one conserved intron. In order for a duplication to be classified as RT the genes must be non-syntenic and 1) one gene must be intronless and the other one not or 2) the genes both have fewer than 3 introns, their intron counts are different, and there are no conserved introns between them. Here we assume that intron gain is a relatively rare event in the timescale studied [32], and newly gained introns in retrogenes are not positionally conserved. We further classified segmental duplicates as distant (interspersed) or tandem by searching for each duplicate's paralog within 5 genes in both the 5' and 3' directions. If paralogs were found within this window the duplicate was classified as tandem.

2.5 Computing Pairwise dN/dS

We calculated synonymous and non-synonymous divergence between pairs of dupli-cates (paralogs) according to the Yang-Nielsen method [33], using the 'yn00' pro-gram, which is a part of PAML.

2.6 Determining Rate Asymmetry

Using a method similar to that proposed in [23], we assigned duplicate pairs appropri-ate outgroup genes and performed a maximum-likelihood relative rate analysis on the resulting triplets using HYPHY [34]. The outgroup gene selected for each pair was required to be orthologous to the pair, and belong to the closest species to the pair's species. The evolutionary model used was that proposed in [35], where each branch has two parameters associated with it: dN and dS. The normalized rate asymmetry between two genes was defined as: $R = |(p1 - p2)| / (p1 + p2)$, where p1 and p2 are the parameters associated with each of the duplicates' branches. In addition we calculate a derived value, ω, defined as the ratio dN/dS and taken to be a measure of selective constraint on a given branch. Thus we can obtain three measures of normalized asymmetry between duplicates: RN, RS, and Rω, which correspond to asymmetry in dN, dS, and ω, respectively. We excluded duplications with pairwise dS or dN lower than 0.00001 or dS greater than 5 due to saturation effects.

2.7 Detection of Disrupted Flanking Regions

In order to locate genes with disrupted flanking regions we examined the syntenic relationship between duplicates and their outgroup gene. We define paralogs as hav-ing direct synteny if the gene immediately adjacent to each paralog is orthologous to the outgroup gene's neighbor (using the same BLASTP criteria as for local synteny). This definition is distinct from local synteny – local synteny is based on any collinear orthologous neighbors among the five flanking genes whereas direct synteny relies on the nearest gene only. If both genes share direct synteny with the outgroup gene then conservation of direct synteny is inferred. If one gene shares direct synteny and the other does not, then a disruption is inferred. We excluded cases where neither gene has synteny with the outgroup.

3 Results

3.1 Prevalence of Gene Duplication Types

Of the 4,386 mammalian duplication events we inferred, 1,593 form 1:2 ortholog groups derived from a single duplication event, while the remaining duplications are "one to many" events (1:N) where the genes belong to a larger duplication cluster. We analyzed the 1:2 and 1:N duplications separately.

We counted the total number of duplications of each type (distant segmental, tandem, and RT) among the 1:2 and 1:N groups. RT, responsible for 49% of the 1:2, and 41% and 1:N duplicated genes, is the most common type of duplication event in these mam-malian genomes (Figure 1). The most striking difference between the 1:2 duplicates

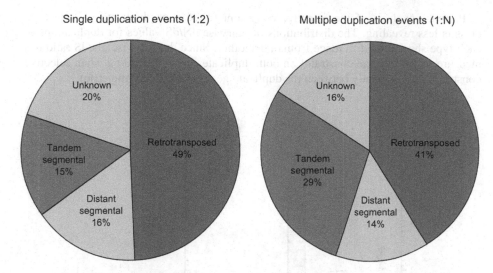

Fig. 1. Prevalence of each duplication type divided into two classes based on whether or not the duplicates are the result of exactly one duplication. These sets are restricted to genes with intact ORFs that are annotated as protein-coding genes by Ensembl.

and the 1:N duplicates is the difference in the proportion of tandem duplications. Among these clusters tandem duplications are significantly more highly represented (p < 0.0001, X^2 test).

3.2 Pairwise Analysis of Duplicate Age and Divergence

We calculated synonymous and non-synonymous divergence between pairs of duplicates and used these estimates to calculate the degree of purifying selection on the gene pair (dN/dS). Synonymous divergence (dS) reveals a striking difference in the age profile of the duplication types (Figure 2). Consistent with observations of bursts of RT in mammalian lineages [36], retrotransposed duplicates show a highly skewed distribution of dS values. Additionally, their tight clustering in the very low dS range suggests that most of these events occurred relatively recently. In contrast, tandem duplications show dS values that are more evenly distributed over time, are less tightly clustered around very low values, and have a higher mean dS. The observed higher mean dS of tandem duplications is consistent with tandem duplications occurring more uniformly over time than retrotranspositions. Distant segmental duplications show a very interesting pairwise dS profile. It is both more uniform and greater than the other dS profiles, suggesting that these events occur at a more stable rate than retrotranspositions, and are older than tandem duplications. The shifted age profile of these distant segmental duplicates raises the possibility that many of them are the result of translocated tandem duplicates. The differences in dS distributions between duplication types demonstrates that the mechanism of duplication plays a large role in determining whether and for how long a genome retains detectable duplicate gene copies.

Pairwise analysis of the selective constraint (dN/dS) on duplicate pairs in each class is less revealing. The distributions of pairwise dN/dS values for duplications of each type show little difference from one another. Since the pairwise dN/dS ratio is a measure of the average constraint on both duplicates, it is misleading when selective constraint differs greatly between the duplicate genes (they are asymmetric).

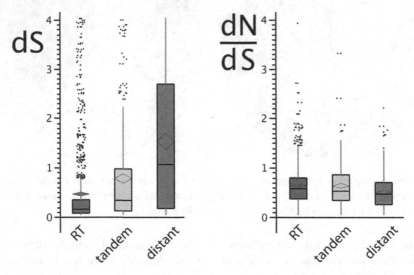

Fig. 2. Distributions of pairwise synonymous divergence (dS) and constraint (dN/dS) on duplicate pairs for each duplication type. Boxes represent interquartile range with the horizontal line being the median; diamonds span a 95% confidence interval around the mean assuming normality. The vertical lines span the extents of 95% of a normal distribution fit to the data.

3.3 Tandem Duplications Show Lower Asymmetry in Selective Constraint

While pairwise analysis of the duplicate genes gives insight to the probable age of the duplication event, it reveals little about the behavior of each individual copy relative to the ancestral state. In order to gain better insight to the evolution of members of a duplicate pair we identified an orthologous outgroup gene for each 1:2 duplication event, yielding 1,388 gene triplets (two paralogs and an ortholog). For each triplet we performed a maximum-likelihood analysis and used a significance value of 0.05 to reject the null model of symmetric evolution. Using this threshold for statistically significant asymmetry we found that tandem duplications are less frequently asymmetric in selective constraint (ω) than either RT gene pairs or distant segmental gene pairs ($p < 0.001$, Fisher's Exact Test) (Figure 3a). Note that ω is a branch-specific parameter analogous to the pairwise average dN/dS computed earlier; it is the ratio dN/dS along a particular branch.

In order to assess the degree of rate asymmetry in each duplication class we computed a measure of normalized rate asymmetry for each duplication event [23]. Retrotransposed and distant segmental duplications exhibit similar magnitudes of asymmetry in selective constraint (Figure 3b) while tandem duplications show significantly lower asymmetry in constraint ($p < 0.05$, Tukey-Kramer HSD Test). This trend parallels that of the asymmetry frequencies, which show that tandem duplicates are less likely to be evolving asymmetrically than the other duplicate types.

Within the retrotransposed class we wanted to confirm that it is the retrotransposed member of a gene pair that is more likely to show a relaxation of constraint. Those genes that are intronless and non-syntenic are labeled as retrogenes. This allowed determination of the retrogene in 1,043 out of the 1,589 RT events. These retrogenes are significantly more likely to exhibit higher levels of dN, dS and ω than their parental genes (Fisher's exact test, $p < 0.0001$), with dN and ω showing more of a difference. As a group the retrogenes have a median ω of 0.7 whereas their parental genes have a median ω of only 0.03. A value of omega near 0.7 implies neutral selection in the species studied [37], which is consistent with reports that most retroduplicates are processed pseudogenes [38].

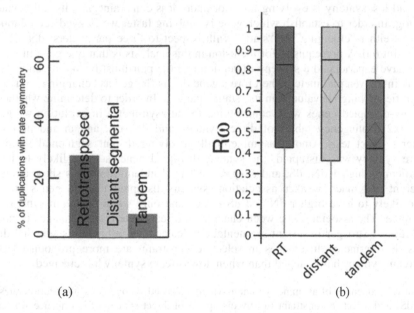

(a)   (b)

Fig. 3. (a) Proportion of 1:2 duplications showing statistically significant asymmetry in branch-specific selective constraint (ω) for each class. (b) Distribution of asymmetry values in branch-specific selective constraint (ω) for each class. Boxes represent interquartile ranges with diamonds spanning a 95% confidence interval around the mean assuming normality; vertical lines span 95% of a normal distribution fitted to the data. See methods for computation of constraint asymmetry, Rω.

3.4 Disruptions in Flanking Regions of Distant Segmental Duplicates Are Associated with Greater Asymmetry in dN and Relaxed Selective Constraint

The difference we observe in the fate of distant and tandem segmental duplicates led us to test the hypothesis that disruptions in the intergenic DNA surrounding duplicates would correspond to changes in the course of that duplicate's protein evolution.

We examined direct synteny in the 372 segmental duplications in the dataset (See Methods). Distant segmental duplicates are more likely to be evolving asymmetrically at the protein and DNA levels when one of the genes has lost direct synteny with the outgroup (Table 1). This association is significant regardless of whether the disruption occurred upstream or downstream of the duplicate. However, the probability of observing asymmetric constraint (ω), is only significantly higher for duplicates that experience upstream disruptions (p=0.021 vs. p=0.113). This suggests that changes in the 5' flanking DNA of a gene may have a greater impact on that gene's functional importance than changes in the 3' flanking DNA. Notably, none of these associations were observed in tandem duplications. It appears that tandem duplications produce paralogs which are apathetic to changes in direct flanking gene order. This may be due to the influence of parental regulatory elements on the entire tandem array [39].

Having observed that distant duplicates are more likely to evolve asymmetrically when one gene has lost direct synteny with the outgroup, we wanted to see if the gene that had lost synteny is evolving faster or under less constraint than its fully syntenic paralog. In order to establish which gene is evolving faster we flagged each duplicate gene as either "changing" or "static" with respect to three parameters: dN, dS, and constraint (ω). A prerequisite for inclusion in this analysis is that one gene in the pair must have experienced a synteny disruption and the pair must be significantly asymmetric in a given parameter. Therefore a gene that is flagged as "changing" must have a significantly higher value than its "static" paralog. In order to determine whether the synteny-disrupted genes were contributing to the asymmetry in evolution we generated 2x2 contingency tables of direct synteny and dN, dS, and ω and performed Fisher's Exact test. Among asymmetrically evolving distant segmental duplicates where synteny was disrupted, the synteny-disrupted gene is more likely to have a significantly higher dN, dS, and ω (p < 0.0001). Tandem duplicates show a slightly different and much weaker association: synteny-disrupted tandem genes are only more likely to have higher dN and dS not ω, and only when upstream synteny was disrupted. The association between changes in direct synteny and changes in protein evolution is stronger in distant segmental duplicates than in tandem segmental duplicates. Furthermore, differences in selective constraint are more pronounced when upstream synteny has changed than when downstream synteny has changed.

Table 1. Frequencies of asymmetry in non-synonymous and synonymous substitution rates (dN and dS) and selective constraint (ω) by disruption of direct synteny. Significance was established via Fisher's exact test: *: p<0.05, **: p<0.01, ***: p < 0.001.

Duplication type	N	Frequency of asymmetry		
		dN	dS	ω
Distant, 5' syntenic	22	50%	50%	9%
Distant, 5' disrupted	56	79%*	73%*	34%*
Distant, 3' syntenic	21	47%	38%	14%
Distant, 3' disrupted	65	78%**	80%***	31%^{ns}
Tandem, 5' syntenic	37	54%	43%	8%
Tandem, 5' disrupted	71	65%^{ns}	49%^{ns}	11%^{ns}
Tandem, 3' syntenic	35	54%	43%	9%
Tandem, 3' disrupted	78	68%^{ns}	51%^{ns}	12%^{ns}

4 Discussion

4.1 Prevalence of Duplication Mechanisms

Our observation of the strong contribution of retrotransposition to mammalian genomes is consistent with other recent observations of high rates of retrotransposition [36][40][41]. By some estimates, processed pseudogenes (resulting from retrotransposition events) outnumber non-processed pseudogenes (segmental duplicates) by a ratio of 7:1 in mouse and 3:1 in human [40]. While many of these retrotransposed duplicates may ultimately suffer inactivating mutations and devolve into pseudogenes, some of them are clearly under purifying selective constraint and have been incorporated into important aspects of mammalian biology [42][11]. Very recent evidence also indicates that retrocopies may give rise to functional siRNAs [12][13]. While the complete extent of the contribution of retrotransposed duplicates to new, functional, mammalian genes remains unknown, the sheer number of these events suggests careful consideration of this mechanism and its impact on mammalian evolution is warranted. The abundance of RT duplicates indicates that the majority of gene duplications produce paralogs that are functionally distinct at birth.

4.2 Pairwise dS Differs across Duplication Types

Several factors may be responsible for the large variation in dS distributions we find across the duplication types. Under the molecular clock assumption, the differences in dS could be explained simply in terms of duplicate age. The lower dS of retrotransposed duplicates may be due to recent bursts of retrotransposition in mammalian genomes combined with the affect of older retrogenes accumulating disablements and becoming processed pseudogenes at a greater rate than segmental duplicates. The difference in pairwise dS between distant segmental and tandem duplicates may also be explained by differences in age.

Several studies have noted that intra-chromosomal duplicates show lower sequence divergence than inter-chromosomal duplicates, lending support to the idea that distant duplicates may be the result of tandem duplication followed by chromosomal rearrangement [43] The homogenizing effect of gene conversion may contribute to this phenomenon. Gene conversion retards the divergence of tandem paralogous DNA sequences. Our observation of lower dS among tandem duplicates is consistent with gene conversion having a greater homogenizing effect on local duplicates than on distant ones [44]. However, our data indicates similar levels of average divergence across all duplication types – we observe that tandem duplicates are evolving at more similar rates to each other, but at higher rates than the static copies in RT and distant segmental duplicates.

Pairwise dN/dS distributions are similar across duplication mechanisms implying that the average selective constraint on pairs of duplicates is similar regardless of the mechanism that generated the duplicate pair. Hence from the perspective of pairwise dN/dS, an accelerating retrogene and its highly constrained parent appear similar to two tandem duplicates that are moderately constrained. If two proteins are allowed to evolve symmetrically at a greater rate, what are the functional implications? Assuming they are both functional, there are two possibilities: Either most protein mutations are functionally neutral, or separate proteins can complement each other via the sub-functionalization of protein domains. The relative contributions of neutrality and subfunctionalization to the symmetric evolution of duplicate proteins is unknown.

4.3 Duplicate Rate Asymmetry

We established that the retrogene in each asymmetric pair was almost always evolving more quickly than its parent. Only in 33 out of 292 cases did the reverse occur. It has been suggested that these represent rare cases where the retrogene has been placed into a regulatory context that renders it more important than the parental gene [23]. It is also possible that retrogenes are biased towards deposition in regulatory regions due to higher chromatin accessibility, thus having an impact on the expression of nearby genes. However, in all of these cases the parental gene also has significantly higher dS than the retrogene, implying that the apparent acceleration in parental protein evolution may simply be due to differences in local mutation rates.

While relatively infrequent in general, asymmetry in selective constraint (ω) is more frequently associated with relocated duplicates than with duplicates in the same genomic neighborhood, consistent with a study that found similar rates of nonsynonymous evolution of linked genes [45] and another that found increased constraint asymmetry in relocated rodent duplicates [23]. This raises an important question: Why do tandem genes evolve at more similar rates? Several studies have shown that genes that are local to one another are often co-expressed and share regulatory information [46][47][48]. Other studies have shown that functionally-related genes tend to be co-expressed [49][50]. Is the constraint similarity of local duplicates due to their functional similarity or shared proximal regulatory modules? We tried to resolve this question by looking at relocated duplicates in more detail. One of our criteria for distant segmental duplicates is that they must be next to two collinear orthologous genes or one on either side. Consequently, these duplicates belong to duplicated segments at least three genes in length and are likely to share some promoter sequences with their paralogs. We showed that disruptions directly adjacent to these relocated duplicates are associated with higher frequencies of rate asymmetry. Furthermore, asymmetry in constraint is slightly more probable in duplicates that have experienced disruptions of upstream synteny as opposed to downstream synteny. These disruptions may correspond to indels or rearrangements in the genes' promoter regions, abruptly altering their expression profiles. In contrast, the association is much weaker and not statistically significant for tandem duplicates – there, the buffering effect of a multi-gene regulatory context may provide robustness in the face of such disruptions. The similarity in constraint asymmetry between directly syntenic relocated duplicates and tandem duplicates suggests that loss of co-regulation may cause the evolution of distal protein sequences to become uncoupled.

References

1. Ohno, S.: Evolution by Gene Duplication. Springer, New York (1970)
2. Britten, R.J.: Divergence Between Samples of Chimpanzee and Human DNA Sequences Is 5%, Counting Indels. Proc. Natl. Acad. Sci. U S A 99, 13633–13635 (2002)

3. Fujiyama, A., Watanabe, H., Toyoda, A., Taylor, T.D., et al. (17 co-authors).: Construction and Analysis of a Human-chimpanzee Comparative Clone Map. Science 295, 131–134 (2002)
4. Fortna, A., Kim, Y., MacLaren, E., Marshall, K., et al. (16 co-authors).: Lineage-specific Gene Duplication and Loss in Human and Great Ape Evolution. PLoS Biol. 2, e207 (2004)
5. Bailey, J.A., Eichler, E.E.: Primate Segmental Duplications: Crucibles of Evolution, Diversity and Disease. Nat. Rev. Genet. 7(7), 552–564 (2006)
6. Demuth, J.P., De Bie, T., Stajich, J.E., Cristianini, N., Hahn, M.W.: The Evolution of Mammalian Gene Families. PLoS One 1, E85 (2006)
7. Panopoulou, G., Hennig, S., Groth, D., Krause, A., Poustka, A.J., Herwig, R., Vingron, M., Lehrach, H.: New Evidence for Genome-wide Duplications at the Origin of Vertebrates Using an Amphioxus Gene Set and Completed Animal Genomes. Genome Res. 13, 1056–1066 (2003)
8. Dehal, P., Boore, J.L.: Two Rounds of Whole Genome Duplication in the Ancestral Vertebrate. PLoS Biol. 3, e314 (2005)
9. Esnault, C., Maestre, J., Heidmann, T.: Human Line Retrotransposons Generate Processed Pseudogenes. Nat. Genet. 24, 363–367 (2000)
10. Harrison, P.M., Zheng, D., Zhang, Z., Carriero, N., Gerstein, M.: Transcribed Processed Pseudogenes in the Human Genome: an Intermediate Form of Expressed Retrosequence Lacking Protein-coding Ability. Nucleic Acids Res. 33, 2374–2383 (2005)
11. Vinckenbosch, N., Dupanloup, I., Kaessmann, H.: Evolutionary Fate of Retroposed Gene Copies in the Human Genome. Proc. Natl. Acad. Sci. U S A 103, 3220–3225 (2006)
12. Tam, O.H., Aravin, A.A., Stein, P., Girard, A., Murchison, E.P., Cheloufi, S., Hodges, E., Anger, M., Sachidanandam, R., Schultz, R.M., Hannon, G.J.: Pseudogene-derived Small Interfering RNAs Regulate Gene Expression in Mouse Oocytes. Nature 453, 534–538 (2008)
13. Watanabe, T., Totoki, Y., Toyoda, A., Kaneda, M., Kuramochi-Miyagawa, S., Obata, Y., Chiba, H., Kohara, Y., Kono, T., Nakano, T., Surani, M.A., Sakaki, Y., Sasaki, H.: Endogenous siRNAs from Naturally Formed dsRNAs Regulate Transcripts in Mouse Oocytes. Nature 453, 539–543 (2008)
14. Hughes, A.L.: The Evolution of Functionally Novel Proteins After Gene Duplication. Proc. Biol. Sci. 256, 119–124 (1994)
15. Force, A., Lynch, M., Pickett, F.B., Amores, A., Yan, Y.L., Postlethwait, J.: Preservation of Duplicate Genes by Complementary, Degenerate Mutations. Genetics 151, 1531–1545 (1999)
16. Serluca, F.C., Sidow, A., Mably, J.D., Fishman, M.C.: Partitioning of Tissue Expression Accompanies Multiple Duplications of the Na+/K+ ATPase Alpha Subunit Gene. Genome Res. 11, 1625–1631 (2001)
17. Adams, K.L., Cronn, R., Percifield, R., Wendel, J.F.: Genes Duplicated by Polyploidy Show Unequal Contributions to the Transcriptome and Organ-specific Reciprocal Silencing. Proc. Natl. Acad. Sci. U S A 100, 4649–4654 (2003)
18. Wagner, A.: Decoupled Evolution of Coding Region and Mrna Expression Patterns After Gene Duplication: Implications for the Neutralist-selectionist Debate. Proc. Natl. Acad. Sci. U S A 97, 6579–6584 (2000)
19. Gu, Z., Nicolae, D., Lu, H.H., Li, W.H.: Rapid Divergence in Expression Between Duplicate Genes Inferred from Microarray Data. Trends Genet. 18, 609–613 (2002)
20. Castillo-Davis, C.I., Hartl, D.L., Achaz, G.: Cis-regulatory and Protein Evolution in Orthologous and Duplicate Genes. Genome Res. 14, 1530–1536 (2004)

21. Conant, G.C., Wagner, A.: Asymmetric Sequence Divergence of Duplicate Genes. Genome Res. 13, 2052–2058 (2003)
22. Lynch, M., Katju, V.: The Altered Evolutionary Trajectories of Gene Duplicates. Trends Genet. 20, 544–549 (2004)
23. Cusack, B.P., Wolfe, K.H.: Not Born Equal: Increased Rate Asymmetry in Relocated and Retrotransposed Rodent Gene Duplicates. Mol. Biol. Evol. 24, 679–686 (2007)
24. Zhang, Z., Kishino, H.: Genomic Background Predicts the Fate of Duplicated Genes: Evidence from the Yeast Genome. Genetics 166, 1995–1999 (2004)
25. Birney, E., Andrews, D., Caccamo, M., Chen, Y., et al. (51 co-authors).: Ensembl 2006. Nucleic Acids Res. 34, d556–d561 (2006)
26. Enright, A.J., Dongen, S.V., Ouzounis, C.A.: An Efficient Algorithm for Large-scale Detection of Protein Families. Nucleic Acids Res. 30, 1575–1584 (2002)
27. Saitou, N., Nei, M.: The Neighbor-joining Method: a New Method for Reconstructing Phylogenetic Trees. Mol. Biol. Evol. 4, 406–425 (1987)
28. Thompson, J.D., Higgins, D.G., Gibson, T.J.: Clustal W: Improving the Sensitivity of Progressive Multiple Sequence Alignment Through Sequence Weighting, Position-specific Gap Penalties and Weight Matrix Choice. Nucleic Acids Res. 22, 4673–4680 (1994)
29. Poptsova, M.S., Gogarten, J.P.: Branchclust: A Phylogenetic Algorithm for Selecting Gene Families. Bmc Bioinformatics 8, 120 (2007)
30. Rogozin, I.B., Sverdlov, A.V., Babenko, V.N., Koonin, E.V.: Analysis of Evolution of Exon-intron Structure of Eukaryotic Genes. Brief Bioinform. 6, 118–134 (2005)
31. Edgar, R.C.: Muscle: a Multiple Sequence Alignment Method with Reduced Time and Space Complexity. Bmc Bioinformatics 5, 113 (2004)
32. Babenko, V.N., Rogozin, I.B., Mekhedov, S.L., Koonin, E.V.: Prevalence of Intron Gain Over Intron Loss in the Evolution of Paralogous Gene Families. Nucleic Acids Res. 32, 3724–3733 (2004)
33. Yang, Z., Nielsen, R., Goldman, N., Pedersen, A.M.: Codon-substitution Models for Heterogeneous Selection Pressure at Amino Acid Sites. Genetics 155, 431–449 (2000)
34. Pond, S.L.K., Frost, S.D.W., Muse, S.V.: Hyphy: Hypothesis Testing Using Phylogenies. Bioinformatics 21, 676–679 (2005)
35. Muse, S.V.: Estimating Synonymous and Nonsynonymous Substitution Rates. Mol. Biol. E 13, 105–114 (1996)
36. Marques, A.C., Dupanloup, I., Vinckenbosch, N., Reymond, A., Kaessmann, H.: Emergence of Young Human Genes After a Burst of Retroposition in Primates. Plos Biol. 3, e357 (2005)
37. Emerson, J.J., Kaessmann, H., Betran, E., Long, M.: Extensive Gene Traffic on the Mammalian X Chromosome. Science 303, 537–540 (2004)
38. Zhang, Z., Harrison, P.M., Liu, Y., Gerstein, M.: Millions of Years of Evolution Preserved: a Comprehensive Catalog of the Processed Pseudogenes in the Human Genome. Genome Res. 13, 2541–2558 (2003)
39. Mijalski, T., Harder, A., Halder, T., Kersten, M., Horsch, M., Strom, T.M., Liebscher, H.V., Lottspeich, F., Angelisde, M.H., Beckers, J.: Identification of Coexpressed Gene Clusters in a Comparative Analysis of Transcriptome and Proteome in Mouse Tissues. Proc. Natl. Acad. Sci. U S A 102, 8621–8626 (2005)
40. Zhang, Z., Carriero, N., Gerstein, M.: Comparative Analysis of Processed Pseudogenes in the Mouse and Human Genomes. Trends Genet. 20, 62–67 (2004)
41. Pan, Z., Zhang, L.: Quantifying the Major Mechanisms of Recent Gene Duplications in the Human and Mouse Genomes: a Novel Strategy to Estimate Gene Duplication Rates. Gen. Biol. 8, r158 (2007)

42. Bradley, J., Baltus, A., Skaletsky, H., Royce-Toll, M., Dewar, K., Page, D.C.: An X-to-autosome Retrogene Is Required for Spermatogenesis in Mice. Nat. Genet. 36, 872–876 (2004)
43. Friedman, R., Hughes, A.L.: The Temporal Distribution of Gene Duplication Events in a Set of Highly Conserved Human Gene Families. Mol. Biol. Evol. 20, 154–161 (2003)
44. Padhukasahasram, B., Marjoram, P., Nordborg, M.: Estimating the Rate of Gene Conversion on Human Chromosome 21. Am. J. Hum. Genet. 75, 386–397 (2004)
45. Williams, E.J., Hurst, L.D.: The Proteins of Linked Genes Evolve at Similar Rates. Nature 407, 900–903 (2000)
46. Lercher, M.J., Blumenthal, T., Hurst, L.D.: Coexpression of Neighboring Genes in Caenorhabditis Elegans Is Mostly Due to Operons and Duplicate Genes. Genome Res. 13, 238–243 (2003)
47. Cohen, B.A., Mitra, R.D., Hughes, J.D., Church, G.M.: A Computational Analysis of Whole-genome Expression Data Reveals Chromosomal Domains of Gene Expression. Nat. Genet. 26, 183–186 (2000)
48. Kikuta, H., Laplante, M., Navratilova, P., Komisarczuk, A.Z., et al. (22 co-authors).: Genomic Regulatory Blocks Encompass Multiple Neighboring Genes and Maintain Conserved Synteny in Vertebrates. Genome Res. 17, 545–555 (2007)
49. Lercher, M.J., Urrutia, A.O., Hurst, L.D.: Clustering of Housekeeping Genes Provides a Unified Model of Gene Order in the Human Genome. Nat. Genet. 31, 180–183 (2002)
50. Singer, G.A.C., Lloyd, A.T., Huminiecki, L.B., Wolfe, K.H.: Clusters of Co-expressed Genes in Mammalian Genomes Are Conserved by Natural Selection. Mol. Biol. Evol. 22, 767–775 (2005)

Estimating the Relative Contributions of New Genes from Retrotransposition and Segmental Duplication Events during Mammalian Evolution

Jin Jun[1], Paul Ryvkin[2], Edward Hemphill[3], Ion Măndoiu[1], and Craig Nelson[3]

[1] Computer Science & Engineering Department, University of Connecticut, Storrs,
CT, 06269, USA
{jinjun,ion}@engr.uconn.edu

[2] Genomics & Computational Biology Graduate Group, University of Pennsylvania,
Philadelphia, PA 19104, USA
pry@mail.med.upenn.edu

[3] Genetics & Genomics Program, Department of Molecular & Cell Biology,
University of Connecticut, Storrs, CT 06269, USA
{edward.hemphill_iii,craig.nelson}@uconn.edu

Abstract. Gene duplication has long been recognized as a major force in genome evolution and has recently been recognized as an important source of individual variation. For many years the origin of functional gene duplicates was assumed to be whole or partial genome duplication events, but recently retrotransposition has also been shown to contribute new functional protein coding genes and siRNA's. Here we present a method for the identification and classification of retrotransposed and segmentally duplicated genes and pseudogenes based on local synteny. Using the results of this approach we compare the rates of segmental duplication and retrotransposition in five mammalian genomes and estimate the rate of new functional protein coding gene formation by each mechanism. We find that retrotransposition occurs at a much higher and temporally more variable rate than segmental duplication, and gives rise to many more duplicated sequences over time. While the chance that retrotransposed copies become functional is much lower than that of their segmentally duplicated counterparts, the higher rate of retrotransposition events leads to nearly equal contributions of new genes by each mechanism.

1   Introduction

The impact of changes in gene copy number on both evolution and human health are under increasing scrutiny. While the creation of new genes and the modulation of gene copy-number via duplication has long been recognized as an important mechanism for the evolution of lineage-specific traits [14], a number of recent studies have suggested that variation in gene family size may be even more widespread than previously appreciated [7] and that gene copy number variation

C.E. Nelson and S. Vialette (Eds.): RECOMB-CG 2008, LNBI 5267, pp. 40–54, 2008.

between individuals may account for differences in disease predisposition within populations [18].

Three primary mechanisms of gene duplication have been described: whole genome duplication [9,31], segmental duplication [3,23], and retrotransposition [11,35]. Whole genome duplication has been important to the evolution of many lineages [31], but it is a relatively rare event. Unlike whole genome duplication events, segmental duplications occur continuously and have contributed significantly to the divergence of gene content between mammalian genomes. Duplication by retrotransposition also occurs quite frequently, but because these new retrotransposed gene copies lack the flanking regulatory material of the parental gene, they have long been believed to give rise primarily to non-functional pseudogenes [16,25]. Recent studies however, have indicated the presence of many apparently functional retrocopies in various mammalian genomes, challenging traditional perspectives on the relevance of this event to genome evolution [17,21,27,32]. Very recently retrotransposition has also been shown to contribute siRNA's [28,33].

In this study we compare the rates of new gene formation by segmental duplication (SD) and retrotransposition (RT) in five eutherian genomes. We show that, while genes arising from SD events are up to six times more likely to remain functional than those arising from RT events, the number of RT events is nearly ten times that of SD events, resulting in roughly equal quantitative contributions of new genes by each duplication mechanism. Our analysis further shows that duplicate genes generated by each mechanism are under similar levels of constraint on their protein coding regions and that silent site substitution profiles of RT duplicate copies are consistent with bursts of retrotransposition during mammalian evolution, while segmental duplication appears to occur at a more stable rate.

2 Methods

2.1 Dataset

Protein sequences for the five species analyzed (human, chimp, mouse, rat and dog) were obtained from Ensembl (release 37) [8]. For genes with multiple alternative transcripts we developed a collapsed gene model that incorporates all potential exons of that gene. Resulting exon coordinates were used to obtain a representative protein sequence used for subsequent homology assignment and dN/dS computations. Ensembl protein family annotations served as a starting point for our analysis. Over all five species, there were 17,341 Ensembl families comprising 113,543 genes. Excluding families with members on unassembled contigs (no reliable synteny information) and families with more than 50 Ensembl genes (due to the excessive computation time required to generate multiple alignments) resulted in 8,872 gene families containing 53,733 genes.

Pseudogenes were identified using Pseudopipe [36] seeded with known transcripts from Ensembl release 37. Over all five species, 17,226 pseudogenes (14,189 processed pseudogenes and 3,037 non-processed pseudogenes) were detected.

Each pseudogene was added to one of the 8,872 Ensembl gene families. This process resulted in super-families consisting of both protein coding genes and related pseudogenes.

2.2　Identification of RT and SD Events

Within each super-family a local synteny level was computed for all pairwise combinations of super-family members. Local synteny is defined as homology of upstream and downstream neighboring genes. For each pair, we checked homology between the 3 nearest up- and downstream neighboring Ensembl annotated genes. Homology between neighbors was defined by a BlastP [1] score of 50 or more and sequence similarity over 80% of corresponding protein sequences. After this analysis, for every pair (g_i, g_j) of family members we obtained two numbers $0 \le n_u^{ij}, n_d^{ij} \le 3$ representing the homology upstream and downstream neighbors. A synteny level $s_{i,j}$ of **2** was assigned to every pair of genes or pseudogenes that had homologous neighbors on both sides, up and down (i.e., whenever $n_u^{ij}, n_d^{ij} \ge 1$). When one side lacked homologous neighbors, we assigned a synteny level $s_{i,j}$ of **1** only if the other side had at least two homologous neighbors; otherwise (i.e., when $n_u^{ij} + n_d^{ij} \le 1$) we assigned a synteny level $s_{i,j}$ of **0**.

Local synteny levels were used in a two-stage clustering algorithm (see Algorithm 1) to identify syntenic ortholog/paralog clusters. In our algorithm, for a set X of genes and pseudogenes, $Sp(X)$ denotes the set of species represented in X. For a set S of species, $LCA(S)$ denotes the last common ancestor in the phylogenetic tree. In the first stage, we used a single-linkage clustering algorithm to obtain core clusters by merging pairs of genes and pseudogenes with local synteny level of 2, predicted to be either orthologs or paralogs resulting from SD events which preserve up and downstream neighbors. In the second stage, we merged pairs of core clusters if every member of one cluster had synteny level of 1 to every member of the other cluster. Any two non-overlapping clusters from this two-stage clustering algorithm are mutually non-syntenic. Second stage clusters spanning a phylogenetically contiguous subset of the species represented in larger clusters from the same super-family represent putative descendants of RT events or SD events that have lost local synteny. Since retrotransposed gene copies generally lack introns due to their RNA-intermediate nature, we distinguish between these possibilities using intron content conservation scores as described below.

Within each cluster produced by the two-stage clustering algorithm there may be successive segmental duplication events. We use UPGMA (Unweighted Pair Group Method with Arithmetic mean) [26] to find these successive SD events. For input to UPGMA we compute the distance between two members g_i and g_j as the Pearson's correlation coefficient between the two vectors, $(n_u^{ik} + n_d^{ik})_k$ and $(n_u^{jk} + n_d^{jk})_k$, i.e. sums of upstream and downstream homologous neighbors with remaining genes g_k in the cluster. Given the UPGMA gene trees, we counted the inner nodes as SD events when two subtrees from such an inner node are in a species-subset relationship. If two subtrees from an inner node had disjoint species sets, this node was considered as a speciation event (Fig. 1).

Algorithm 1. Two-Stage Clustering Algorithm

Input: Family of genes and pseudogenes $F = \{g_1, g_2, \ldots, g_N\}$ with species
   information and pairwise synteny levels $s_{i,j}$

Initialization:
   $C \leftarrow \emptyset$
   $U \leftarrow \{g_1, g_2, \ldots, g_N\}$
(Stage1) Single-linkage clustering with synteny level 2:
   While $U \neq \emptyset$ **do**
      Select an arbitrary member g_i of U
      $U \leftarrow U \setminus \{g_i\}; C_{open} \leftarrow \{g_i\}$
      While there exists $g_j \in U$ with synteny 2 to a member of C_{open}, **do**
         $U \leftarrow U \setminus \{g_i\}; C_{open} \leftarrow C_{open} \cup \{g_i\}$ // Add g_j to core cluster
      $C \leftarrow C \cup C_{open}$
(Stage2) Merging of clusters with high average pairwise synteny:
   While there is a (C_l, C_m) where SYNTENIC_TEST (C_l, C_m) is true, **do**
      $C \leftarrow C \setminus \{C_l, C_m\}$
      $C \leftarrow C \cup \{C_l \cup C_m\}$
Return C

SYNTENIC_TEST(A, B)
If $Sp(A)$ and $Sp(B)$ are subsets of different lineages, i.e.
   $LCA(Sp(A)) \neq LCA(Sp(A \cup B))$ and $LCA(Sp(B)) \neq LCA(Sp(A \cup B))$, **then**
   If $s_{i,j} = 1$ for every pair $g_i \in A$, $g_j \in B$ **then return true**
Else, if $LCA(Sp(A)) = LCA(Sp(A \cup B))$ **then**
   $A' \leftarrow$ set of genes/pseudogenes of A of species descending from $LCA(Sp(B))$
   If $s_{i,j} = 1$ for every pair $g_i \in A'$, $g_j \in B$ **then return true**
Else, return false

We distinguish between putative descendants of RT events or SD events that have lost local synteny using intron conservation scores between descendant genes and pseudogenes. The intron conservation rate between two paralogous genes was calculated as the ratio of the number of shared introns divided by the total number of intron positions from the protein/intron alignment between two genes (based upon the method of [20]. An event was identified as an RT duplication if the average intron conservation rate to paralogs outside the cluster was below 1/3.

2.3   Event Assignment to Tree Branches and Evidence of Function

We use parsimony to assign each inferred duplication event to a specific branch of the 5-species tree. We assign each event to the tree branch corresponding to the exact set of species spanned by the descendant genes of the detected duplication event, which we refer to as *assigned* events. *Intact* events are defined as those duplication events that have no apparent disruption (e.g. in stop codons) of the protein coding reading frame and an Ensembl annotated gene in each of the

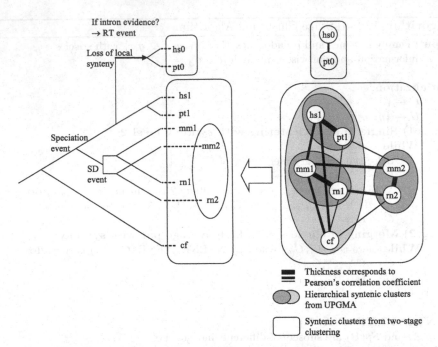

Fig. 1. Inferring SD and RT events using local synteny and hierarchical clustering. This example shows how SD and RT events are inferred from a super-family having 9 members: 2 members per each species except for dog, from the results of our clustering algorithms (on the right side) to corresponding events (on the left side). By using two-stage clustering algorithm, two syntenic clusters are formed, shown as hollow rounded rectangles. Loss of introns in one cluster suggests that the loss of synteny was due to an RT event. UPGMA builds hierarchical clusters within each syntenic cluster and speciation and SD events are inferred based on species sets.

species spanned by the cluster. *Functional* events are defined by the clusters of putative protein coding genes with average dN/dS ratio below 0.5 over all pairs of genes within the cluster. Pairwise dN and dS measures were estimated using the YN00 program of PAML [34].

3   Results

3.1   Lineage Distribution of Duplication Events

Events giving rise to clusters of genes with no conservation of synteny relative to "parental" genes and low inter-cluster intron conservation rates were classified as *RT events*, while events giving rise to clusters of genes with high local synteny to parental genes were classified as *SD events*. Events corresponding to gene clusters with indeterminate intron conservation or local synteny to parental genes were classified as *ambiguous*. This analysis resulted in the classification of

	3 internal branches	Whole tree
SD **functional** / assigned events	**148** / 301 = 49.17%	1,649
RT **functional** / assigned events	**187** / 2,349 = 7.96%	12,078

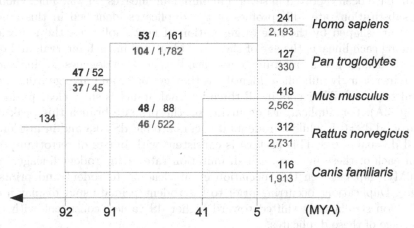

Fig. 2. Numbers of gene duplication events from segmental duplication (above the line) and retrotransposition (below the line). Numbers represent the assigned SD or RT events on each branch. Numbers typeset in bold on three internal branches are counts of functional events, defined in this study as intact events that yield clusters with average dN/dS ratio below 0.5 over pairs of homologous Ensembl genes. For three internal branches, fractions of the functional events over the total assigned events are shown, e.g. **53**/161 for SD events on primate branch. Evolutionary ages are based on [30].

a total of 2,035 SD events, 12,507 RT events, and 2,742 ambiguous events. Using parsimony to assign non-ambiguous events to branches of the species tree resulted in 52 SD and 45 RT events on the branch leading to primates and rodents (the in-group), 161 SD and 1,782 RT events on the primate branch leading to humans and chimps, and 88 SD and 522 RT events on the rodent branch leading to mice and rats (Fig. 2). Gene duplication events for the root and terminal branches of the tree were also counted, but were not used for further analysis due to the difficulty in estimating the degree of purifying selection on very recent duplication on the terminal branches and the age of duplications on the root. 386 SD and 429 RT events could not be reliably assigned to specific branches of the tree using parsimony and were also omitted from further analysis.

Duplication event counts on the three internal branches of the tree reveal an excess of RT events over SD events along all but the deepest branches of the tree, suggesting an average rate of RT copy formation 3-10 times higher than that of SD copy formation (Fig. 2). Deviation from this ratio along the in-group branch may be the result of a period of relative inactivity of retrotransposition compounded with the difficulty of detecting the products of old RT events not under purifying selective pressure [11].

3.2 Rates of Duplication

Rates of retrotransposition vary significantly over time and bursts of retrotransposition have been reported in several mammalian lineages [11,35]. The synonymous substitution rate (dS) profiles of the duplicates identified in this study (Fig. 3) are shaped by the rate of generation of new duplicates, the mutation rates along each lineage, the age of the genes identified in each interval, and our ability to identify genes uniformly along each lineage. Pseudogenes, for instance, become increasingly difficult to identify as they get older and diverge from their original sequence. RT events in all three internal branches show clear peaks in dS (Fig. 3A). For duplications occurring on the primate branch this peak occurs around dS=0.1, while in rodents it occurs around dS=0.3 and in in-groups around dS=0.6 ~ 0.8. This pattern is consistent with bursts of retrotransposition in each of these lineages, a high mutation rate in the rodent lineage, and the 36Myr gap between the speciation events leading to rodent and primate lineages. Duplications occurring prior to the rodent/primate split display a dS distribution significantly shifted toward higher dS values, consistent with the greater age of these duplicates.

Segmental duplications show similar patterns in dS but a more uniform distribution of dS values than RT duplicates (Fig. 3B and C), suggesting that segmental duplication is a more uniform process that occurs at less variable rates than retrotransposition. It is interesting to note that the inferred age distribution of segmental duplication events is more uniform than that of the RT

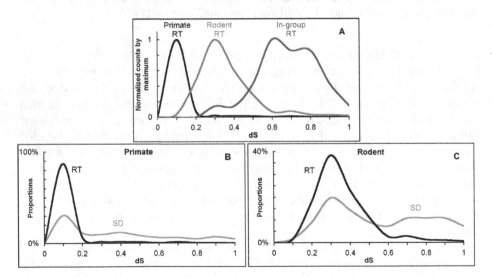

Fig. 3. Histograms of average dS over pairs of Ensembl genes and pseudogenes. (A) for clusters resulting from RT events on the primate, rodent, and the in-group branch leading to primates and rodents, (B) for clusters resulting from SD events and RT events on the primate lineages and (C) on the rodent lineages.

duplicates but is not perfectly flat, suggesting that there may be some variation in the rate of segmental duplication over evolutionary time.

3.3   Functional Preservation Rates

It is probable that young duplicate genes may escape inactivation for some time despite lacking any apparent function. Since Ensembl gene predictions rely upon the presence of an intact coding region rather than any evidence of selection pressure upon the sequence, the gene clusters resulting from intact duplication events should be comprised of both functional genes and duplicates that are not functional, but have escaped inactivation. Evidence of purifying selection is often used as evidence for function, and the ratio of synonymous to non-synonymous changes (dN/dS) in the protein-coding region of a gene is a convenient way of estimating this selective pressure [13]. For example, dN/dS ratio < 0.5 has been used as stringent functionality criteria between retrotransposed genes and their parental genes [6]. Also Torrents et al. showed that there is a clear discrimination between dN/dS ratios of pseudogenes and those of functional genes, supporting the use of dN/dS ratios as evidence of function [29]. Here we compute dN/dS ratios between all pairs of descendants from each duplication event. This pairwise approach is computationally rapid, is independent of precise reconstruction of the entire gene tree, and allows for the detection of functionalized descendant clusters of a duplication event that are not constrained relative to the parental genes.

Analysis of the dN/dS ratios of clusters derived from duplication events is quite revealing. Fig. 4A compares clusters of RT duplication event descendants with intact protein coding reading frames (intact) and clusters of RT duplicates with inactivated reading frames (inactivated). Aggregate dN/dS values of a significant portion of intact clusters overlap with the dN/dS values of inactivated clusters in the region of the graph where dN/dS is greater than ∼ 0.5. Assuming that the vast majority of inactivated clusters (clusters whose members have inactivating mutations in their protein coding regions) are not under purifying selection for protein coding function, those intact clusters that fall into this range are unlikely to encode functional proteins, despite lacking any clearly inactivating mutation. By inference, those clusters that display significantly lower aggregate dN/dS values (< 0.5) than inactivated clusters are likely to be under stabilizing selection for protein coding function.

Panels B through D of Fig. 4 compare dN/dS values of duplicate clusters derived from RT and SD events on each of the three internal branches of the mammalian tree. In the oldest internal branch of the tree (in-group) very few clusters generated by either duplication mechanism can be detected that are not under some degree of purifying selection pressure. This is probably due to the difficulty in identifying very old non-functional sequences. Such sequences are expected to drift away from their parental sequence making identification increasingly difficult with advanced age. Clusters derived from duplication events along the rodent branch have a bimodal distribution of dN/dS ratio resulting from RT and SD events that gave rise to putatively functional gene copies (aggregate dN/dS values < 0.5), and clusters with no clear evidence of stabilizing

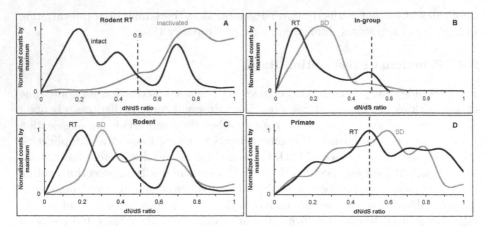

Fig. 4. (A) Histograms of average dN/dS ratio over pairs of Ensembl genes for clusters resulting from intact RT events and average dN/dS ratio over pairs of genes and pseudogenes for clusters resulting from inactivated RT events on the rodent lineage. Histograms of average dN/dS ratio over pairs of Ensembl genes for clusters resulting from intact SD events and RT events on the (B) in-group branch leading to primates and rodents, (C) rodent, and (D) primate.

selective pressure. Duplication events along the primate branch gave rise to clusters with more uniformly distributed aggregate dN/dS values spanning the entire range of measurements. This is likely to be a reflection of the relatively short period of time these new genes have been under purifying selection and is consistent with the relatively low dS values of duplicates detected along this branch (Fig. 3B).

3.4   Distribution of Duplication Events within the Mammalian Tree

The total number of RT and SD duplication events detected in this study is illustrated in Fig. 2. Along each branch the number of events giving rise to clusters with evidence of purifying selective pressure on their protein coding regions is in bold typeset, while the total number of events detected is in denominators. From these numbers it is clear that that we detect far more RT events than SD events, but that far fewer of these events give rise to functional protein coding genes than their SD counterparts. Analysis of the internal branches individually reveals possible differences in the relative probability of these events giving rise to functional genes in different lineages. In the most basal branch shared by rodents and primates there is a slight excess of functional SD events over functional RT events, while the two mechanisms appear to contribute equal numbers of functional events in the rodent lineage. The primate and rodent branches show similar rates of assigned SD events, but in primates fewer of these events give rise to functional descendants (Table 1). A decreased rate of functionalization is apparent in the RT events on the primate lineage. Despite an RT event rate nearly twice that seen in rodents, the number of functional RT events in primates is only ∼ 25% greater than that in rodents.

Table 1. Rates of duplication events for rodent and primate lineages.

Events per million yrs	SD events			RT evnets		
	Assigned	Intact	Functional	Assigned	Intact	Functional
Rodents	1.76	1.56	0.96	10.4	1.42	0.92
Primates	1.87	1.31	0.62	20.7	3.41	1.21

4 Discussion

4.1 Identification and Characterization of Gene Duplications During Mammalian Evolution

Identifying gene duplication events and placing them in a phylogenetic framework depends upon sensitive identification of duplicate copies, reliable clustering of orthologs, and differentiating between lineage specific gene loss events and more recent duplications. To identify groups of duplicated sequences we combine Ensembl gene predictions with Pseudopipe pseudogene identification. Combining predicted genes and pseudogenes in our gene families significantly reduces the complexity of placing duplication events on the phylogenetic tree; gene loss events are represented by pseudogenes and need not be inferred. Of course, this approach is less effective as pseudogenes age and become more difficult to detect deep in the tree. Undetected gene loss events deeper in the tree may lead to mis-assignment of some duplication events to younger branches and a consequent underestimation of the age of these gene families. But using local synteny to help classify duplication events appears to work relatively well for the species analyzed in this study.

Once duplicated genes have been identified and assigned to large gene families, clusters of orthologs within those families must be constructed to infer the time of the duplication event that gave rise to each cluster. Our clustering algorithm uses both the protein-coding information embedded in the Ensembl gene family assignments, and the local genome structure surrounding duplicate copies, to differentiate between DNA and RNA based duplications and to order successive segmental duplication events. This method is effective because random insertion of a retrocopied cDNA into the genome is very unlikely to recreate any significant synteny with orthologs or paralogs (data not shown). The very low false-positive rate associated with measures of local synteny means that genes that share synteny with paralogs are almost certainly the result of segmental duplications regardless of intron content. Therefore this method is unlikely to misclassify RT duplicates as segmental duplications. Segmental duplications, however, can lose synteny to their paralogs over time [10,19], which may result in some segmental duplications being mis-assigned to the RT class. To account for this we use conventional intron content criteria [32] to further discriminate between non-syntenic DNA based duplications and RT duplicate copies. Duplicate pairs that maintain synteny with their paralog are most likely DNA based, while non-syntenic paralogs with significant intron loss are likely RT duplicates. Comparison with other studies identifying RT duplicates in mammalian genomes suggest that using synteny criteria in addition to intron based criteria improves the reliability of RT duplicate classification and that duplications characterized as RT duplicates on the basis of intron content alone may in fact be SD duplicates.

While the gradual degradation of synteny can create problems for placing duplication events on a phylogenetic tree, it conversely enables the differentiation of successive segmental duplication events. Gene families generated by rounds of segmental duplication can be difficult to classify into definitive orthologous groups using protein-coding sequences alone. By examining flanking gene content, however, orthologous groups of paralogs can often be clearly resolved and iterative DNA based duplications placed on the phylogenetic tree. As a result we can see that while synteny decays over time, dN/dS values may also decrease, reflecting the prolonged influence of stabilizing selection.

Detection of duplicate genes will always depend strongly on the depth and quality of genome annotation. This fact is reflected in our results in the highest number of duplicates detected in the two most well annotated genomes in the study, human and mouse. While it is difficult to predict how many duplicates have been missed in current genome annotations, estimates of duplication rates from the most well-annotated genomes are now judged to be quite accurate [4,22,32]. The consistency of these estimates across the tree suggests that the number of duplications events is not highly variable between these species, but definitive demonstration of that finding must await further annotation (see also [5]).

4.2 Rates of Duplication

Lineage specific gene duplication, by retrotransposition or segmental duplication, is a major force in the evolution of differences between genomes. Thousands of new genes have been born over the course of mammalian evolution, and while not all of these new genes live, they provide significant quantities of raw material for species-specific evolution and account for many of the known differences between closely related mammalian genomes [5]. Retrotransposition, in particular, appears to be peppering the genome with large numbers of duplicate retrocopies that can act as insertional mutagens [12], new duplicate genes [32], and siRNA's [28,33]. Analysis of retrotransposon activity during vertebrate evolution shows strong peaks of activity [15] and it is therefore not surprising that RT duplication of genes shows similar peaks in birth rates. Segmental duplications, however, are not expected to be dependant on retrotransposon duplication machinery and appear to occur at a more stable rate. Consistent with these expectations, the age profiles of the segmental duplications identified in our study are more broadly distributed than the RT age profiles, but interestingly, they are not perfectly uniform over time and may indicate of bursts of segmental duplication activity in the evolutionary history of these genomes (see also [2,24]).

4.3   The Fate of Newly Duplicated Genes

At the moment a newly duplicated gene is born it is presumed to be an exact copy of the duplicated portion of the parental gene (cDNA for retrocopies;and introns, exons, and flanking material for segmental duplicates). Over time, however, mutation, coupled with selection, leads to the divergence of the new copy's sequence from its parent/paralog. The progressive aging of a duplicate is revealed in its dS profile, as we move deeper on the tree, dS values between duplicate pairs become progressively larger, reflecting the age of the duplications. If a new duplicate is functional, purifying selection will serve to remove deleterious non-synonymous mutations from the population, and the ratio of non-synonymous to synonymous changes (dN/dS) will diverge from that of non-functional copies. Full resolution of the degree of purifying selective pressure however, takes time, and estimating this pressure on young duplicates can be difficult. Indeed, we find significant separation between putative functional and non-functional de-scendants of a duplication event in populations of genes that have had sufficient time for this difference to become apparent (see the rodent branch Fig. 4A and C). For the young primate branch the divergence between functional and non-functional descendants is less clear. At virtually all time-points, however, there are duplicates that have not yet been inactivated, but also show no evidence of purifying selection on their protein coding sequence. Whether this is the result of copies evading inactivation simply due to chance, or the reflection of some other phenomenon is unknown. We also observe the converse phenomenon, old copies that appear to have dN/dS ratios consistent with purifying selection, but inactivating mutations in their protein-coding region. This could be the result of recent inactivating mutations after long periods of purifying selection, or the result of purifying selection acting on fragments of the original protein coding sequence.

While the general effects of time, mutation, and selective pressure discussed above apply to all new duplicates, we wondered if RT duplicates and SD du-plicates would show different degrees of purifying selective pressure. Interest-ingly, in age-matched populations of segmental and retrotransposed duplicates, there is no dramatic difference in selection pressure on genes born by these two mechanisms (Fig. 4). What is most clearly different between these two populations is the *proportion* of copies that show evidence of purifying selec-tive pressure. Of the duplication events assigned to the branches leading to primates and rodents, only about six percent (150/2,304) of RT events give rise to duplicates showing evidence of purifying selection, while forty percent (101/249) of SD events appear to generate functional descendants (Fig. 2). The very high rate of RT events coupled with the very low rate of functionaliza-tion of gene copies generated by these events, and the lower rate of SD events with much higher rate of descendant gene functionalization, results in nearly equal contributions of new genes to eutherian genomes by each of these two mechanisms.

References

1. Altschul, S.F., Gish, W., Miller, W., Myers, E.W., Lipman, D.J.: Basic local alignment search tool. J. Mol. Biol. 215(3), 403–410 (1990)
2. Bailey, J.A., Eichler, E.E.: Primate segmental duplications: crucibles of evolution, diversity and disease. Nat. Rev. Genet. 7(7), 552–564 (2006)
3. Bailey, J.A., Gu, Z., Clark, R.A., Reinert, K., Samonte, R.V., Schwartz, S., Adams, M.D., Myers, E.W., Li, P.W., Eichler, E.E.: Recent segmental duplications in the human genome. Science 297(5583), 1003–1007 (2002)
4. International Human Genome Sequencing Consortium. Finishing the euchromatic sequence of the human genome. Nature 431(7011), 931–945 (2004)
5. Demuth, J.P., De Bie, T., Stajich, J.E., Cristianini, N., Hahn, M.W.: The evolution of mammalian gene families. PLoS ONE 1, e85 (2006)
6. Emerson, J.J., Kaessmann, H., Betran, E., Long, M.: Extensive gene traffic on the mammalian x chromosome. Science 303(5657), 537–540 (2004)
7. Fortna, A., Kim, Y., MacLaren, E., Marshall, K., Hahn, G., Meltesen, L., Brenton, M., Hink, R., Burgers, S., Hernandez-Boussard, T., Karimpour-Fard, A., Glueck, D., McGavran, L., Berry, R., Pollack, J., Sikela, J.M.: Lineage-specific gene duplication and loss in human and great ape evolution. PLoS Biol. 2(7), E207 (2004)
8. Hubbard, T., Andrews, D., Caccamo, M., Cameron, G., Chen, Y., Clamp, M., Clarke, L., Coates, G., Cox, T., Cunningham, F., Curwen, V., Cutts, T., Down, T., Durbin, R., Fernandez-Suarez, X.M., Gilbert, J., Hammond, M., Herrero, J., Hotz, H., Howe, K., Iyer, V., Jekosch, K., Kahari, A., Kasprzyk, A., Keefe, D., Keenan, S., Kokocinsci, F., London, D., Longden, I., McVicker, G., Melsopp, C., Meidl, P., Potter, S., Proctor, G., Rae, M., Rios, D., Schuster, M., Searle, S., Severin, J., Slater, G., Smedley, D., Smith, J., Spooner, W., Stabenau, A., Stalker, J., Storey, R., Trevanion, S., Ureta-Vidal, A., Vogel, J., White, S., Woodwark, C., Birney, E.: Ensembl 2005. Nucleic Acids Res. 33, 447–453 (2005)
9. Hurley, I., Hale, M.E., Prince, V.E.: Duplication events and the evolution of segmental identity. Evol. Dev. 7(6), 556–567 (2005)
10. Huynen, M.A., Bork, P.: Measuring genome evolution. Proc. Natl. Acad. Sci. USA 95(11), 5849–5856 (1998)
11. Marques, A.C., Dupanloup, I., Vinckenbosch, N., Reymond, A., Kaessmann, H.: Emergence of young human genes after a burst of retroposition in primates. PLoS Biol. 3(11), e357 (2005)
12. Mills, R.E., Bennett, E.A., Iskow, R.C., Devine, S.E.: Which transposable elements are active in the human genome? Trends Genet. 23(4), 183–191 (2007)
13. Nekrutenko, A., Makova, K.D., Li, W.H.: The k(a)/k(s) ratio test for assessing the protein-coding potential of genomic regions: an empirical and simulation study. Genome Res. 12(1), 198–202 (2002)
14. Ohno, S.: Evolution by gene duplication. Allen and Unwin, London (1970)
15. Ohshima, K., Hattori, M., Yada, T., Gojobori, T., Sakaki, Y., Okada, N.: Whole-genome screening indicates a possible burst of formation of processed pseudogenes and alu repeats by particular l1 subfamilies in ancestral primates. Genome Biol. 4(11), R74 (2003)
16. Petrov, D.A., Hartl, D.L.: Patterns of nucleotide substitution in drosophila and mammalian genomes. Proc. Natl. Acad. Sci. USA 96(4), 1475–1479 (1999)
17. Potrzebowski, L., Vinckenbosch, N., Marques, A.C., Chalme, F., Jègou, B., Kaessmann, H.: Chromosomal gene movements reflect the recent origin and biology of therian sex chromosomes. PLoS Biol. 6(4), e80 (2008)

18. Redon, R., Ishikawa, S., Fitch, K.R., Feuk, L., Perry, G.H., Andrews, T.D., Fiegler, H., Shapero, M.H., Carson, A.R., Chen, W., Cho, E.K., Dallaire, S., Freeman, J.L., Gonzalez, J.R., Gratacos, M., Huang, J., Kalaitzopoulos, D., Komura, D., MacDonald, J.R., Marshall, C.R., Mei, R., Montgomery, L., Nishimura, K., Okamura, K., Shen, F., Somerville, M.J., Tchinda, J., Valsesia, A., Woodwark, C., Yang, F., Zhang, J., Zerjal, T., Armengol, L., Conrad, D.F., Estivill, X., Tyler-Smith, C., Carter, N.P., Aburatani, H., Lee, C., Jones, K.W., Scherer, S.W., Hurles, M.E.: Global variation in copy number in the human genome. Nature 444(7118), 444–454 (2006)

19. Rocha, E.P.: Inference and analysis of the relative stability of bacterial chromosomes. Mol. Biol. Evol. 23(3), 513–522 (2006)

20. Rogozin, I.B., Wolf, Y.I., Sorokin, A.V., Mirkin, B.G., Koonin, E.V.: Remarkable interkingdom conservation of intron positions and massive, lineage-specific intron loss and gain in eukaryotic evolution. Curr. Biol. 13(17), 1512–1517 (2003)

21. Sakai, H., Koyanagi, K.O., Imanishi, T., Itoh, T., Gojobori, T.: Frequent emergence and functional resurrection of processed pseudogenes in the human and mouse genomes. Gene. 389(2), 196–203 (2007)

22. She, X., Cheng, Z., Zollner, S., Church, D.M., Eichler, E.E.: Mouse segmental duplication and copy number variation. Nat. Genet. (2008)

23. She, X., Jiang, Z., Clark, R.A., Liu, G., Cheng, Z., Tuzun, E., Church, D.M., Sutton, G., Halpern, A.L., Eichler, E.E.: Shotgun sequence assembly and recent segmental duplications within the human genome. Nature 431(7011), 927–930 (2004)

24. She, X., Liu, G., Ventura, M., Zhao, S., Misceo, D., Roberto, R., Cardone, M.F., Rocchi, M., Green, E.D., Archidiacano, N., Eichler, E.E.: A preliminary comparative analysis of primate segmental duplications shows elevated substitution rates and a great-ape expansion of intrachromosomal duplications. Genome Res. 16(5), 576–583 (2006)

25. Shemesh, R., Novik, A., Edelheit, S., Sorek, R.: Genomic fossils as a snapshot of the human transcriptome. Proc. Natl. Acad. Sci. USA 103(5), 1364–1369 (2006)

26. Sneath, P.H.A., Sokal, R.R.: Numerical Taxonomy. W.H. Freeman and Company, San Francisco (1973)

27. Svensson, O., Arvestad, L., Lagergren, J.: Genome-wide survey for biologically functional pseudogenes. PLoS Comput. Biol. 2(5), e46 (2006)

28. Tam, O.H., Aravin, A.A., Stein, P., Girard, A., Murchison, E.P., Cheloufi, S., Hodges, E., Anger, M., Sachidanandam, R., Schultz, R.M., Hannon, G.J.: Pseudogene-derived small interfering RNAs regulate gene expression in mouse oocytes. Nature 453(7194), 534–538 (2008)

29. Torrents, D., Suyama, M., Zdobnov, E., Bork, P.: A genome-wide survey of human pseudogenes. Genome Res. 13(12), 2559–2567 (2003)

30. Ureta-Vidal, A., Ettwiller, L., Birney, E.: Comparative genomics: genome-wide analysis in metazoan eukaryotes. Nat. Rev. Genet. 4(4), 251–262 (2003)

31. Van de Peer, Y., Taylor, J.S., Meyer, A.: Are all fishes ancient polyploids? J. Struct. Funct. Genomics 3(1-4), 65–73 (2003)

32. Vinckenbosch, N., Dupanloup, I., Kaessmann, H.: Evolutionary fate of retroposed gene copies in the human genome. Proc. Natl. Acad. Sci. USA 103(9), 3220–3225 (2006)

33. Watanabe, T., Totoki, Y., Toyoda, A., Kaneda, M., Kuramochi-Miyagawa, S., Obata, Y., Chiba, H., Kohara, Y., Kono, T., Nakano, T., Surani, M.A., Sakaki, Y., Sasaki, H.: Endogenous siRNAs from naturally formed dsRNAs regulate transcripts in mouse oocytes. Nature 453(7194), 539–543 (2008)
34. Yang, Z.: Paml: a program package for phylogenetic analysis by maximum likelihood. Comput. Appl. Biosci. 13(5), 555–556 (1997)
35. Zhang, Z., Carriero, N., Gerstein, M.: Comparative analysis of processed pseudogenes in the mouse and human genomes. Trends Genet. 20(2), 62–67 (2004)
36. Zhang, Z., Carriero, N., Zheng, D., Karro, J., Harrison, P.M., Gerstein, M.: Pseudopipe: an automated pseudogene identification pipeline. Bioinformatics 22(12), 1437–1439 (2006)

Discovering Local Patterns of Co-evolution

Yifat Felder[1],[**] and Tamir Tuller[1,2,*,**]

[1] School of Computer Science
Tel Aviv University
{felderyi,tamirtul}@post.tau.ac.il
[2] Department of Molecular Microbiology and Biotechnology
Tel Aviv University

Abstract. Co-evolution is the process in which a set of orthologs exhibits a similar or correlative pattern of evolution. Co-evolution is a powerful way to learn about the functional interdependencies between sets of genes and cellular functions, about their complementary and backup relations, and more generally, for answering fundamental questions about the evolution of biological systems.

Orthologs that exhibit strong signal of co-evolution in part of the evolutionary tree may show mild signal of co-evolution in other parts of the tree. The major reasons for this phenomenon are noise in the biological input, genes that gain or lose functions, and the fact that some measures of co-evolution relate to rare events such as positive evolution. Previous works in the field dealt with the problem of finding sets of genes that co-evolved along an entire underlying phylogenetic tree, without considering the fact that often co-evolution is local.

In this work, we describe a new set of biological problems that are related to finding patterns of local co-evolution. We discuss their computational complexity and design algorithms for solving them. These algorithms outperform other bi-clustering methods as they are designed specifically for solving the set of problems mentioned above. We use our approach to trace the co-evolution of fungal and Eukaryotic genes at a high resolution across the different parts of the corresponding phylogenetic trees. Our analysis shows that local co-evolution is a wide-scale phenomenon.

Keywords: Co-evolution, evolution rates, gene copy number, gene deletion and duplication, functional ontology, bi-clustering, systems biology.

1 Introduction

Co-evolution is the process by which a set of orthologs exhibits a similar or a correlative pattern of evolution. Co-evolution can be measured in various ways, the most common are similarity in absolute evolutionary rate (ER[1]) [12, 17, 20],

---

[*] Corresponding author.
[**] Y.F. and T.T contributed equally to this work.
[1] In this work we used dN/dS as an estimate of evolutionary rate.

C.E. Nelson and S. Vialette (Eds.): RECOMB-CG 2008, LNBI 5267, pp. 55–71, 2008.
© Springer-Verlag Berlin Heidelberg 2008

correlative ER [9, 14], and similarity in the pattern of protein presence in the proteomes of a set of organisms [16, 18, 27].

Detecting co-evolving sets of orthologs is an important matter since physically interacting proteins [9, 14] and functionally related proteins [7, 8, 17, 20] tend to co-evolve. Thus, an appropriate analysis of co-evolving genes can lead to a better understanding of the evolution of various cellular processes and gene modules (*e.g.* see [15]).

The most famous approach for detecting co-evolution is the phylogenetic profiles [16, 18, 27]. It searches groups of orthologs with similar phyletic patterns. The main disadvantage of this approach is the fact that it totally ignores the topology of the organisms' evolutionary tree. A similar measure is the Propensity for Gene Loss (PGL) in evolution [7, 8, 11]. Genes with lower PGL have lower ER and tend to be essential for the organism viability. It has been proven recently [8] that orthologs with correlative PGL tend to be functionally related.

Another related measure for evolutionary distance is the difference between the average ER of pairs of orthologs [12, 17, 20]. Using this measure Marino *et al.* showed that there is a strong connection between gene evolutionary rates and function [17].

All previous approaches for detecting co-evolution have not considered the fact that gene modules can exhibit strong patterns of co-evolution in some parts of the evolutionary tree while exhibiting a very weak signal of co-evolution in other periods of their evolution. There are a few main reasons for this phenomenon.

First, evolving genes may gain or lose functions (see *e.g.* [10]); loss or gain of a new function can move an orthologous from one co-evolving module to another one. Second, the analyzed biological data may be noisy or partial in some portions of an evolutionary tree while it can have higher quality in other parts. In such cases, searching sets of orthologs with similar evolution along the *entire* phylogenetic tree may result in high false negative rates. Third, there are co-evolutionary problems that are local by definition. For example, genes tend to undergo positive selection in a small fraction of their history (see *e.g.* [2]). Thus, if we define co-evolution as a process in which a set of orthologs undergo positive selection together, we should not expect that such type of co-evolution will span over the entire phylogenetic tree.

The goal of this work is to study the *Local Co-Evolutionary* problem. Namely, given a phylogenetic tree and an evolutionary pattern of orthologous sets along the evolutionary tree we aim to find sub-sets of orthologs with similar evolutionary patterns along subtrees of the evolutionary tree (see Figure 1 *C.*). We formalize a new set of *Local Co-Evolutionary* problems, study their computational hardness and describe algorithms and heuristics for solving them. Simulation study show that these algorithms give much better performances than popular bi-clustering algorithms for gene expression. Finally, we generate three relevant biological datasets which include ER and gene Copy Number (CN) of thousands of orthologs across evolutionary trees with dozens of nodes, and use our method for analyzing them.

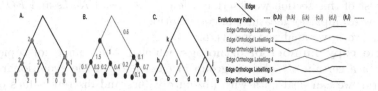

Fig. 1. *A.* A hypothetic example of a node orthologous labelling which includes gene copy number in each node of the evolutionary tree. *B.* A hypothetic example of an edge orthologous labelling which includes ER along each edge of the evolutionary tree. *C.* The goal of the local co-evolutionary problem is to find large sets of orthologs that have similar pattern of evolution across large subtrees of the evolutionary tree.

In each of the three datasets, we found hundreds of orthologous sets that exhibit local co-evolution. Large fractions of these sets were functionally enriched and fitted our knowledge regarding the evolution of the studied organisms. However, as our approach suggests a new set of tools for analyzing co-evolution, the resolution and the abundance of our analysis are significantly higher than the results reported in previous studies.

2   Definitions and Preliminaries

Let $T = (V, E)$ be a tree, where V and E are the tree *nodes* and tree *edges* respectively. In this work we consider rooted binary phylogenetic trees (*i.e.* the degree of each node in the tree is 1, 2, or 3), and all the trees described in this work are species trees. A node of degree 1 is named a *leaf*, a node with degree 3 is named an *internal node*, and the root has degree 2. A tree T' is a subtree of T if it is a connected subgraph in T. We denote such a relation by $T' \subseteq T$. Note that by the above definition an internal node of a tree T can be a leaf in the subtree $T' \subseteq T$.

A *Node Orthologous Labelling* (*NOL*) of a tree T, is a set of labelling (real numbers) for each of the nodes in T, an *Edge Orthologous Labelling* (*EOL*) for a tree T, is a set of labellings for each of the edges in T (see Figure 1). An Orthologous Labelling (*OL, i.e.* a *NOL* or *EOL*) of a tree is named the evolutionary pattern along the tree.

Let S denote a set of *OL*s in T, and let S' be a subset of S. Let $D_c(S', T')$ denote a measure for co-evolution along a subtree, $T' \subseteq T$. Such measures return a real positive number which reflects how similar is the co-evolution of the *OL*s from S' along the subtree T' (0 reflects an identical evolution). Formally, we deal with versions of the following problem:

Problem 1. Local Co-Evolution
Input: A phylogenetic tree, $T = (V, E)$, a set of *NOL*s or *EOL*s, $S = [S_1, .., S_m]$, three natural numbers, n', m', a real number, d, and a measure of co-evolution, $D_c(\cdot, \cdot)$.

Question: Is there a subtree $T' = (V', E') \subseteq T$ with $|E'| = n'$, and a subset $S' \subseteq S$ with $|S'| = m'$, such that $D_c(S', T') \leq d$?

In the rest of this section we describe a few examples of *NOLs* and *EOLs*, and give a few examples of measures of co-evolution.

In this work, we analyzed one *NOL*:

(1) **Gene copy number of orthologs**, which is the number of copies of a gene from a certain orthologous group in each node of the evolutionary tree. In general, we can deal both with absolute values and discrete values of gene copy numbers. In the discrete case, we are only interested whether a certain orthologous appears or not in each node of the evolutionary tree and not in the number of times it appears, while in the absolute value we do consider the number of times each orthologous appears in each node of the evolutionary tree.

We also analyzed two *EOLs*:

(1) **Evolutionary rate (ER)**. In this work we used the non-synonymous substitution rate, dN, divided by the synonymous substitution rate, dS (*i.e.* dN/dS) as an estimator of the ER. We examined absolute, discrete, and relative values of dN/dS. The absolute case is dN/dS (a positive real number) without additional processing. In the discrete case, we only consider three possibilities: ER > 1 (positive selection, $dN/dS > 1$), ER ≈ 1 (neutral selection, $dN/dS \approx 1$), or ER < 1 (purifying selection, $dN/dS < 1$). In the relative case, we perform an additional normalization of the ERs of each orthologous group by comparing them to the ERs of other orthologous groups. This is done by computing for each edge of the tree the rank of the ER of an orthologous group among the ERs of all orthologous group.

(2) **Change in orthologous gene Copy Numbers (CN)** along the tree edges. In this case, we can check the exact changes or only the direction of the changes (*i.e.* if the copy number increases, decreases, or does not change along an edge).

We analyzed the following measures of co-evolution (Figure 2)[2]:

(1) $\mathbf{D_{c1}}(\mathbf{S}' = [\mathbf{S_1'}, \mathbf{S_2'}, .., \mathbf{S_f'}], \mathbf{T}')$ is the maximal L_1 norm between all the pairs of $S_1', S_2', .., S_f'$ along the evolutionary subtree T'. D_{c1} measures the similarity of the absolute values of the *OLs* (see Figure 2 A.). Thus, orthologs that have similar ERs along each branch of T' will have a significantly low D_{c1}.

(2) $\mathbf{D_{c2}}(\mathbf{S}' = [\mathbf{S_1'}, \mathbf{S_2'}, .., \mathbf{S_f'}], \mathbf{T}') = 1 - |r|$, where r denotes the minimal Spearman correlation among all pairs of the *OL* of $S_1', S_2', .., S_f'$ along the edges or nodes of T'. Orthologs can differ in their average ER but exhibit similar fluctuations in their ER (see Figure 2 B.). D_{c1} can not discover such pattern of co-evolution but D_{c2}, as it finds sets of orthologs with correlative pattern of evolution, is suitable for this task.

(3) $\mathbf{D_{c3}}(\mathbf{S}' = [\mathbf{S_1'}, \mathbf{S_2'}, .., \mathbf{S_f'}], \mathbf{T}' = (\mathbf{E}', \mathbf{V}')) =$ $|E'| - |\{e \in E' : (S_{1,e}' = \ell) \wedge (S_{2,e}' = \ell) \wedge .. \wedge (S_{f,e}' = \ell)\}|$ where ℓ is a certain labelling. This measure is used for finding a large subtree and a set of orthologs with identical labelling along most of this subtree (see Figure 2 C.).

---

[2] We usually give examples that are related to ERs but with the appropriate changes all the measures can be implemented on *NOLs* and on labellings that are related to CNs.

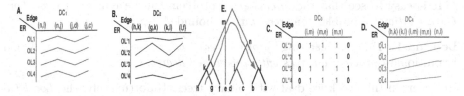

Fig. 2. Illustration of the four measures of co-evolution ($A.\ D_{c1}$, $B.\ D_{c2}$, $C.\ D_{c3}$ and $D.\ D_{c4}$) along a hypothetical evolutionary tree ($E.$).

In this work, we used this measure for finding subtree where a set of orthologs undergo positive selection (*i.e.* $ER > 1$) together. To this end, we first performed a two levels discretization of the ERs; one discrete level was assigned to the ERs above 1 and the second discrete level was assigned to the ERs below 1.

(4) $\mathbf{D_{c4}(S', T')}$: In the case of $D_{c4}(S', T')$, we want to find a path along the evolutionary tree (*i.e.* T' is a path), and a set of OLs, S', that have similar monotonic/non-monotonic decreasing/increasing evolutionary pattern along the path (see Figure 2 $D.$). $D_{c4}(S', T') = d$ denotes that the maximal number of components of an orthologous labelling, $S_i \in S'$ that should be changed for fitting it to the pattern that the path induces, is less than d. This measure can be useful for discovering modules of orthologs that exhibit together acceleration or deceleration in their ER due to speciation.

3 Hardness Issues

This section deals with hardness issues which are related to the *Local Co-Evolutionary* problem. We show that some versions of the problem are NP-hard, but in practice it seems that the *Local Co-Evolutionary* problem has a shorter running time than the *bi-clustering* problem which is highly used in the context of gene expression analysis. Furthermore, we show that there are versions of *Local Co-Evolutionary* problem that have a fixed-parameter tractable (FPT) or that are polynomial. Due to lack of space, most of the proofs in this section are deferred to the full version of this paper.

A *bi-clustering* is a subset of genes and a subset of conditions with the property that the selected genes are co-expressed (according to some measure of co-expression) in the selected conditions. Some versions of the *bi-clustering* problem are NP-hard (see for example [5, 6]). Let $D_b(S_g, S_c)$ denote a measure of co-expression of a set of genes, S_g, across a set of conditions, S_c. Formally, the *bi-clustering* problem is defined as follows:

Problem 2. Bi-clustering
Input: A set M of k vectors of length ℓ (each vector is related to one gene, and each component of the vectors is related to one condition), a measure of co-expression, $D_b(\cdot, \cdot)$, two natural numbers, p, q, and positive real number c.
Question: Does the input include a set of p genes, S_g, and a set of q conditions, S_c, such that $D_b(S_g, S_c) < c$?

It is easy to see that the *bi-clustering* problem can be reduced to the *Local Co-Evolutionary* problem on trees with unbounded degree.

Lemma 1. *The* bi-clustering *problem can be reduced to the* Local Co-Evolutionary *problem on trees with unbounded degree.*

However, in this work we deal with binary trees. Unfortunately, the *Local Co-Evolutionary* problem is NP-hard also for binary trees. We prove it by reduction from the *Balanced Complete Bipartite Subgraph* problem that is known to be NP-Complete [21]:

Problem 3. Balanced Complete Bipartite Subgraph
Input: A bipartite graph, $G = (V_1, V_2, E)$, and a positive integer k.
Question: Are there two disjoint subset $V_1' \subseteq V_1, V_2' \subseteq V_2$ such that $|V_1'| = |V_2'| = k$ and for all $v_1 \in V_1', v_2 \in V_2', (v_1, v_2) \in E$.

Theorem 1. *The* Local Co-Evolution *on binary trees is NP-hard.*

Proof. We will prove theorem 1 by reduction from the *Balanced Complete Bipartite Subgraph* problem.

Given an input $< G = (V_1, V_2, E), k >$ to the *Balanced Complete Bipartite Subgraph* problem we will generate the following input to the *Local Co-Evolution* problem, $< T, S, n', m', d, D_c >$:

- T is a tree with $|V_1| + |V_2|$ edges, an edge for each $v \in V_1 \cup V_2$.
- $S = [S_1, .., S_{|V_1|+|V_2|}]$ includes $|V_1| + |V_2|$ vectors of length $|V_1| + |V_2|$ (*i.e.* $\forall_i |S_i| = |V_1| + |V_2|$). A vector for each $v \in V_1 \cup V_2$, where $\forall_{i,j} S_{i,j} = \delta((i, j) \in E)$ (*i.e.* $|S|$ is the distance matrix of G).
- $n' = |V_1| + |V_2|$, $m' = k$, $d = |V_1| + |V_2| - k$.
- As a distance measure we use $D_c = D_{c3}$.

$\Longrightarrow$ Suppose G contains a balanced complete bipartite subgraph of size k, (V_1', V_2') (*i.e.* the answer to the *Balanced Complete Bipartite Subgraph* problem with $< G, k >$ is YES). By definition each pair of vertices in the V_1' share k neighbors (the vertices in V_2'), thus, all the k vectors of evolutionary patterns that are related to the V_1' contain k common '1's (the vertices that are in V_2'), *i.e.* $d = |V_1| + |V_2| - k$, and thus the answer to the *Local Co-Evolution* problem is YES.

$\Longleftarrow$ Suppose the answer to the *Local Co-Evolution* problem with $< T, S, n' = |V_1| + |V_2|, m' = k, d = |V_1| + |V_2| - k, D_{c3} >$ is YES. This means that there is a set of k edge orthologous labellings that share k positions with '1' along the tree T, thus G includes two sets of k vertices each, such that all the vertices in one set (that is related to the k edge orthologous labellings) are connected to all the vertices in the second set (that is related to the k positions), *i.e.* balanced complete bipartite subgraph of size k.

Practically, due to the following lemma, it seems that the running time of the co-evolutionary problem should be shorter than the running time of the bi-clustering problem.

Lemma 2. *[22] The number of subtrees in a tree with n nodes is about 1.48^n.*

For example, suppose that we are dealing with n conditions in the case of the bi-clustering problem or tree of size n in the case of the *Local Co-Evolutionary* problem. The number of subgroups of n conditions is 2^n, while, by lemma 2, the number of connected subtrees of a tree with n nodes is only about 1.48^n. In practice, this can make a big difference, for example, if $n = 20$ there are less than $2,542$ connected subtrees while more than 10^6 subgroups of n conditions.

Similarly, considering a greedy procedure, there are at most n' ways to expand by one edge/node a subtree of size n', while in the general bi-clustering case, there are $n - n'$ such possibilities. This can make a big difference if $n' = o(1)$.

Finally, there are versions of the co-evolutionary problem that are polynomial. For example, if we search exact co-evolution along paths (*i.e.* D_{c4}), or more generally along subtrees with k leaves[3] or k nodes. In this case, if k is constant, the number of such subtrees is polynomial.

Lemma 3. *The number of subtrees with k leaves in a tree with n nodes is less than $\binom{n}{k}$.*

Lemma 4. *The number of subtrees with k nodes in a tree with n nodes is less than $(n-1) \cdot 2^k \cdot (k-1)!$.*

We say that an evolutionary pattern along a subtree T' is *supported* by a set of orthologous labellings, S', if there is $s \in S'$ with that evolutionary pattern along T'. Suppose that each component of an OL can have one of α possibilities. In this case, the number of possible evolutionary patterns that are supported by at least one of the OLs, in a tree with n nodes and when the input includes $|S|$ OLs is less than $min(|S| \cdot 1.48^n, \sum_{k=1}^{n}(n-1) \cdot (k-1)! \cdot \alpha^k)$. The left component, is due to the fact that each of the 1.48^n subtrees can have maximum of S different evolutionary patterns, since each of the $|S|$ labelling induces only one pattern. The right components is an enumeration of all the possible patterns on all possible subtrees with k nodes ($0 < k \le n$).

Thus, the problem of finding sets of OLs with the same pattern of evolution along a subtrees of an evolutionary tree is a fixed-parameter tractable (FPT), which is exponential in the size of the tree, n. If we are interested in paths, *i.e.* subtrees with $k = 2$ leaves, this problem is polynomial since by lemma 3 there are $O(n^2)$ paths, where each path can have at most $|S|$ evolutionary pattern, a total of $|S| \cdot n^2$ evolutionary patterns.

4   Methods

4.1   Heuristics and Algorithms

This section includes a brief description of the algorithms we developed for finding local patterns of co-evolution. We designed two main algorithms. The first

---

[3] Note that subtree with k leaves can includes $O(n)$ nodes. One such example is when the input tree is a path of length $n/2$ (two leaves, and $n - 2$ internal nodes), and $n/2 - 2$ additional leaves that are connected to each of these $n - 2$ internal nodes of the path, and when we seek co-evolution along paths (i.e. $k = 2$).

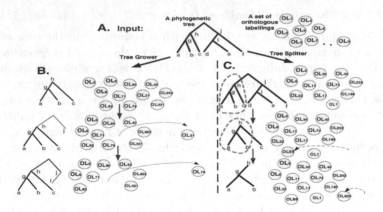

Fig. 3. An illustration of the two algorithms. *A*. The input. *B*. The *Tree Grower* algorithm. *C*. The *Tree Splitter* algorithm.

algorithm, *Tree Grower* starts with set of orthologs with similar patterns of evolution along small subtrees, and expands this initial trees while possibly decreasing the set of orthologs (Figure 3 *B*.). The second algorithm, *Tree Splitter*, first finds sets of orthologs with similar pattern of evolution along the entire input tree, and recursively cuts edges from the initial tree while possibly increasing the sets of orthologs (Figure 3 *C*.). Detailed description of the two algorithms will appear in the full version of this paper.

The Tree Grower Algorithm. The first stage of the *Tree Grower* algorithm included generating a collection of sets of *OL*s (seeds) that have a high co-evolutionary score along a small subtree (*e.g.* a subtree with around $\log(n)$ nodes or edges). The set of seeds was generated by the FPT procedure that we described in the previous sections, or by implementing K-means [23] on the *OL*s that are induced along each of the small subtrees. Next, the *Tree Grower* procedure greedily grew solutions with larger subtrees that may have less *OL*s than in the initial seeds.

Let $f_c(|E|, |S|)$ denote the running time for computing $D_c(S, T)$. In the most general case, the running time of the *Tree Grower* on an input tree $T = (V, E)$, a set of *OL*s, S, and initial set of seeds of size $|H|$ is $O((|E| + |S|) \cdot |S| \cdot |E| \cdot |H| \cdot f_c(|E|, |S|))$.

The Tree Splitter Algorithm. In this case, by the FPT procedure and by K-means we first generated a set of clusters of *OL*s along the entire input phylogenetic tree. Next, in each stage the *Tree Splitter* algorithm cut edges from the subtree that was related to each cluster while greedily increasing the size of the set of *OL*s that is related to the cluster.

Let K denote the initial number of clusters, the running time of *Tree Splitter* is $|K| \cdot |S| \cdot |E| \cdot f_c(|E|, |S|)$. The *Tree Splitter* algorithm is usually faster than the *Tree Grower*.

4.2   P-Values and GO Enrichments

P-Values: upper bound, lower bound, and exact empirical p-values.
As the statistical nature of the problems mentioned in this work is not clear we
used few empirical p-values for evaluating co-evolving sets of OLs .

Empirical p-values for a co-evolving set of m' OLs over subtrees of size n',
when the input includes m OLs along a tree of size n, was computed by the
following permutation test: 1) Generate 1000 permutated versions of the input,
each permutated versions is a result of $O(n \cdot m)$ single permutations of the OLs
of the original input. 2) Implement the algorithms for finding co-evolving set on
these random inputs. 3) Compute the fraction of times the algorithms found a
co-evolving set with larger properties (m' and n') than the original one.

A raw but much faster empirical bounds on the p-value can be computed
based on the empirical estimation of the probability that two OLs have distance
score (co-evolutionary score) less than d on a tree with similar properties as T'
(*e.g.* the same topology or the same number of edges/nodes). Let p denote this
empirical probability. Thus, since in a set of m' OLs there are at least $m'/2$
independent pairs and no more than $\binom{m'}{2}$ pairs of OLs, the upper and lower
bounds on the p-value are $p^{m'/2}$ and $p^{\binom{m'}{2}}$ respectively.

GO-enrichment. GO enrichment of the co-evolving sets were computed using
the GO ontology of *S. cerevisiae* (downloaded from the Saccharomyces genome
database[4]) and *H. Sapiens* (downloaded from EBI - BioMart[5]). We used the
algorithm of Grossmann *et al.* [19] for detecting over-represented GO terms.

4.3   Implementation

The software for the algorithms (*Tree Grower* and *Tree Splitter*) was written in
C++, and the implementation run on regular PCs (Pentium M, 1400MHz with
512MB of RAM, and with Windows XP).

4.4   The Biological Inputs

We analyzed three biological datasets: 1) ER of $1,372$ orthologous sets ($12,348$
genes) along the phylogenetic tree of nine yeasts (Figure 4 *A*.); we named this
dataset the *yeast ER* dataset. 2) Gene copy number of $6,227$ orthologous sets
($56,043$ genes) along the same phylogenetic tree of the nine yeasts (Figure 4
A.); we named this dataset the *yeast CN* dataset. 3) gene copy number of $4,851$
orthologous sets ($33,957$ genes) along the phylogenetic tree of seven eukaryotes
(Figure 4 *B*.); we named this dataset the *eukaryote* dataset.

The preparation of these biological inputs included dozens of steps which are
depicted in Figure 5. In the case of the evolutionary rates datasets (Figure 5 A.),
the major stages included identifying the phylogenetic tree, generating sets of

---

[4] http://www.yeastgenome.org/
[5] http://www.biomart.org/

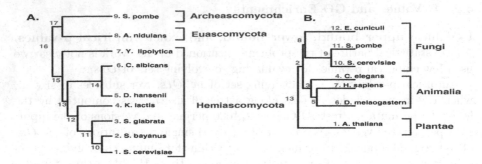

Fig. 4. The phylogenetic trees of the analyzed biological datasets: *A*. The yeast dataset, *B*. The eukaryote dataset.

orthologs without paralogs, aligning these sets, using maximum likelihood for reconstructing the ancestral genes of these orthologs (the sequences at the internal nodes of the phylogenetic tree), and using these orthologs and ancestral genes for computing the dN/dS values along each branch of the phylogentic tree. These dN/dS can be used as $EOLs$ or can undergo normalization or descritization (as we described in section 2).

In the case of gene copy numbers datasets (Figure 5 B.), the initial steps were similar. However, in this case we used only the sets of orthologs that exhibit at

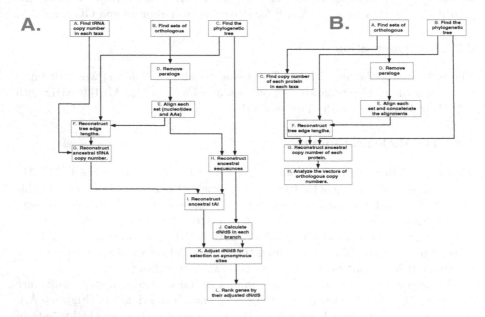

Fig. 5. The various steps of the preprocessing of the biological inputs. *A*. The yeast ER, *B*. The copy number datasets.

least one change in their corresponding gene copy number along the phylogenetic tree. By maximum likelihood, we reconstructed the ancestral copy number for each of these sets. The result of these steps is a set $NOLs$ that can be further processed or can be translated to $EOLs$ (as we described in section 2).

Due to lack of space, the exact details regarding the preparation of the biological inputs are deferred to the full version of this paper.

5   Experimental Results

5.1   Synthetical Inputs

For evaluating the performances of our algorithms we designed the following simulation:

1) We generated random trees with 12 - 52 nodes by random hierarchical clustering of the trees' leaves, and generated random sets of 1000 - 3000 OLs that are related to these trees. The labellings were sampled from the uniform distribution $U[0,3]$.

2) In these random inputs, we "planted" solutions, which are OLs (with 100 - 300 orthologs) that have high co-evolutionary score in large subtree (*e.g.* 5 - 20 nodes) of the input tree. We added additive noise with uniform distribution $U[-0.15, 0.15]$ to each component of the "planted" solutions.

3) We implement the two algorithms, *Tree Grower* and *Tree Splitter*, on these inputs.

4) We compared the performances of the algorithms to two popular bi-clustering algorithms (SAMBA [6] and the algorithm of Cheng and Church ($C\&C$) [3]). To this end, we used two measures of performances: False Positive (FP) rate, which is the fraction of orthologs (OFP) or tree branches (BFP) in the output that are not part of a "planted" solution, and False Negative (FN) rate, which is the fraction of orthologs (OFN) or tree branches (BFN) in the "planted" solution that do not appear in output.

Figure 6 includes a summary of the simulation study. As can be seen, the performances of our algorithms are very good and far exceed the performances of the competing bi-clustering algorithms. For example, when considering *all* the synthetical inputs, the average OFN, OFP, BFN, and BFP of the *Tree Splitter* are 0.002, 0.25, 0.07, and 0.14 respectively. For comparison, the average OFN, OFP, BFN, and BFP of the algorithm of $C\&C$ are 0.52, 0.76, 0.16, and 0.61 respectively. This result motivates designing algorithms that are specific for solving the *co-evolutionary* problem, instead of using general bi-clustering algorithms.

Finally, our simulation showed that there are many inputs where *Tree Splitter* outperform the *Tree Grower* algorithm. However, there are cases where the *Tree Grower* gave better results. Thus, we employed both algorithms in the biological analysis.

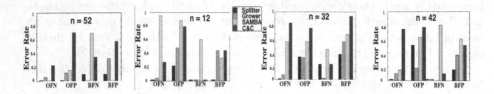

Fig. 6. Simulation study of the algorithms. The figure depicts the average OFP, OFN, BFP, and BFN of the two algorithms (the *Tree Grower* and the *Tree Splitter*), and two bi-clustering algorithms (SAMBA and *C&C*) for different sizes of input trees (n is the number of nodes in the input trees). For each size of input trees we averaged the error rates of 100 simulations.

5.2 Biological Inputs: Results and Discussion

In this section, we describe our main biological findings. The full lists of all the co-evolving sets that were found along with their local co-evolutionary patterns, and their functional enrichments will appear in the full version of the paper. A summary of the results appears in Figure 7A. The fact that 10% - 56% of the co-evolving sets that we found are functionally enriched is very encouraging, as it demonstrates that our measures and algorithms are capable of detecting real biological phenomena.

The biological datasets describe the evolution of diverse sets of organisms and *OLs*, along different time ranges (see Figure 7 B.). The Eukaryote dataset includes both multicellular and unicellular organisms and describes evolution along 1642 million years. The yeasts are unicellular organisms that appeared 837 million years ago (see [1, 11] for the divergence times of the different phylogenetic groups). The yeast ER dataset includes conserve *OLs* that have exactly one ortholog in each organism while the yeast CN dataset includes *OL* with varying number of ortholog in each organism (see section 4.4). Our analysis shows that

A.

Measure Dataset	Dc₁	Dc₂	Dc₃	Dc₄
Yeast ER	258 (87)	1394 (139)	12	58 (14)
Yeast CN, EOL	382 (106)	236 (71)	---	83 (50)
Yeast CN, NOL	190 (56)	239 (63)	---	94 (59)
Eukaryote CN, EOL	114 (56)	176 (70)	--	34 (14)
Eukaryote CN, NOL	94 (45)	170 (74)	--	32 (18)

B.

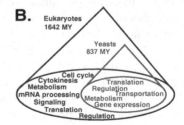

Fig. 7. A. Summary of the biological results. The number of local co-evolving groups and the number of enriched co-evolving groups (in brackets) that were found in each of the biological datasets according to each of the co-evolution measures. B. A global view at the co-evolving functions (GO groups) in the yeast and the Eukaryote datasets, and the appearance time of each of the analyzed biological groups.

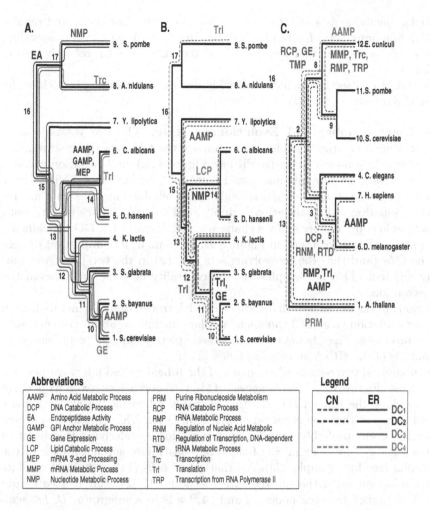

Fig. 8. A detailed description of the co-evolving sets of *OLs*, in the three biological datasets, that were enriched with metabolic and regulatory GO functions: *A*. Yeast ER, *B*. Yeast *CN*, *C*. Eukaryote *CN*. We marked regions in the trees where we detected co-evolving sets of *OLs* that are enriched with the aforementioned GO functions (see the abbreviation list in the middle of the figure). We used different colors (see the legend above) to distinguish between the different types of co-evolution. Dashed lines correspond to *CN* based co-evolution (*EOL* or *NOL*), and continues lines correspond to ER based co-evolution (*EOL*).

there are cellular processes, such as metabolism and regulation, that exhibit co-evolution in all the datasets. Figure 7 B. depicts the enriched GO functions that were found in the co-evolving sets of *OLs* in each of the datasets.

Figure 8 includes an intensive view on the co-evolution of the biological processes that are related to metabolism and regulation in the three biological datasets. The figure depicts the regions in the evolutionary trees where we detected co-evolving sets of OLs that are enriched with metabolic and regulatory GO functions. This figure also includes information on the corresponding measures of co-evolution that were used for detecting each of the co-evolving sets of OLs.

The rest of this section includes a few highlights of our finding in each of the biological datasets.

Yeast Copy Number and Evolutionary Rate. The two yeast datasets are interesting since they enable us to compare the two types of co-evolution: co-evolution via similar/correlative ER (see Figure 8 A.), and evolution via similar/correlative gene copy number (see Figure 8 B.). Many metabolic cellular functions (*e.g.* metabolism of amino acids), and cellular functions that are related to regulation (*e.g.* translation) exhibit local co-evolutionary pattern both via changes in copy number and via changes in ER. Though the GO enrichments that appear in Figure 8 A. and in Figure 8 B. are similar, it is important to note that the OLs (and thus the co-evolving sets of OLs) in the two cases are completely different. This fact emphasizes the centrality of these processes in the yeast evolution.

Previous works in the field that have dealt with translation and metabolism in the yeast evolution (*e.g.* [24]) came into similar conclusions about the centrality of these processes. They, however, have used completely different techniques (*e.g.* the analysis of the tRNA adaptation index [24]).

We discovered two regions where many of the fungal genes underwent positive selection (see Figure 8 A.). The larger set of OLs (554 orthologs) exhibits positive selection along the branch (11, 12), this probably following the whole genome duplication event that has occurred at this bifurcation [28]. This whole genome duplication event probably served as a driving force underlying this burst of positive selection, by relaxing the functional constraints acting on each of the gene copies (see for example [25]). Another set of OLs (112 orthologs) exhibits positive selection along the subtree with the nodes 13, 14, and 15 (see Figure 8 A.). The branch between nodes 13 and 14, leads to a subgroup (*D. hansenii* and *C. albicans*) that evolved a modified version of the genetic code [4], and the branch between nodes 13 and 15 leads to *Y. lipolytica* (which is a sole member in one of the three taxonomical clusters of the *Saccharomycotina* [26]).

Eukaryote Copy Number. As mentioned, this biological dataset gives a wider evolutionary view than the yeasts' datasets. Cellular processes that are related to metabolism, signaling, and mRNA processing exhibit co-evolution patterns along this dataset (see Figures 7 B. and 8). One striking phenomenon is that many of these co-evolving sets (87%) exhibited co-evolution (according to all the measures of co-evolution) along the subtrees of the *Animalia* and *Fungi*, and excluding the subtree of the *Plantae*. We believe that this phenomenon proves that many gene modules changed their functionality after the split between the *Plantae* and the two other groups (*Animalia* and *Fungi*).

Cases where homolog protein complexes in *Plantae* and *Animalia* have a rather different functions in these two organism groups were reported in the past. For example, the *COP9 signalosome*, a repressor of photomorphogenesis in *Plantae*, regulates many developmental processes in *Animalia* [13]. Our analysis, however, may suggest that this is a wide scale phenomenon.

Co-Evolution of Cellular Functions. The functional enrichments of the co-evolving *OL*s can teach us about functional interdependencies between cellular functions and about the co-evolution of cellular functions. We found many sub-trees where sets of *OL*s that are enriched with various GO functions exhibited co-evolution. In many cases the relations between the different GO functions seemed trivial. For example, *Translation* and *Gene expression*, that exhibited a copy number based co-evolution in the yeasts (Figure 8 *B*.) subtree that is under internal node 12, are two biological processes that relate to producing proteins or RNAs from the corresponding genes (DNA sequences).

However, there are more intriguing cases. For example, *Translation* and *Amino acid metabolic process* that exhibited a copy number based co-evolution in the Eukaryotes (Figure 8 *C*.) in the subtree that included nodes 1, 2, 3, 4, 5, and 8. The link between these two processes is not immediate and is probably not direct. A possible explanation is that the evolution of the metabolism of various Amino Acids (*AA*) changed the composition of the *AA* pool in the yeast cell. These changes were followed by a corresponding evolution of the translation machinery (*e.g.* the yeast tRNA copy numbers).

6   Conclusions

This work introduced a set of local co-evolutionary problems. As some of these problems are NP-hard, we suggested two heuristics for solving them. We demonstrated the biological significance of the local co-evolutionary problems through the analysis of three biological datasets. We found that more than 90% of the co-evolving sets of *OL*s that we found indeed exhibit local co-evolution (*i.e.* co-evolution along *part* of the phylogenetic tree). This fact confirms that our approach is desirable.

As a future work we intend to extend this work in three directions. First, in this work we described two heuristics for solving co-evolutionary problems, these heuristics gave very encouraging results in the simulation study. However, as we believe that better algorithms are within reach. We plan to spend more time designing faster and more accurate algorithms for solving these problems. A related open problem is to find approximation algorithms for solving at least some of the co-evolutionary problems mentioned in this work.

Second, due to lack of space, we focused in this work on four versions of the *Local Co-Evolutionary* problem. However, we intend to use the concept that was described here for solving both more specific queries (*e.g.* finding co-evolving sets of *OL* along a subtree that includes at least one leaf) and more general ones (*e.g.* a *joint* analysis of ER and copy number of orthologs across a phylogenetic tree).

Third, generating biological inputs for local co-evolutionary problems is a non-trivial task (see section 4.4) as it includes dozens of preprocessing steps that should be performed properly. We plan to use our approach for studying co-evolution across the entire tree of life. To this end, we intend to generate the phylogenetic tree and the *OLs* of hundreds of organisms (Archaea, Bacteria, and Eukaryota), and to analyze this input by our approach.

Acknowledgment

We would like to thank Prof. Eytan Ruppin and Prof. Martin Kupiec for helpful discussions. T.T. was supported by the Edmond J. Safra Bioinformatics program at Tel Aviv University and the Yeshaya Horowitz Association through the Center for Complexity Science.

References

1. Benton, M.J., Donoghue, P.C.J.: Paleontological evidence to date the tree of life. Mol. Biol. Evol. 24(1), 26–53 (2007)
2. Berbee, M., Taylor, J.: Systematics and evolution. In: McLaughlin, D., McLaughlin, E., Lemke, P. (eds.) The Mycota, vol. VIIB, pp. 229–245. Springer, Berlin (2001)
3. Cheng, Y., Church, G.M.: Biclustering of expression data. In: Proc. 8th Int. Conf. Intell. Syst. Mol. Biol., pp. 93–103 (2000)
4. Dujon, B., Sherman, D., Fischer, G., Durrens, P., Casaregola, S., et al.: Genome evolution in yeasts. Nature 430, 35–44 (2004)
5. Ben-Dor, A., et al.: Discovering local structure in gene expression data: The order-preserving submatrix problem. J. Comput. Biol. 10(3-4), 373–384 (2003)
6. Tanay, A.: Discovering statistically significant biclusters in gene expression data. Bioinformatics 18, S136–144 (2002)
7. Barker, D., et al.: Predicting functional gene links using phylogenetic-statistical analysis of whole genomes. PLoS Comput. Biol. 1, 24–31 (2005)
8. Barker, D., et al.: Constrained models of evolution lead to improved prediction of functional linkage from correlated gain and loss of genes. Bioinformatics 23(1), 14–20 (2007)
9. Juan, D., et al.: High-confidence prediction of global interactomes based on genome-wide coevolutionary networks. PNAS 105(3), 934–939 (2008)
10. Ober, D., et al.: Molecular evolution by change of function. alkaloid-specific homospermidine synthase retained all properties of deoxyhypusine synthase except binding the eif5a precursor protein. J. Biol. Chem. 278(15), 12805–12812 (2003)
11. Krylov, D.M., et al.: Gene loss, protein sequence divergence, gene dispensability, expression level, and interactivity are correlated in eukaryotic evolution. Genome Res. 13(10), 2229–2235 (2003)
12. Wall, D.P., et al.: Functional genomic analysis of the rate of protein evolution. Proc. Natl. Acad. Sci. U.S.A. 102(15), 5483–5488 (2005)
13. Oron, E., et al.: Genomic analysis of cop9 signalosome function in drosophila melanogaster reveals a role in temporal regulation of gene expression. Mol. Syst. Biol. 3, 108 (2007)
14. Pazos, F., et al.: Correlated mutations contain information about protein-protein interaction. J. Mol. Biol. 271, 511–523 (1997)
15. Wapinski, I., et al.: Natural history and evolutionary principles of gene duplication in fungi. Nature 449, 54–65 (2007)

16. Wu, J., et al.: Identification of functional links between genes using phylogenetic profiles. Bioinformatics 19, 1524–1530 (2003)
17. Marino-Ramirez, L., et al.: Co-evolutionary rates of functionally related yeast genes. Evolutionary Bioinformatics, 2295–2300 (2006)
18. Bowers, P.M., et al.: Prolinks: a database of protein functional linkages derived from coevolution. Genome Biology 5, R35(2004)
19. Grossmann, S., et al.: An improved statistic for detecting over-represented gene ontology annotations in gene sets. In: Apostolico, A., Guerra, C., Istrail, S., Pevzner, P.A., Waterman, M. (eds.) RECOMB 2006. LNCS (LNBI), vol. 3909, pp. 85–98. Springer, Heidelberg (2006)
20. Chena, Y., et al.: The coordinated evolution of yeast proteins is constrained by functional modularity. Trends in Genetics 22(8), 416–419 (2006)
21. Garey, M.R., Johnsons, D.S.: Computers and Interactability: A Guide to the Theory of NP-Completeness, p. 196. Freeman, New York (1979)
22. Knudsen, B.: Optimal multiple parsimony alignment with affine gap cost using a phylogenetic tree. In: Benson, G., Page, R.D.M. (eds.) WABI 2003. LNCS (LNBI), vol. 2812, pp. 433–446. Springer, Heidelberg (2003)
23. MacQueen, J.B.: Some methods for classification and analysis of multivariate observations. In: Proceedings of 5-th Berkeley Symposium on Mathematical Statistics and Probability, vol. 1, pp. 281–297. University of California Press, Berkeley (1967)
24. Man, O., Pilpel, Y.: Differential translation efficiency of orthologous genes is involved in phenotypic divergence of yeast species. Nature Genetics 39, 415–421 (2007)
25. Ohno, S.: Evolution by gene duplication. Springer, Heidelberg (1970)
26. Scannell, D.R., Butler, G., Wolfe, K.H.: Yeast genome evolution-the origin of the species. Yeast 24(11), 929–942 (2007)
27. Snel, B., Huynen, M.A.: Quantifying modularity in the evolution of biomolecular systems. Genome Res. 14(3), 391–397 (2004)
28. Wolfe, K.H., Shields, D.C.: Molecular evidence for an ancient duplication of the entire yeast genome. Nature 387(6634), 708–713 (1997)

Ancestral Reconstruction by Asymmetric Wagner Parsimony over Continuous Characters and Squared Parsimony over Distributions

Miklós Csűrös

Department of Computer Science and Operations Research
University of Montréal
C.P. 6128, succ. Centre-Ville, Montréal, Québec, H3C 3J7, Canada
csuros@iro.umontreal.ca

Abstract. Contemporary inferences about evolution occasionally involve analyzing infinitely large feature spaces, requiring specific algorithmic techniques. We consider parsimony analysis over numerical characters, where knowing the feature values at terminal taxa allows one to infer ancestral features, namely, by minimizing the total number of changes on the edges using continuous-valued distance measures. In particular, we show that ancestral reconstruction is possible in linear time for both an asymmetric linear distance measure (Wagner parsimony) over continuous-valued characters, and a quadratic distance measure over finite distributions. The former can be used to analyze gene content evolution with asymmetric gain and loss penalties, and the latter to reconstruct ancestral diversity of regulatory sequence motifs and multi-allele loci. As an example of employing asymmetric Wagner parsimony, we examine gene content evolution within Archaea.

1 Introduction

Phylogenetic studies commonly operate with molecular sequence data, where homologous characters take values over a finite space. When working with characters such as numbers of paralogs within homologous gene families, allele frequencies, sequence length polymorphisms, or DNA sequence motif distributions, the analysis of theoretically infinite feature spaces becomes necessary [1]. In such situations, one can resort to parsimony criteria to infer ancestral states, or score candidate phylogenies by minimizing the total change of the feature in question. Change is quantified by using different types of distance measures which are appropriate for the study. A popular parsimony criterion for features that can be ordered linearly is the so-called Wagner parsimony [2,3] in which change is penalized simply by the absolute value of the numerical difference on an edge. Another criterion used sometimes is the minimization of squared distance between the numerical values [4].

Wagner parsimony has been used to infer the evolution of gene family size. Change in the family size, however, is not always equally likely in both directions,

C.E. Nelson and S. Vialette (Eds.): RECOMB-CG 2008, LNBI 5267, pp. 72–86, 2008.

as losses may be more frequent than gains, or vice versa. We propose a modification of the original Wagner parsimony criterion for such situations, where increases and decreases are penalized linearly, but with different penalty factors. We discuss the resulting optimization problem, and show how to compute the parsimony score, as well as the ancestral states in linear time, regardless of the actual values at the terminal taxa. We also show that squared parsimony over finite distributions can be computed efficiently, by performing the minimization in each coordinate separately, without considering the restriction to the probability simplex.

We demonstrate the utility of asymmetric Wagner parsimony by an analysis of gene content evolution in Archaea.

2   Algorithmic Results

2.1   Problem Statement

Consider the following general parsimony framework, introduced by Sankoff and Rousseau [5]. Let $\mathcal{T} = (\mathcal{V}, \mathcal{E})$ be a rooted tree that represents a phylogeny, with node set $\mathcal{V}$ and edge set $\mathcal{E}$. The set of tree leaves is denoted by $\mathcal{L}$. It is assumed that every non-leaf node has at least two children. Each node $u \in \mathcal{V}$ is associated with a *label* $\xi[u] \in \mathcal{X}$ where $\mathcal{X}$ is the space of possible labels. The focus of this study is the case when $\mathcal{X}$ is a numerical infinite space such as $\mathcal{X} = \mathbb{R}^d$ or $\mathcal{X} = \{0, 1, 2, \dots\}$. The label space is equipped with a *change weight* function $\Delta \colon \mathcal{X}^2 \mapsto [0, \infty)$. (Classically, Δ is a proper distance metric, but we will consider asymmetric functions, as well.) We are interested in the following problem.

General parsimony labeling problem. Given the tree T, label space $(\mathcal{X}, \Delta)$, and fixed assignments $\xi[u]$ at the leaves $u \in \mathcal{L}$, find $\xi[v]$ for all inner nodes $v \in \mathcal{V} \setminus \mathcal{L}$ that minimize the total change

$$\sum_{uv \in \mathcal{E}} \Delta(\xi[u] \to \xi[v]).$$

The problem in this form was introduced in [5] as a Steiner tree problem [6] with a distance metric Δ. Some specific cases of the general problem have been extensively studied. The case of nonnegative integers $\mathcal{X} = \mathbb{N}$ and $\Delta(y \to x) = |y - x|$, is known as Wagner parsimony that can be solved in linear time [2,3]. The case $\mathcal{X} = \mathbb{R}$ and $\Delta(y \to x) = (y - x)^2$ is known as squared parsimony, which also has a linear-time solution [4]

The parsimony labeling problem is encountered in phylogenetic studies when one wants to estimate the ancestral state of some feature that is represented by the labels [7,8]. An unknown phylogeny can also be inferred by searching for the topology $\mathcal{T}$ over the leaf set $\mathcal{L}$ that minimizes the parsimony score [1].

Features in question may be (continuous-valued) allele frequencies, in which case squared-parsimony is in fact equivalent to likelihood maximization under a Brownian motion model [4,1]. Wagner parsimony has been used to infer the evolution of sequence length polymorphisms [9], genome size [10], and gene family size.

2.2 General Solution by Dynamic Programming

The general parsimony problem has a solution by dynamic programming, as elucidated in the pioneering paper of Sankoff and Rousseau [5]. The key idea is to define the *subtree weight functions* $f_u(x)\colon x \in \mathcal{X} \mapsto [0,\infty]$ for each node $u \in \mathcal{V}$, which give the minimum weight within the subtree $\mathcal{T}_u$ rooted at u when $\xi[u] = x$. For leaves, $f_u(x) = 0$ if $x = \xi[u]$; otherwise, $f_u(x) = \infty$. For an inner node u, the following recursion holds.

$$f_u(y) = \sum_{v \in \text{children}(u)} \min_{x \in \mathcal{X}} \Big(\Delta(y \to x) + f_v(x) \Big). \tag{1}$$

For every edge $uv \in \mathcal{E}$, define the *stem weight functions*

$$h_v(y) = \min_{x \in \mathcal{X}} \Big(\Delta(y \to x) + f_v(x) \Big), \tag{2}$$

so that

$$f_u(y) = \sum_{v \in \text{children}(u)} h_v(y). \tag{3}$$

The minimum total weight is then $\min_y f_{\text{root}}(y)$, and the optimal labeling can be determined by backtracking. For a finite label space, the general solution takes $O(|\mathcal{X}|^2)$ time on each edge. For an infinite space, it is not immediately clear how the minimization can be done in practice. Luckily, it is possible to compute f and h efficiently in many important cases [2,4,5].

2.3 Asymmetric Wagner Parsimony

Often, the labels represent features that are more easily lost than gained [11,7]. Gene content evolution, in particular, is characterized by frequent gene loss, which may be properly captured in parsimony methods by penalizing gains more than losses [12]. We define the *asymmetric Wagner parsimony* problem as that of general parsimony labeling when

$$\mathcal{X} \subseteq \mathbb{R} \quad \text{and} \quad \Delta(y \to x) = \begin{cases} \gamma(x - y) & \text{if } y < x; \\ \lambda(y - x) & \text{if } x < y, \end{cases}$$

where $\gamma, \lambda > 0$ are gain and loss penalty factors, respectively. The pivotal observation for an algorithmic solution is given by the following lemma; the claim is illustrated in Figure 1.

Lemma 1. *For every non-leaf node* $u \in \mathcal{V} \setminus \mathcal{L}$, *the subtree weight function is a continuous, convex, piecewise linear function. In other words, there exist* $k \geq 1$, $\alpha_0 < \alpha_1 < \cdots < \alpha_k$, $x_1 < x_2 \cdots < x_k$, *and* $\phi_0, \ldots, \phi_k \in \mathbb{R}$ *that define* f_u *in the following manner.*

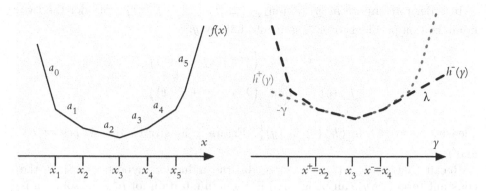

Fig. 1. Illustration of Lemma 1. **Left:** for asymmetric Wagner parsimony, the subtree weight function f is always piecewise linear with slopes $a_0, \ldots, a_k$ ($k = 5$ here). **Right:** the stem weight function h is determined by the two auxiliary functions h^+ and h^-, which are obtained by "shaving off" the steep extremities of f, and replacing them with slopes of $-\gamma$, and λ, respectively.

$$
f_u(x) = \begin{cases}
\phi_0 + \alpha_0 x & \text{if } x \leq x_1; \\
\phi_1 + \alpha_1(x - x_1) & \text{if } x_1 < x \leq x_2; \\
\ldots \\
\phi_{k-1} + \alpha_{k-1}(x - x_{k-1}) & \text{if } x_{k-1} < x \leq x_k; \\
\phi_k + \alpha_k(x - x_k) & \text{if } x_k < x,
\end{cases}
\tag{4}
$$

where $\phi_1 = \phi_0 + \alpha_0 x_1$ and $\phi_{i+1} = \phi_i + \alpha_i(x_{i+1} - x_i)$ for all $0 < i < k$. Moreover, if u has d children, then $a_0 = -d\gamma$ and $a_k = d\lambda$.

Proof. The proof proceeds by induction over the tree in a postorder traversal, following the recursion of (1). By the definition of Δ, if v is a leaf, then

$$
h_v(x) = \begin{cases}
\gamma(\xi[v] - x) & \text{if } x \leq \xi[v]; \\
\lambda(x - \xi[v]) & \text{if } \xi[v] < x.
\end{cases}
\tag{5}
$$

Base case. If all d children of u are leaves, then (3) and (5) imply that (4) holds with some $k \leq d$, $\alpha_0 = -d\gamma$ and $\alpha_k = d\lambda$. For a more precise characterization, let $\mathcal{C}$ be the set of children of u, and consider the set of leaf labels $\mathcal{S} = \{\xi[v] : v \in \mathcal{C}\}$. Then $k = |\mathcal{S}|$, and $\{x_1, \ldots, x_k\} = \mathcal{S}$. Furthermore, for all $i = 1, \ldots, k$, $\alpha_i = t_i \lambda - (d - t_i)\gamma$ with $t_i = \sum_{v \in \mathcal{C}} \{\xi[v] \leq x_i\}$, where $\{\cdot\}$ denotes the indicator for the event within the braces; i.e., t_i is the number of children that carry a label that is not larger than x_i. Finally, $\phi_0 = \gamma \sum_{v \in \mathcal{C}} \xi[v]$.

Induction step. Assume that u is an inner node at which (4) holds for every non-leaf descendant. Let v be a non-leaf child of u. By the induction hypothesis, $f_v(x)$ is a piecewise linear function as in (4) with some parameters $(\alpha_i : i = 0, \ldots, k)$, and $(x_i : i = 1, \ldots, k)$.

In order to compute $h_v(y) = \min_{x \in \mathcal{X}}\Big(\Delta(y \to x) + f_v(x)\Big)$, consider the two minimization problems over $\mathcal{X}$ split into half by y:

$$h_v^+(y) = \min_{x \in \mathcal{X}; x > y}\Big(\gamma(x - y) + f_v(x)\Big)$$

$$h_v^-(y) = \min_{x \in \mathcal{X}; x \leq y}\Big(\lambda(y - x) + f_v(x)\Big).$$

Clearly, $h_v(y) = \min\Big\{h_v^+(y), h_v^-(y)\Big\}$. Figure 1 illustrates the shapes of h^+ and h^-.

Recall that $\alpha_0 < \alpha_1 < \cdots < \alpha_k$ by the induction hypothesis. Since the constant term $(-\gamma y)$ can be ignored in the minimization for h^+, the solution is determined by the point $x^+ = x_j$ with $j = \min\{i : \alpha_i + \gamma \geq 0\}$. In particular,

$$h_v^+(y) = \begin{cases} \gamma \cdot (x^+ - y) + f_v(x^+) & \text{if } y < x^+; \\ f_v(y) & \text{if } y \geq x^+. \end{cases}$$

In a similar manner, let $x^- = x_j$ with $j = \min\{i : \alpha_i - \lambda \geq 0\}$. Then

$$h_v^-(y) = \begin{cases} f_v(y) & \text{if } y < x^-; \\ \lambda \cdot (y - x^-) + f_v(x^-) & \text{if } y \geq x^-. \end{cases}$$

Notice that the induction hypothesis implies that $x^+ \leq x^-$, since $\alpha_0 + \gamma < 0$ and $\alpha_k - \lambda > 0$ hold. By the definition of x^+ and x^-, it is also true that $h_v^+(y) \leq f_v(y)$ if $y < x^+$, and that $h_v^-(y) \leq f_v(y)$ if $y \geq x^-$. Hence, h_v is a piecewise linear function in the form

$$h_v(y) = \begin{cases} \gamma \cdot (x^+ - y) + f_v(x^+) & \text{if } y < x^+; \\ f_v(y) & \text{if } x^+ \leq y < x^-; \\ \lambda \cdot (y - x^-) + f_v(x^-) & \text{if } y \geq x^-. \end{cases} \qquad (6)$$

The formula also shows that when $\xi[u] = y$, the best labeling for v is either $x = x^+$ for $y < x^+$ (i.e., net gain on edge uv), or $x = x^-$ for $y \geq x^-$ (i.e., net loss), or else $x = y$ (no change).

Equations (5) and (6) show that $h_v(y)$ is always a continuous, convex, piecewise linear function with slopes $(-\gamma)$ on the extreme left and λ on the extreme right. Consequently, $f_u(y) = \sum_{v \in \text{children}(u)} h_v(y)$ is also a continuous, convex, piecewise linear function, with slopes $(-d\gamma)$ on the left and $d\lambda$ on the right. Hence, the induction hypothesis holds for u. □

The proof provides the recipe for implementing the dynamic programming of (1). The algorithm has to work with piecewise linear functions as in (4), parametrized by the set of slopes $(\alpha_i : i = 0, \ldots, k)$, breakpoints $(x_i : i = 1, \ldots, k)$ and shift ϕ_0. The parameters are naturally sorted as $\alpha_0 < \alpha_1 < \cdots < \alpha_k$ and $x_1 < x_2 < \cdots < x_k$, and can be thus stored as ordered arrays. The algorithm is sketched as follows.

W1 DYNAMIC PROGRAMMING FOR ASYMMETRIC WAGNER
W2 initialize $h_u(\cdot)$ and $f_u(\cdot)$ as null at each node $u \in \mathcal{V}$
W3 **for** all nodes u in postorder traversal
W4 **if** u is a leaf
W5 **then** set $h_u(x)$ as in (5)
W6 **else**   ▷ $h_v(x)$ *is computed for all children* v *already*
W7   compute $f_u(x) = \sum_{v \in \text{children}(u)} h_u(x)$
W8   **if** u is not the root then compute $h_u(y)$ by (6)
W9 find the minimum of $f_{\text{root}}(x)$
W10 backtrack for the optimal labeling if necessary

Theorem 1. *For a tree $\mathcal{T}$ of height h and $n = |\mathcal{V}|$ nodes, asymmetric Wagner parsimony can be solved in $O(n \min\{h, D\} \log d_{\max})$ time where D is the number of different leaf labels and $d_{\max}$ is the maximum arity.*

Proof. First, notice that the breakpoints at each f_u and h_u are exactly the set of different leaf labels in the subtree rooted at u, with at most D elements. Line W5 takes $O(1)$ time at each leaf. In Line W8, a binary search for x^+ and x^- takes $O(\log k)$ time if there are k breakpoints. In Line W7, piecewise linear functions need to be summed, which can be done by straightforward modification of well-known linear-time merging algorithms for ordered lists [13]. In order to sum the piecewise linear functions, the breakpoints must be processed in their combined order, and the intermediate slopes need to be computed. The procedure takes $O(k \log d)$ time, if the node has d children, and there are a total of k breakpoints at the children's stem weight functions. Now, $k \leq D$, and, thus, every node can be processed in $O(D \log d_{\max})$ time. The $O(nh \log d_{\max})$ bound comes from the fact that k is bounded by the number of leaves in the subtree. The total computing time for nodes that are at the same distance from the root is then $O(n \log d_{\max})$. By summing across all levels, we get $O(nh \log d_{\max})$ computing time. □

Remark. Lemma 1 and its proof show that there is an optimal solution where every non-leaf node carries a label that appears at one of the leaves. Accordingly, it is enough to keep track of $f_u(x)$ only where x takes one of the leaf label values. Adapting Sankoff's general parsimony algorithm over the discrete finite label space defined by the D label values of interest yields an $O(nD^2)$ algorithm.

2.4 Squared Parsimony

In certain applications, node labels are distributions such as allele frequencies [14], or probabilistic sequence motifs [15]. Suppose, for example, that we identified homologous regulatory sequence motifs in some genomes related by a known phylogeny. A particular instance of the motif is a DNA oligomer $s_1 s_2 \cdots s_\ell$ with a fixed length ℓ. From the set, we compile sequence motifs describing each terminal node by the labels $\xi_{is}[u]$, which give the relative frequency of each nucleotide s at motif position $i = 1, \ldots, \ell$. From the node labels, we would like to infer the compositional distribution of the motif at ancestral nodes. In a recent example, Schwartz and coworkers [15] examined the evolution of splicing signals in eukaryotes. The authors deduced that the 5' splice site and the branch site were degenerate in the earliest eukaryotes, in agreement with previous studies by Irimia and coworkers [16]. These findings are intriguing as they hint at the prevalence of alternative splicing in the earliest eukaryotes. Schwartz et al. [15] reconstructed the diversity of ancestral splicing signals by using a squared change penalty $\Delta(\mathbf{y} \to \mathbf{x}) = \sum_{i=1}^{\ell} \sum_{s=A,C,G,T} (x_{is} - y_{is})^2$. An equivalent sum-of-squares penalty was suggested by Rogers [14] in a different context, where $i = 1, \ldots, \ell$ would stand for genetic loci and s would index possible alleles at each locus. Since the positions can be handled separately, we consider the problem of general parsimony at a given position. Specifically, we assume that the labels are distributions over a finite set $\mathcal{A} = \{1, 2, \ldots, r\}$. The change penalty is defined by

$$\Delta(\mathbf{y} \to \mathbf{x}) = \sum_{s \in \mathcal{A}} (x_s - y_s)^2.$$

The case of a binary alphabet $r = 2$ was shown to be solvable in linear time by Maddison [4]. The algorithm is stated for the general parsimony problem with $\mathcal{X} = \mathbb{R}$ and $\Delta(y \to x) = (y - x)^2$. While Maddison's algorithm is trivially extended to any dimension with $\mathcal{X} = \mathbb{R}^r$ and $\Delta(y \to x) = \sum_i (y_i - x_i)^2$, the extension to distributions with $r > 2$ is not immediately obvious. In [15], the distributions were discretized to an accuracy of 0.02, and then solved on the corresponding grid by using Sankoff's dynamic programming. Notice that there are 23426 such discretized distributions, and dynamic programming over a finite alphabet takes quadratic time in the alphabet size. Here we show that Maddison's algorithm can be carried out at each coordinate independently, as the computed solution is automatically a distribution.

Squared Parsimony for a Continuous Character

For a discussion, we restate the result of [4].

Lemma 2. *In the general parsimony problem with $\mathcal{X} = \mathbb{R}$ and $\Delta(y \to x) = (y - x)^2$, subtree weight functions are quadratic. In other words, at each non-leaf node u, there exist $\alpha, \mu, \phi \in \mathbb{R}$ such that*

$$f_u(x) = \alpha(x - \mu)^2 + \phi. \tag{7}$$

Proof. We will use the simple arithmetic formula that

$$\sum_{i=1}^{d} \alpha_i (x - \mu_i)^2 = \alpha(x - \bar{\mu})^2 + \alpha\big(\mu^{(2)} - (\bar{\mu})^2\big) \tag{8}$$

with

$$\alpha = \sum_{i=1}^{d} \alpha_i, \quad \bar{\mu} = \frac{\sum_{i=1}^{d} \alpha_i \mu_i}{\sum_{i=1}^{d} \alpha_i}, \quad \mu^{(2)} = \frac{\sum_{i=1}^{d} \alpha_i \mu_i^2}{\sum_{i=1}^{d} \alpha_i}.$$

The proof proceeds by induction over the tree in a postorder traversal, following the recursion structure of Eq. (1).

Base case. Let u be an inner node with d children $\{v_1, \ldots, v_d\}$ that are all leaves. By (1),

$$f_u(y) = \sum_{i=1}^{d} (y - \xi[v_i])^2.$$

Hence (8) applies with $\alpha_i = 1$ and $\mu_i = \xi[v_i]$. Specifically, (7) holds with $\mu = \sum_{i=1}^{d} \xi[v_i]/d$.

Induction step. Suppose that u is an inner node with d children $\{v_1, \ldots, v_d\}$, which are all either leaves, or inner nodes for which (7) holds. Let $v = v_i$ be an arbitrary child node. If v is a leaf, then $h_v(y) = (y - \xi[v])^2$. If v is an inner node with $f_v(x) = \alpha(x - \mu)^2 + \phi$, then

$$h_v(y) = \min_x \big((y - x)^2 + \alpha(x - \mu)^2 + \phi \big)$$

$$= \min_x \left\{ (\alpha + 1) \left(x - \frac{y + \alpha\mu}{\alpha + 1} \right)^2 \right\} + \frac{\alpha}{\alpha + 1} (y - \mu)^2 + \phi$$

$$= \frac{\alpha}{\alpha + 1} (y - \mu)^2 + \phi.$$

Notice that the best labeling at v is achieved with $x = \frac{y + \alpha\mu}{\alpha + 1}$.

Consequently, the stem weight function can be written as $h_{v_i}(x) = \alpha_i(x - \mu_i)^2 + \phi_i$ for every child v_i with some $\alpha_i, \mu_i, \phi_i \in \mathbb{R}$. By (3),

$$f_u(x) = \sum_{i=1}^{d} \big(\alpha_i(x - \mu_i)^2 + \phi_i \big) = \alpha(y - \bar{\mu})^2 + \phi,$$

where $\phi = \alpha \big(\mu^{(2)} - (\bar{\mu})^2 \big) + \sum_{i=1}^{d} \phi_i$, and $\alpha, \bar{\mu}, \mu^{(2)}$ are as in (8). Therefore, (7) holds at u. $\square$

The proof of Lemma 2 shows how the parameters α and μ need to be computed in a postorder traversal. Namely, for every node u, the following recursions hold for the parameters $\alpha = \alpha_u$ and $\mu = \mu_u$ of (7).

$$\alpha_u = \begin{cases} \text{undefined} & \text{if } u \text{ is a leaf;} \\ \sum_{v \in \text{children}(u)} \beta_v & \text{otherwise;} \end{cases} \tag{9a}$$

$$\mu_u = \begin{cases} \xi[u] & \text{if } u \text{ is a leaf;} \\ \dfrac{\sum_{v \in \text{children}(u)} \beta_v \mu_v}{\sum_{v \in \text{children}(u)} \beta_v} & \text{otherwise;} \end{cases} \tag{9b}$$

where

$$\beta_v = \begin{cases} 1 & \text{if } v \text{ is a leaf;} \\ \dfrac{\alpha_v}{\alpha_v + 1} & \text{otherwise.} \end{cases} \tag{9c}$$

Squared Parsimony for Distributions

Suppose that the nodes are labeled with finite distributions over a set $\mathcal{A} = \{1, 2, \ldots, r\}$. Accordingly, we write $\xi_i[u]$ with $i = 1, \ldots, r$ for the i-th probability value at each node u. Node labelings are scored by the square parsimony penalty: $\Delta(\mathbf{y} \to \mathbf{x}) = \sum_{i=1}^{r}(y_i - x_i)^2$, where $\mathbf{y}$ and $\mathbf{x}$ are distributions over $\mathcal{A}$, i.e., points of the $(r - 1)$-dimensional simplex in $\mathbb{R}^r$ defined by $0 \leq \xi_i[u]$ for all i, and $\sum_{i=1}^{r} \xi_i[u] = 1$. Suppose that one carries out the minimization coordinate-wise, for each i separately, without making particular adjustments to ensure that the ancestral labels also define a distribution. By Lemma 2, such an independent ancestral reconstruction finds the subtree weight functions of the form $f_{u,i}(x) = \alpha_u(x - \mu_{u,i})^2 + \phi_{u,i}$ in each coordinate i. (Equations (9a) and (9c) show that α_u and β_u are determined by the tree topology alone, and are thus the same in each coordinate.)

Theorem 2. *The coordinate-wise independent ancestral reconstruction produces the optimal solution for distributions.*

Proof. Let $f_{u,i}(x)$ denote the subtree weight function for coordinate i at node u. Clearly, $\sum_{i=1}^{r} f_{u,i}(x_i)$ is a lower bound on the true subtree weight functionbreak $f_u(x_1, \ldots, x_r)$ for the distributions. Consequently, it is enough to show that the solution by coordinate-wise reconstruction leads to valid distributions. From Equation (9b), if u is an inner node, then $\sum_{i=1}^{r} \mu_{u,i} = \sum_{v \in \text{children}(u)} \frac{\beta_v}{\alpha_u} \sum_{i=1}^{r} \mu_{v,i}$. As $\sum_{i=1}^{r} \mu_{u,i} = 1$ holds at every leaf u, the equality holds at all nodes by induction. It is also clear that $\mu_{u,i} \geq 0$ is always true, since β_v is never negative. In particular, the optimal labelings at the root define a distribution with $\xi_i[\text{root}] = \mu_{\text{root},i}$.

In the proof of Lemma 2, we showed that if the parent of an inner node v is labeled by $\mathbf{y} = (y_1, \ldots, y_r)$, then the optimal labeling at v is $\xi_i[v] = x_i = \frac{y_i + \alpha_v \mu_{v,i}}{\alpha_v + 1}$. Now, $\sum_{i=1}^{r} x_i = \frac{\sum_i y_i + \alpha_v \sum_i \mu_{v,i}}{\alpha_v + 1} = 1$ if $\sum_i y_i = 1$ holds. Since the independent ancestral reconstructions produce a distribution at the root, the backtracking procedure produces a distribution at every inner node v. $\square$

3   Gene Content Evolution in Archaea

We applied asymmetric Wagner parsimony to the analysis of gene content evolution in Archaea. We note that parsimony-based analysis has its well-known shortcomings, such as the underestimation of gene loss, and the imposition of uniformity across lineages and genes, which may be avoided with sophisticated probabilistic methods [17,18]. Nevertheless, parsimony may give important insights by providing a conservative estimate of ancestral gene content, and by underlining some general idiosyncrasies without much procedural difficulty.

Makarova and coauthors [19] delineated homologous gene families across 41 completely sequenced and annotated archaeal genomes. They analyzed some characteristic features of archaeal genome evolution, and extrapolated the gene composition of the last archaeal common ancestor, or LACA. The analysis relied on so-called phyletic profiles, which are binary patterns of family presence-absence, in conjunction with parsimony-based ancestral reconstruction algorithms [20]. In our analysis, we used the available information on the number of paralogs within different genomes.

3.1 Data and Methods

Data was downloaded from ftp://ftp.ncbi.nih.gov/pub/wolf/COGs/arCOG. The data set defines 7538 families (so-called archaeal clusters of orthologous genes, or *arCOGs*) in 41 genomes. Figures 2 and 3 show the organisms and their phylogenetic relationships. The abbreviations are those used in [19] and the arCOG database: the Appendix lists the organism names and the abbreviations. The archaeal phylogeny is based on the one used by Makarova et al. (Figure 7 in [19]) for inferring gene content evolution, using additional considerations to partially resolve certain polytomies. Namely, we assume the monophyly of the Pyrococcus genus within Thermococcales [21], and the monophyly of Methanomicrobia excluding Halobacteriales [22], as depicted in Figure 3.

In order to perform the analysis, an adequate gain and loss penalization needed to be chosen. The ratio between the two penalty factors influences how much of the reconstructed history is dominated by gene loss [12]. Since the inference depends only on the ratio of the gain and loss penalties, we set $\lambda = 1$, and performed the reconstruction at different gain penalties γ. We selected a gain penalty of $\gamma = 1.6$, matching the estimate of [19] the closest. The reconstruction results in a LACA genome of 984 families and 1106 genes, which is similar in the corresponding statistics to such extant archaea as *Methanopyrus kandleri* (Metka; 1121 arCOGs with 1336 genes) and *Cenarchaeum symbiosum* (Censy; 918 arCOGs with 1296 genes).

3.2 Results

Gene content at LACA. The reconstructed set of ancient families contains 96 families inferred as present, and 107 as absent in contradiction with [19]. The two reconstructions qualitatively give a very similar picture, pointing to a LACA genome complexity comparable to the simplest free-living prokaryotes such as Mycoplasma. Table 1 shows a summary of the functional categorization for the inferred primordial gene families. Among the gene families present in LACA, 91 (9%) included more than one gene. The majority of these families (77 of 91) have closer homologs among Bacteria than among Eukaryota, which would be expected if Archaea emerged from a bacterial lineage. These multi-gene families are indicative of ancestral adaptations: notable cases include reverse gyrase (2 paralogs), hinting at a hyperthermophilic LACA, and various genes implicated in pyruvate oxidation that has a pivotal importance in archaeal metabolism [21].

Losses and gains of families. Figures 2 and 3 show further details of the ancestral reconstruction. Using asymmetric Wagner parsimony, it was possible to

Table 1. Ancestral gene content at LACA. Columns: (a) arCOG functional category code, (b) functional category description, (c) LACA families with more than one member, (d) total number of families at LACA.

Cat[a]	Description[b]	Multi[c]	Fam[d]
	Information storage and processing		
J	Translation	4	153
K	Transcription	6	59
L	Replication	7	57
	Cellular processes and signaling		
D	Cell cycle control	3	5
V	Defense mechanisms	3	19
T	Signal transduction mechanisms	2	8
M	Cell wall, membrane and envelope biogenesis	7	23
N	Cell motility	1	5
U	Intracellular trafficking and secretion	1	11
O	Posttranslational modification, protein turnover, chaperones	5	41
	Metabolism		
C	Energy production and conversion	10	77
G	Carbohydrate transport and metabolism	6	37
E	Amino acid transport and metabolism	14	101
F	Nucleotide transport and metabolism	3	46
H	Coenzyme transport and metabolism	3	70
I	Lipid transport and metabolism	2	23
P	Inorganic ion transport and metabolism	1	45
Q	Secondary metabolites biosynthesis, transport and catabolism	1	23
R,S	Poorly characterized or unknown	12	197
	Total	91	984

postulate expansions and reductions within gene families, in addition to the families' appearance and elimination. Numerous losses, just as in the reconstruction of [19], are associated with symbiotic lifestyles (Censy and Naneq). Our studies also agree on examples of significant losses coupled with major gains in Thermococcales (node 7) and Thermoplasmales (node 9), hinting at unusually dynamic genomes. Our reconstructions of lineage-specific changes, however, often differ numerically, as illustrated in Table 2. Namely, Wagner parsimony tends to postulate fewer genes at inner nodes, and family gains on deep branches also tend to be lower. Our reconstruction seems more conservative, and at times even more plausible. For instance, we posit major gains in Desulfurococcales and Sulfolobales (nodes 4 and 5) lineages, whereas [19] postulates an extremely large genome for their common ancestor (node 3) instead.

Patterns of diversification. Interestingly, large losses are not always associated with compact genomes: Methanosarcina species (cf. Fig. 3) are among the archaea with the largest genomes, but terminal lineages have disposed of many families to end up with their current gene repertoire. The finding points to different paths of specialization from a versatile ancestor, accompanied by the elimination of redundant functions.

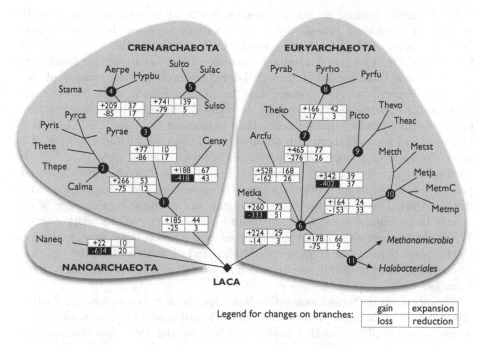

Fig. 2. Changes of gene repertoire in main lineages. On each branch, we computed the number of arCOG families gained and lost, as well as those that were retained but underwent changes in the number of paralogs (i.e., expansions or reductions). The numbers are shown in the small tables, in which darkened cells highlight major losses. Correspondence between numbered nodes and taxonomic groups is given in Table 2. The subtree below node 11 is shown in Figure 3.

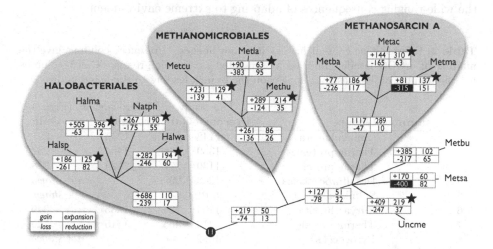

Fig. 3. Gene content evolution within Halobacteriales and Methanomicrobia. Stars highlight substantial expansions (at least half as many as family gains).

On branches leading to major lineages, newly appearing families typically outnumber expanding families by a factor of two to eight. It is not surprising that gains on those branches would be so frequent: the substantial differences in lifestyles are presumably possible only by acquiring genes with adequate new functionalities through lateral transfer or other means of evolutionary innovations. At the same time, terminal branches often display abundant family expansions: in 29 of the 41 terminal lineages, there are less than twice as many newly acquired genes than expanding families. This point is illustrated in Figure 3, showing a detailed reconstruction within a subtree. The most dramatic expansions are seen in Sulfolobales (below node 5 in Fig. 2), Methanosarcina and Halobacteriales (cf. Fig. 3). The branches leading to the progenitors of the same groups are precisely those with the most gains inferred in this study. The abundance of expansions is not a simple consequence of relatively large genome sizes, since expansions are frequent even in relative terms. Within Halobacteriales, 7.5–18% of families expanded on terminal branches; on the terminal branches of *M. hungatei* (Methu) and *M. acetivorans* (Metac), more than 12% of families did, in contrast with an overall average of 5.7% on terminal branches.

The observed patterns exemplify adaptations to new environments. Such an adaptation may be prompted by the acquisition of new functions, with ensuing series of gene duplications that lead to sub-functionalization, and, thus, specialization. A further scrutiny of such scenarios, is unfortunately difficult, because a substantial number of lineage-specific expansions are within poorly characterized families. In the most extreme case of *H. marismortui* (Halma), for example, 126 (31%) of 396 expanding families are poorly characterized. The top arCOG functional categories represented by the remaining expansions are C (energy: 35 families), E (amino acid metabolism: 33), K (transcription: 26), and T (signal transduction: 25). The functional variety of lineage-specific expansions illustrates the wide-ranging consequences of adapting to extreme environments.

Table 2. Inferred gene content history in major linages. "Presence" columns give the number of arCOG families inferred at the listed taxonomic groups. "Gain" columns list the number of families that appear on the branch leading to the listed nodes.

		This study		Makarova et al. (Fig. 7)	
Node number	Group	Presence	Gain	Presence	Gain
1	Crenarchaeota	1148	185	1245	291
2	Thermoproteales	1339	266	1404	237
3	Thermoprotei	1139	77	2128	928
4	Desulfurococcales	1263	209		*not shown*
5	Sulfolobales	1801	741		*not shown*
6	Euryarchaeota	1194	224	1335	349
7	Thermococcales	1413	465	1715	720
8	Pyrococcus	1562	166		*not shown*
9	Thermoplasmales	1134	342	1474	643
10	"Class I" methanogens	1205	164	1563	415

4   Conclusion

When small data sets need to be analyzed, or reasonable assumptions for probabilistic analysis are not available, parsimony is a well-justified method of choice. Even in phylogenetic reconstructions, parsimony may enjoy an advantage over sophisticated likelihood methods, as it enables the faster exploration of the search space by quick scoring of candidate phylogenies [1]. The present work augments the set of parsimony tools available for the analysis of numerical evolutionary characters in a range of applications, including the analysis of gene content, regulatory motifs, and allele frequencies.

References

1. Felsenstein, J.: Inferring Phylogenies. Sinauer Associates, Sunderland (2004)
2. Farris, J.S.: Methods for computing Wagner trees. Syst. Zool. 19, 83–92 (1970)
3. Swofford, D.L., Maddison, W.P.: Reconstructing ancestral states using Wagner parsimony. Math. Biosci. 87, 199–229 (1987)
4. Maddison, W.P.: Squared-change parsimony reconstructions of ancestral states for continuous-valued characters on a phylogenetic tree. Syst. Zool. 40, 304–314 (1991)
5. Sankoff, D., Rousseau, P.: Locating the vertices of a Steiner tree in arbitrary metric space. Math. Program. 9, 240–246 (1975)
6. Hwang, F.K., Richards, D.S.: Steiner tree problems. Networks 22, 55–89 (1992)
7. Cunningham, C.W., Omland, K.E., Oakley, T.H.: Reconstructing ancestral character states: a critical reappraisal. Trends Ecol. Evol. 13, 361–366 (1998)
8. Pagel, M.: Inferring the historical patterns of biological evolution. Nature 401, 877–884 (1999)
9. Witmer, P.D., Doheny, K.F., Adams, M.K., Boehm, C.D., Dizon, J.S., Goldstein, J.L., Templeton, T.M., Wheaton, A.M., Dong, P.N., Pugh, E.W., Nussbaum, R.L., Hunter, K., Kelmenson, J.A., Rowe, L.B., Brownstein, M.J.: The development of a highly informative mouse simple sequence length polymorphism (SSLP) marker set and construction of a mouse family tree using parsimony analysis. Genome Res. 13, 485–491 (2003)
10. Caetano-Anollés, G.: Evolution of genome size in the grasses. Crop. Sci. 45, 1809–1816 (2005)
11. Omland, K.E.: Examining two standard assumptions of ancestral reconstructions: repeated loss of dichromatism in dabbling ducks (Anatini). Evolution 51, 1636–1646 (1997)
12. Koonin, E.V.: Comparative genomics, minimal gene sets and the last universal common ancestor. Nat. Rev. Microbiol. 1, 127–136 (2003)
13. Cormen, T.H., Leiserson, C.E., Rivest, R.L., Stein, C.: Introduction to Algorithms, 2nd edn. MIT Press, Cambridge (2001)
14. Rogers, J.S.: Deriving phylogenetic trees from allele frequencies. Syst. Zool., 52–63 (1984)
15. Schwartz, S., Silva, J., Burstein, D., Pupko, T., Eyras, E., Ast, G.: Large-scale comparative analysis of splicing signals and their corresponding splicing factors in eukaryotes. Genome Res. 18, 88–103 (2008)
16. Irimia, M., Penny, D., Roy, S.W.: Coevolution of genomic intron number and splice sites. Trends Genet. 23, 321–325 (2007)

17. Csűrös, M., Miklós, I.: A probabilistic model for gene content evolution with duplication, loss, and horizontal transfer. In: Apostolico, A., Guerra, C., Istrail, S., Pevzner, P.A., Waterman, M. (eds.) RECOMB 2006. LNCS (LNBI), vol. 3909, pp. 206–220. Springer, Heidelberg (2006)
18. Iwasaki, W., Takagi, T.: Reconstruction of highly heterogeneous gene-content evolution across the three domains of life. Bioinformatics 23, i230–i239 (2007)
19. Makarova, K.S., Sorokin, A.V., Novichkov, P.S., Wolf, Y.I., Koonin, E.V.: Clusters of orthologous genes for 41 archaeal genomes and implications for evolutionary genomics of archaea. Biology Direct 2, 33 (2007)
20. Mirkin, B.G., Fenner, T.I., Galperin, M.Y., Koonin, E.V.: Algorithms for computing evolutionary scenarios for genome evolution, the last universal common ancestor and dominance of horizontal gene transfer in the evolution of prokaryotes. BMC Evol. Biol. 3, 2 (2003)
21. Fukui, T., Atomi, H., Kanai, T., Matsumi, R., Fujiwara, S., Imanaka, T.: Complete genome sequence of the hyperthermophilic archaeon Thermococcus kodakaraensis KOD1 and comparison with Pyrococcus genomes. Genome Res. 15, 352–363 (2005)
22. Brochier, C., Forterre, P., Gribaldo, S.: An emerging phylogenetic core of Archaea: phylogenies of transcription and translation machineries converge following addition of new genome sequences. BMC Evol. Biol. 5, 36 (2005)

A   Species Names and Abbreviations

The following organisms are included in the study.

Aerpe *Aeropyrum pernix*, **Arcfu** *Archaeoglobus fulgidus*, **Calma** *Caldivirga maquilingensis* IC-167, **Censy** *Cenarchaeum symbiosum*, **Halma** *Haloarcula marismortui* ATCC 43049, **Halsp** Halobacterium species strain NRC-1, **Halwa** *Haloquadratum walsbyi*, **Hypbu** *Hyperthermus butylicus*, **Metac** *Methanosarcina acetivorans*, **Metba** *Methanosarcina barkeri fusaro*, **Metbu** *Methanococcoides burtonii* DSM 6242, **Metcu** *Methanoculleus marisnigri* JR1, **Methu** *Methanospirillum hungatei* JF-1, **Metja** *Methanocaldococcus jannaschii*, **Metka** *Methanopyrus kandleri*, **Metla** *Methanocorpusculum labreanum* Z, **Metma** *Methanosarcina mazei*, **MetmC** *Methanococcus maripaludis* C5, **Metmp** *Methanococcus maripaludis* S2, **Metsa** *Methanosaeta thermophila* PT, **Metst** *Methanosphaera stadtmanae*, **Metth** *Methanothermobacter thermoautotrophicus*, **Naneq** *Nanoarchaeum equitans*, **Natph** *Natronomonas pharaonis*, **Picto** *Picrophilus torridus* DSM 9790, **Pyrab** *Pyrococcus abyssi*, **Pyrae** *Pyrobaculum aerophilum*, **Pyrca** *Pyrobaculum calidifontis* JCM 11548, **Pyrfu** *Pyrococcus furiosus*, **Pyrho** *Pyrococcus horikoshii*, **Pyris** *Pyrobaculum islandicum* DSM 4184, **Stama** *Staphylothermus marinus* F1, **Sulac** *Sulfolobus acidocaldarius* DSM 639, **Sulso** *Sulfolobus solfataricus*, **Sulto** *Sulfolobus tokodaii*, **Theac** *Thermoplasma acidophilum*, **Theko** *Thermococcus kodakaraensis* KOD1, **Thepe** *Thermofilum pendens* Hrk 5, **Thete** *Thermoproteus tenax*, **Thevo** *Thermoplasma volcanium*, **Uncme** Uncultured methanogenic archaeon.

An Alignment-Free Distance Measure for Closely Related Genomes

Bernhard Haubold[1], Mirjana Domazet-Lošo[1,2], and Thomas Wiehe[3]

[1] Max-Planck-Institute for Evolutionary Biology, Department of Evolutionary Genetics, Plön, Germany
[2] Faculty of Electrical Engineering and Computing, University of Zagreb, Zagreb, Croatia
[3] Institute of Genetics, Universität zu Köln, Cologne, Germany

Abstract. Phylogeny reconstruction on a genome scale remains computationally challenging even for closely related organisms. Here we propose an alignment-free pairwise distance measure, K_r, for genomes separated by less than approximately 0.5 mismatches/nucleotide. We have implemented the computation of K_r based on enhanced suffix arrays in the program kr, which is freely available from guanine.evolbio.mpg.de/kr/. The software is applied to genomes obtained from three sets of taxa: 27 primate mitochondria, eight *Staphylococcus agalactiae* strains, and 12 *Drosophila* species. Subsequent clustering of the K_r values always recovers phylogenies that are similar or identical to the accepted branching order.

1 Introduction

Gene phylogenies do not necessarily coincide with organism phylogenies. This well known observation leads to the idea of reconstructing phylogenies from all available genetic information, that is, from complete genomes. In fact, the study of whole genome phylogenies started as soon as suitable data became available [8]. In spite of much progress since then, the computational obstacles to such analyses are still considerable and a good part of bioinformatics is concerned with solving them [6].

To the uninitiated the reconstruction of genome phylogenies might appear to simply involve the scaling up of available techniques for reconstructing gene phylogenies: compute a multiple sequence alignment and estimate the genealogy from that. However, in the wake of the first genome projects it proved difficult if not impossible to scale existing gene-centered alignment software from input of a few kilo bases to several mega bases. This left two avenues to explore: development of more efficient alignment algorithms and development of alignment-free methods of distance computation.

In the years following publication of the first genomes of free-living organisms, phylogenomics—as the field concerned with reconstructing phylogenies from genomes became known—made great strides on both counts [14]. Alignment algorithms and alignment tools have received most attention as they are useful in many sequence comparison tasks [6]. In contrast, alignment-free sequence comparison has a more narrow applicability, the classical case being phylogeny reconstruction from pairwise distances [3]. The great advantage of this approach is that it obviates the computationally intensive alignment step. In fact, alignment-free distance measures may even be

C.E. Nelson and S. Vialette (Eds.): RECOMB-CG 2008, LNBI 5267, pp. 87–99, 2008.
© Springer-Verlag Berlin Heidelberg 2008

used in the computation of multiple sequence alignments. For example, pairwise distances based on exact word (k-tuple) matches [27] underlie the fast mode of guide tree construction in the popular multiple sequence alignment program `clustalw` [18].

Two classes of methods for alignment-free sequence comparison can be distinguished: (i) methods based on word frequencies, the utility of which may depend on the word length chosen, and (ii) resolution-free methods, where no such parameter choice is necessary [26]. These methods have been applied to, for example, phylogeny reconstruction from γ-protobacterial genomes [5] and the analysis of regulatory sequences in metazoan genomes [17]. One disadvantage of alignment-free methods is that there is generally no model to map their results to evolutionary distances. Models describing the mutation probabilities of homologous nucleotides have been continuously refined since the pioneering work on this topic by Jukes and Cantor in the late 1960's [16,29]. However, a recent study indicates that k-tuple distances may be highly accurate when compared to conventional model-based distances [28].

We have developed a new alignment-free distance measure, which we call K_r. The central idea of our approach is that closely related sequences share longer exact matches than distantly related sequences. In the following we derive K_r, describe its implementation, and demonstrate its utility through simulation. We then apply it to three data sets of increasing size: 27 primate mitochondrial genomes, eight complete genomes of the bacterial pathogen *Streptococcus agalactiae*, which is a leading cause of bacterial sepsis in neonates [24], and the complete genomes of twelve species of *Drosophila* [25]. In each case cluster analysis of K_r values recovers a topology that is close or identical to the accepted phylogeny.

2 Approach and Data

2.1 Definition of K_r

Consider two sequences, $Q =$ TATAC and $S =$ CTCTGG, which we call *query* and *subject*, respectively. For every suffix of Q, $Q[i..|Q|]$, we look up the shortest prefix, $Q[i..j]$, that is absent from S. This special prefix is called a *Shortest Absent Prefix* (SAP) and denoted by q_i. We start by examining the first suffix of our example query, which covers the entire sequence: $Q[1..|Q|] =$ TATAC. Its shortest prefix, $Q[1..1] =$ T, does occur in S and hence we extend it by one position to get $Q[1..2] =$ TA, which is absent from S yielding our first SAP, $q_1 =$ TA. Next we determine the shortest prefix of $Q[2..|Q|] =$ ATAC that is absent from S and find $q_2 =$ A, and so on. Notice that there is no prefix of $Q[5..|Q|] =$ C that is absent from S. In this case we define $q_i = Q[i..|Q| + 1]$; in other words, we pretend that Q (and S) are terminated by a unique sentinel character (\$) to guarantee that q_i exists for all i. Finally we have the SAPs $q_1 =$ TA, $q_2 =$ A, $q_3 =$ TA, $q_4 =$ A, and $q_5 =$ C\$.

Our algorithm is based on the lengths of the SAPs, $|q_i|$. The key insight leading from these lengths to a distance measure is that if Q and S are closely related, they are characterized by many long exact repeats between Q and S. As a consequence, SAP lengths will tend to be greater than if Q and S are only distantly related.

To make this notion rigorous, we define the observed aggregate SAP length

$$A_{\mathrm{o}} = \sum_{i=1}^{|Q|} |q_i|$$

and its expectation, A_{e}, which can be computed either analytically [12] or through shuffling of S. Next, we take the logarithm of $A_{\mathrm{o}}/A_{\mathrm{e}}$ and normalize this quantity by the maximum value it can take to define the index of repetitiveness, I_{r}

$$I_{\mathrm{r}}(Q, S) = \frac{\ln(A_{\mathrm{o}}/A_{\mathrm{e}})}{\ln(\max(A_{\mathrm{o}})/A_{\mathrm{e}})}, \tag{1}$$

where

$$\max(A_{\mathrm{o}}) = \begin{cases} \binom{|Q|+2}{2} - 1 & \text{if } |Q| \le |S| \\ (|Q| - |S| + 1)(|S| + 1) + \binom{|S|+1}{2} - 1 & \text{otherwise.} \end{cases} \tag{2}$$

We therefore have

$$\sim 0 \le I_{\mathrm{r}} \le 1.$$

The ceiling of the I_{r} domain is exact—any pair of identical query and subject sequences are maximally repetitive and have $I_{\mathrm{r}} = 1$. In contrast, the floor is an expectation for reasonably long shuffled sequences of any GC content. The definition of I_{r} presented here extends an earlier version [13] by adding the query/subject distinction and the normalization.

We used simulations to explore the relationship between I_{r} and the number of pairwise mismatches per nucleotide, d. One thousand pairs of 10 kb long sequences with a fixed d were generated and Figure 1 displays d as a function of simulated $\ln(I_{\mathrm{r}})$ values. The shape of the bottom right hand part of the curve tells us that in pairs of similar sequences few mutations have a large effect on I_{r}. We found that the relationship between divergence and I_{r} could conveniently be modeled with the statistical software R [22] using two logistic functions, one covering $\ln(I_{\mathrm{r}}) > -2.78$ and the other covering the rest. Given these two functions, we define the number of pairwise differences based on the I_{r}, d_{r}:

$$d_{\mathrm{r}} = \begin{cases} \frac{0.1380}{1 + e^{(-2.2016 - \ln(I_{\mathrm{r}}))/-0.5307}} & \text{if } \ln(I_{\mathrm{r}}) > -2.78 \\ \frac{0.6381}{1 + e^{(-5.5453 - \ln(I_{\mathrm{r}}))/-1.7113}} & \text{otherwise.} \end{cases} \tag{3}$$

The dashed line in Figure 1 indicates that this model gives a useful approximation of the simulated values shown as dots.

Finally, the number of pairwise mismatches, d_{r}, was converted into our distance measure, K_{r}, using the formula by Jukes and Cantor [16]:

$$K_{\mathrm{r}} = -\frac{3}{4} \ln\left(1 - \frac{4}{3} d_{\mathrm{r}}\right). \tag{4}$$

2.2 Asymmetric Values of K_{r}

In general and depending on which sequence is designated query, the two resulting I_{r} values differ, that is, $I_{\mathrm{r}}(S_1, S_2) \ne I_{\mathrm{r}}(S_2, S_1)$. Direct application of equations (3) and (4)

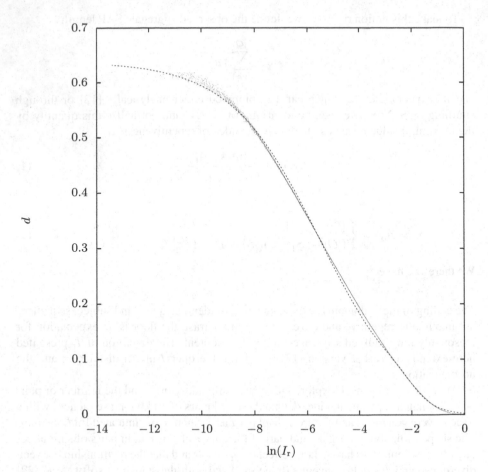

Fig. 1. Simulated (dots) and modeled (dashed) relationship between the number of pairwise differences per site, d, and the index of repetitiveness, I_r. Each dot represents an average of 1000 I_r values calculated from 1000 pairs of sequences characterized by a given value of d. The model relationship is stated in Equation (3).

would translate this inequality into asymmetric matrices of K_r values, which is unacceptable for a metric. In the case of two "ideal" sequences devoid of insertions/deletions and repetitive elements, the difference is due to stochastic placement of mutations along a DNA sequence. However, indels and shared repetitive elements may cause systematic differences between the two possible query/subject configurations.

Figure 2A shows an example in which S_1 has undergone large deletions and as a result is much shorter than S_2, i.e. S_2 is only locally homologous to S_1. In this case $I_r(S_1, S_2) < I_r(S_2, S_1)$. However, regions in S_2 that have no homologue in S_1 are characterized by SAPs that are only as long as expected by chance. We have therefore implemented a global and a local mode for K_r computation. In the global mode all SAPs are included in the analysis. In the local mode the user can set the fraction,

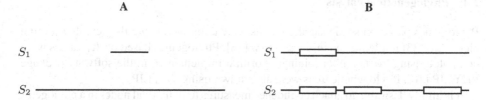

Fig. 2. Sources of asymmetric I_r values. **A:** S_2 is only locally homologous to S_1, in which case $I_r(S_1, S_2) < I_r(S_2, S_1)$; **B:** S_1 contains a lower copy number of a genetic element than S_2, in which case again $I_r(S_1, S_2) < I_r(S_2, S_1)$.

say 0.5, of SAP lengths compatible with randomness that are excluded from the analysis. All applications to real data presented in this paper were computed using the local mode.

Figure 2B illustrates variation in the copy number of a shared element: S_1 contains one copy of the element and S_2 three, which again leads to $I_r(S_1, S_2) < I_r(S_2, S_1)$. Since many mutations are necessary to reverse the effect of a single gene duplication on I_r, we always chose the lower of the two values for the computation of K_r.

2.3 Implementation

Conceptually, SAP lengths are determined in a single bottom-up traversal of a generalized sufix tree [11] containing the forward and reverse strands of the query and subject data sets. Each internal node in this tree, n, is classified as *isQuery* if the subtree rooted on it has leaves referring to positions in the query sequences, and as *isSbjct*, if the subtree rooted on it has leaves referring to positions in the subject sequences. Both properties propagate up the tree. If n *isQuery* and *isSbjct*, its child nodes, c_i, are searched for two relevant cases: First, c_i may be a leaf referring to a query position x. In that case the desired SAP length, $|q_x|$, is the string depth of c_i plus 1. Second, c_i may be an internal node with the property *isQuery* but not *isSbjct*. Then the leaves of the subtree rooted on c_i are looked up and the string depth of c_i plus 1 is the desired length of the SAPs referred to by these leaves.

We based the implementation of the suffix tree traversal on its more space-efficient sister data structure, the enhanced suffix array [2]. For this purpose we used the suffix array library by Manzini and Ferragina [19], as it is fast and space-efficient [21]. In its original form, the library was limited to the analysis of $2^{31} \approx 2 \times 10^9$ characters, which we have re-engineered to a limit of $2^{63} \approx 9 \times 10^{18}$ characters.

Our program for calculating the K_r is called kr. It takes as input a set of FASTA-formatted sequences and returns a distance matrix in PHYLIP [10] format. The program can be accessed via a simple web interface at

$$http://guanine.evolbio.mpg.de/kr/$$

The C source code of kr is also available from this web site under the GNU General Public License.

2.4 Phylogenetic Analysis

Phylogenies based on sequence alignments were computed using the neighbor joining algorithm [23] implemented in `clustalw` [18]. Phylogenies based on K_r values were computed using the neighbor joining algorithm implemented in the software package PHYLIP [10]. Phylogenetic trees were also drawn using PHYLIP.

It is highly desirable to attach confidence measures to individual nodes in a phylogeny. A popular method for achieving this is bootstrap analysis [7]. The central question in any bootstrap analysis is, what is the unit to be sampled with replacement (bootstrapped)? In traditional bootstrap analysis of phylogenies, columns of homologous nucleotides are sampled with replacement from the underlying multiple sequence alignment [9]. This cannot be applied in the context of an alignment-free distance measure such as K_r. Instead, we propose to sample random fragments of 500 bp length with replacement from the original sequences.

Table 1. Primate mitochondrial genomes analyzed in this study

#	Name	Genbank Common Name	Accession
1	*Cebus albifrons*	white-fronted capuchin	NC_002763.1
2	*Chlorocebus aethiops*	African green monkey	NC_007009.1
3	*Chlorocebus pygerythrus*	green monkey	NC_009747.1
4	*Chlorocebus sabaeus*	green monkey	NC_008066.1
5	*Chlorocebus tantalus*	green monkey	NC_009748.1
6	*Colobus guereza*	guereza	NC_006901.1
7	*Cynocephalus variegatus*	Sunda flying lemur	NC_004031.1
8	*Gorilla gorilla*	western Gorilla	NC_001645.1
9	*Homo sapiens*	human	NC_001807.4
10	*Hylobates lar*	common gibbon	NC_002082.1
11	*Lemur catta*	ring-tailed lemur	NC_004025.1
12	*Macaca mulatta*	rhesus monkey	NC_005943.1
13	*Macaca sylvanus*	Barbary ape	NC_002764.1
14	*Nasalis larvatus*	proboscis monkey	NC_008216.1
15	*Nycticebus coucang*	slow loris	NC_002765.1
16	*Pan paniscus*	pygmy chimpanzee	NC_001644.1
17	*Pan troglodytes*	chimpanzee	NC_001643.1
18	*Papio hamadryas*	hamadryas baboon	NC_001992.1
19	*Pongo pygmaeus*	Bornean orangutan	NC_001646.1
20	*Pongo pygmaeus abelii*	Sumatran orangutan	NC_002083.1
21	*Presbytis melalophos*	mitred leaf monkey	NC_008217.1
22	*Procolobus badius*	western red colobus	NC_008219.1
23	*Pygathrix nemaeus*	Douc langur	NC_008220.1
24	*Pygathrix roxellana*	golden snub-nosed monkey	NC_008218.1
25	*Semnopithecus entellus*	Hanuman langur	NC_008215.1
26	*Tarsius bancanus*	Horsfield's tarsier	NC_002811.1
27	*Trachypithecus obscurus*	dusky leaf monkey	NC_006900.1

Table 2. *Streptococcus agalactiae* genomes and the corresponding multilocus sequence types analyzed in this study

# Strain	Accession	Sequence Type
1 18RS21	AAJO01000000	ST19
2 2603V/R	AAJP01000000	ST110
3 515	AAJQ01000000	ST23
4 NEM316	AAJR01000000	ST23
5 A909	AAJS01000000	ST7
6 CJB111	CP000114	ST1
7 COH1	AE009948	ST17
8 H36B	AL732656	ST6

2.5 Data Sets

Three sets of genomes were analyzed: 27 primate mitochondrial genomes (total of 446.23 kb), genomes of eight *S. agalactiae* strains (17.39 Mb), and the genomes of twelve *Drosophila* species (2.03 Gb).

The 27 primate mitochondrial genomes available from Genbank were downloaded and compared without any further editing (Table 1).

The eight *S. agalactiae* genomes previously analyzed by [24] were downloaded from Genbank and subjected to K_r computation without further editing (Table 2). Complete multilocus sequence data for the sequence types corresponding to the these genomes was obtained from mlst.net [1].

The 12 *Drosophila* genomes consisting of up to 14,547 contigs each were downloaded from

http://rana.lbl.gov/drosophila/caf1/all_caf1.tar.gz

Unsequenced regions in these genomes marked by N were removed before K_r analysis, as these generate suffixes with long matching prefixes that distort the K_r.

3 Results

3.1 Clustering of Simulated DNA Sequences

Figure 1 demonstrates that the model underlying the computation of K_r is reasonably exact for divergence $d \leq 0.5$, which roughly corresponds to $\ln(I_r) \geq -8$, or $I_r \geq 0.0003$. In order to explore the utility of K_r for sequence clustering, we simulated a set of 12 DNA sequences of 10 kb with a maximal d of 0.5, that is, by distributing 5000 segregating sites on a random topology generated using the coalescent simulation program ms [15]. The true phylogeny of these sequences is shown in Figure 3A. Neighbor joining analysis of the 66 Jukes-Cantor distances between the dozen simulated sequences yielded the phenogram shown in Figure 3B, which is topologically identical to the true phylogeny. The branch lengths of Figure 3A and B also look almost indistinguishable. However, they differ in numerical detail as illustrated for the edges

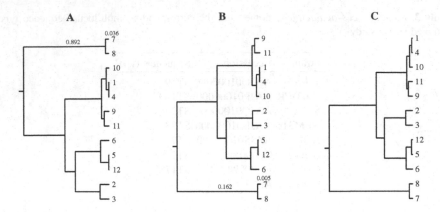

Fig. 3. Reconstructing the phylogeny of 12 simulated sequences. **A:** True phylogeny; **B:** phylogeny based on multiple sequence alignment by clustalw [18]; **C:** phylogeny based on K_r. The small numbers on the edges leading from taxon 7 to the root illustrate branch length differences between phylogenies **A** and **B**.

connecting taxa 7 and 8 to the root. Phylogeny reconstruction based on K_r returned the tree shown in Figure 3C. It is topologically identical to the cluster diagram based on standard pairwise distances (Figure 3B). Again, the branch lengths also look very similar but the diagram reveals small differences such as the distance between taxa 5 and 12, which is larger in the K_r phylogeny than in the other two. We shall see that the K_r measure has a tendency to overestimate terminal branch lengths.

Next we investigate the performance of K_r when applied to real sequences.

3.2 Clustering Primate Mitochondrial Genomes

Figure 4 displays two phylogenies of primate mitochondrial genomes, one based on K_r (A), the other on a multiple sequence alignment by clustalw (B). The two trees share important clades, particularly groups of closely related taxa. For example, the well-known great ape clade (asterisk in Figure 4) is resolved correctly using K_r. In contrast, within the Cercopithecinae (bullet in Figure 4) *Pio hamadryas* ought to cluster with the macaques (Figure 4B) rather than with the green monkeys (*Chlorocebus*, Figure 4A).

3.3 Clustering *Streptococcus agalactiae* Genomes

Tettelin and colleagues analyzed the complete genomes of eight *S. agalactiae* strains and reconstructed their phylogeny by comparing gene content [24]. Surprisingly, they obtained a phylogeny that did not cluster strains 515 and NEM316, even though these belong to the same multilocus sequence type (ST23; Table 2). In our K_r phylogeny of complete genomes these strains again appear as closest neighbors with 100% bootstrap support (Figure 5A). Overall the topology of this phylogeny is similar to a clustalw tree based on multilocus sequence data (Figure 5B). In contrast to the topology, the branch lengths derived from the two methods differ markedly, with the

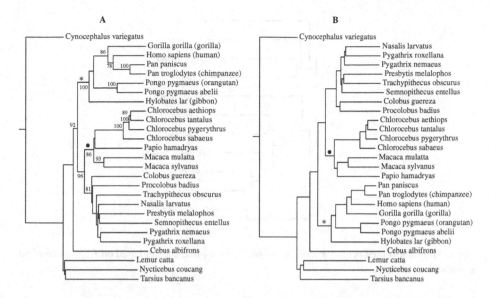

Fig. 4. Phylogeny of 27 primate mitochondrial genomes. The asterisk (∗) marks the ape clade (Hominoidea), the bullet (•) the Cercopithecinae among the old world monkeys (Cercopithecidae). **A**: Distance estimates based on K_r, bootstrap (100 replicates) greater than 75% are shown; **B**: distances based on multiple sequence alignment, all bootstrap values were greater than 95%.

external branches being much longer in the K_r tree. This is not simply a consequence of the K_r tree being computed from whole genomes and the clustalw tree from multilocus sequence data. When we subjected the same multilocus sequence data to K_r

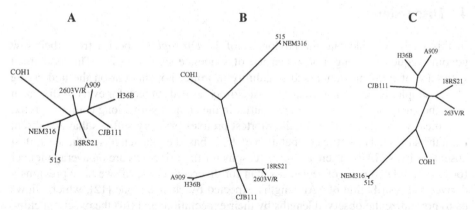

Fig. 5. Phylogenies of eight *Streptococcus agalactiae* strains. **A**: Based on K_r and whole genomes, all bootstrap values (100 replicates) were 100%; **B**: same set of organisms as **A**, but tree based on an alignment of multilocus sequence data using clustalw; **C**: same organisms and data as in **B**, but clustering based on K_r.

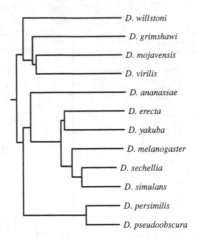

D. willstoni
D. grimshawi
D. mojavensis
D. virilis
D. ananassae
D. erecta
D. yakuba
D. melanogaster
D. sechellia
D. simulans
D. persimilis
D. pseudoobscura

Fig. 6. Midpoint-rooted neighbor-joining tree of 12 *Drosophila* species based on K_r and complete genome sequences

analysis, we obtained the tree shown in Figure 5C. This is topologically similar to the alignment-based tree but has longer terminal branches.

3.4   Clustering *Drosophila* Genomes

Calculating the K_r values for the 12 *Drosophila* species investigated took four days and 18 hours of CPU time on a computer with 64 GB RAM. The resulting phylogeny in Figure 6 has the same topology as the tree computed as part of the *Drosophila* dozen project [25].

4   Discussion

In this study we calculate the phylogeny of 12 *Drosophila* species from their raw genome sequences using a new measure of sequence similarity, K_r. This is defined with ease of implementation and scalability in mind. For this reason the underlying idea is simple: if we compare a query to a closely related subject, for every suffix taken from the query one finds on average a suffix in the subject with a long common prefix. Specifically, we concentrated on the shortest prefixes of query suffixes that are absent from the subject. The entire computation of K_r is based on the lengths of these shortest absent prefixes, SAPs. There are three reasons for this: (i) SAPs are on average longer for closely related pairs of sequences than for divergent pairs; (ii) we have previously derived the distribution of SAP lengths expected by chance alone [12], which allows us to normalize the observed lengths by their expectation; and (iii) the exact matching strategy for distance computation we propose is very quick as it is based on enhanced suffix array traversal [2].

Technicalities aside, our approach is to transform exact match lengths to distances using the Jukes Cantor model [16]. This is the oldest and simplest model of nucleotide evolution. It is clear that its application across species with strong intra-genomic variation in mutation rates as observed in *Drosophila* [20] violates the model assumption of rate uniformity across residues and positions. However, the very large amount of sequence information contained in the *Drosophila* genomes leads to the recovery of the correct clades from K_r in spite of the simplifications of the model.

The trade-off between speed and precision is well known in the field of sequence alignment. For example, clustalw has a slow, accurate and a fast, approximate mode ("quicktree") for guide tree computation. Like our K_r calculation, the fast mode of guide tree reconstruction is based on alignment-free pairwise sequence comparison. However, kr is both faster and more sensitive than the quicktree mode. For example, kr takes half as long as clustalw in quicktree mode to compute the guide tree for the 27 primate mitochondrial genomes. The difference in run time grows to 12-fold for a simulated sample of 27 sequences that are 100 kb long, that is 6 times longer than the primate mitochondrial genomes. In addition, K_r tends to resolve closely related sequences better than the quicktree mode (not shown).

The reason for this sensitivity to small differences in sequence similarity was apparent in the long terminal branches of the phylogeny based on multilocus sequence data (Figure 5C) compared to the alignment-based phylogeny (Figure 5B). This emphasis on recent mutations is already apparent in the simulated relationship between divergence, d, and I_r (Figure 1). The lower right corner of this graph indicates that the addition of few mutations to a pair of identical sequences has a strong effect on the I_r and hence on K_r. This suggests great sensitivity to differences among closely related sequences, leading to the long terminal branches observed in *S. agalactiae* (Figures 5A and C) and *Drosophila* (Figure 6). Sensitivity and speed of execution make kr a promising tool for the computation of guide trees that can be used as input to multiple sequence alignment programs such as clustalw or the more powerful MAVID [4].

The sensitivity of K_r restricts its application to closely related DNA sequences, which is an important limitation of our method. Figure 1 allows us to quantify the range of diversity values for which K_r computations might be attempted: For divergence values greater than 0.5 the relationship between d and I_r becomes increasingly noisy. Under the Jukes-Cantor model of sequence evolution [16] a d-value of 0.5 corresponds to 0.82 substitutions/site. Substitution rates in *Drosophila* genes vary between 11.0×10^{-9} and 27.1×10^{-9}/site/year [20]. If we take the average of these values (19.5×10^{-9}), we arrive at a maximum evolutionary distance of 43.4 million years for our method. This is approximately the divergence time of the *Drosophila* clade analyzed in Figure 6. Taxa with lower substitution rates could, of course, be analyzed to correspondingly greater evolutionary distances, but this rough calculation illustrates the caveat that K_r should only be applied to closely related genomes. Given this proviso, our distance measure gives biologically meaningful results on scales ranging from mitochondrial to metazoan nuclear genomes.

Acknowledgements

We thank Peter Pfaffelhuber, Angelika Börsch-Haubold, and an anonymous reviewer for comments that improved this manuscript.

References

1. Aanensen, D.M., Spratt, B.G.: The multilocus sequence typing network: mlst.net. Nucleic Acids Res. 33(Web Server issue) , W728–W733 (2005)
2. Abouelhoda, M.I., Kurtz, S., Ohlebusch, E.: The enhanced suffix array and its applications to genome analysis. In: Proceedings of the second workshop on algorithms in bioinformatics. Springer, Heidelberg (2002)
3. Blaisdell, B.E.: A measure of the similarity of sets of sequences not requiring sequence alignment. Proceedings of the National Academy of Sciences, USA 83, 5155–5159 (1986)
4. Bray, N., Pachter, L.: MAVID: Constrained ancestral alignment of multiple sequences. Genome Research 14, 693–699 (2004)
5. Chapus, C., Dufraigne, C., Edwards, S., Giron, A., Fertil, B., Deschavanne, P.: Exploration of phylogenetic data using a global sequence analysis method. BMC Evolutionary Biology 5, 63 (2005)
6. Dewey, C.N., Pachter, L.: Evolution at the nucleotide level: the problem of multiple whole-genome alignment. Hum. Mol. Genet. 15(Spec. No. 1), R51–R56 (2006)
7. Efron, B.: Bootstrap methods: another look at the Jackknife. The Annals of Statistics 7, 1–26 (1979)
8. Eisen, J.A.: Phylogenomics: improving functional predictions for uncharacterized genes by evolutionary analysis. Genome Research 8, 163–167 (1998)
9. Felsenstein, J.: Confidence limits on phylogenies: an approach using the bootstrap. Evolution 39, 783–791 (1985)
10. Felsenstein, J.: PHYLIP (Phylogeny Inference Package) version 3.6. Distributed by the author. Department of Genome Sciences, University of Washington, Seattle (2005)
11. Gusfield, D.: Algorithms on Strings, Trees, and Sequences: Computer Science and Computational Biology. Cambridge University Press, Cambridge (1997)
12. Haubold, B., Pierstorff, N., Möller, F., Wiehe, T.: Genome comparison without alignment using shortest unique substrings. BMC Bioinformatics 6, 123 (2005)
13. Haubold, B., Wiehe, T.: How repetitive are genomes? BMC Bioinformatics 7, 541 (2006)
14. Hervé, P., Delsuc, F., Lartillot, N.: Phylogenomics. Annual Review of Ecology, Evolution, and Systematics 36, 541–562 (2005)
15. Hudson, R.R.: Generating samples under a Wright-Fisher neutral model of genetic variation. Bioinformatics 18, 337–338 (2002)
16. Jukes, T.H., Cantor, C.R.: Evolution of protein molecules. In: Munro, H.N. (ed.) Mammalian Protein Metabolism, vol. 3, pp. 21–132. Academic Press, New York (1969)
17. Kantorovitz, M.R., Robinson, G.E., Sinha, S.: A statistical method for alignment-free comparison of regulatory sequences. Bioinformatics 23, i249–i255 (2007)
18. Larkin, M.A., Blackshields, G., Brown, N.P., Chenna, R., McGettigan, P.A., McWilliam, H., Valentin, F., Wallace, I.M., Wilm, A., Lopez, R., Thompson, J.D., Gibson, T.J., Higgins, D.G.: Clustal w and clustal x version 2.0. Bioinformatics 23(21), 2947–2948 (2007)
19. Manzini, G., Ferragina, P.: Engineering a lightweight suffix array construction algorithm. In: Möhring, R.H., Raman, R. (eds.) ESA 2002. LNCS, vol. 2461, pp. 698–710. Springer, Heidelberg (2002)
20. Moriyama, E.N., Gojobori, T.: Rates of synonymous substitution and base composition of nuclear genes in Drosophila. Genetics 130(4), 855–864 (1992)
21. Puglisi, S.J., Smyth, W.F., Turpin, A.H.: A taxonomy of suffix array construction algorithms. ACM Comput. Surv. 39, 4 (2007)
22. R Development Core Team. R: A Language and Environment for Statistical Computing. R Foundation for Statistical Computing, Vienna, Austria (2007) ISBN 3-900051-07-0

23. Saitou, N., Nei, M.: The neighbor-joining method: a new method for reconstructing phylgenetic trees. Molecular Biology and Evolution 4, 406–425 (1987)
24. Tettelin, H., Masignani, V., Cieslewicz, M.J., Donati, C., Medini, D., Ward, N.L., Angiuoli, S.V., Crabtree, J., Jones, A.L., Durkin, A.S., Deboy, R.T., Davidsen, T.M., Mora, M., Scarselli, M., Margarit y Ros, I., Peterson, J.D., Hauser, C.R., Sundaram, J.P., Nelson, W.C., Madupu, R., Brinkac, L.M., Dodson, R.J., Rosovitz, M.J., Sullivan, S.A., Daugherty, S.C., Haft, D.H., Selengut, J., Gwinn, M.L., Zhou, L., Zafar, N., Khouri, H., Radune, D., Dimitrov, G., Watkins, K., O'Connor, K.J., Smith, S., Utterback, T.R., White, O., Rubens, C.E., Grandi, G., Madoff, L.C., Kasper, D.L., Telford, J.L., Wessels, M.R., Rappuoli, R., Fraser, C.M.: Genome analysis of multiple pathogenic isolates of Streptococcus agalactiae: implications for the microbial "pan-genome". Proc. Natl. Acad. Sci. USA 102(39), 13950–13955 (2005)
25. Drosophila 12 Genomes Consortium. Evolution of genes and genomes on the Drosophila phylogeny. Nature 450, 203–218 (2007)
26. Vinga, S., Almeida, J.: Alignment-free sequence comparison—a review. Bioinformatics 19, 513–523 (2003)
27. Wilbur, W.J., Lipman, D.J.: Rapid similarity searches of nucleic acid and protein data banks. Proceedings of the National Academy of Sciences, USA 80, 726–730 (1983)
28. Yang, K., Zhang, L.: Performance comparison between k-tuple distance and four model-based distances in phylogenetic tree reconstruction. Nucleic Acids Res. 36(5), e33 (2008)
29. Yang, Z.: Computational Molecular Evolution. Oxford University Press, Oxford (2006)

Gene Team Tree:
A Compact Representation of All Gene Teams

Melvin Zhang and Hon Wai Leong

School of Computing, National University of Singapore

Abstract. The identification of conserved gene clusters is an important step towards understanding genome evolution and predicting the function of genes. Gene team is a model for conserved gene clusters that takes into account the position of genes on a genome. Existing algorithms for finding gene teams require the user to specify the maximum distance between adjacent genes in a team. However, determining suitable values for this parameter, δ, is non-trivial. Instead of trying to determine a single best value, we propose constructing the *gene team tree (GTT)*, which is a compact representation of all gene teams for every possible value of δ. Our algorithm for computing the GTT extends existing gene team mining algorithms without increasing their time complexity. We compute the GTT for *E. coli* K-12 and *B. subtilis* and show that *E. coli* K-12 operons are recovered at different values of δ. We also describe how to compute the GTT for multi-chromosomal genomes and illustrate using the GTT for the human and mouse genomes.

1 Introduction

Biological evidence suggests that genes which are located close to one another in several different genomes tend to code for proteins that have a functional interaction [Snel et al., 2002]. Such regions are also commonly known as *conserved gene clusters*. Computational studies done in Overbeek et al. [1999] showed that functional dependency of proteins can be inferred by considering the spatial arrangement of genes in multiple genomes. In the study of prokaryotic genomes, the identification of conserved gene clusters is used in predicting operons [Ermolaeva et al., 2001] and detecting horizontal gene transfers [Lawrence, 1999]. Therefore, the identification of conserved gene clusters is an important step towards understanding genome evolution and function prediction.

A popular model of conserved gene clusters is the *gene team* model [Béal et al., 2004]. Current algorithms require the specification of the parameter δ, which is the maximum distance between adjacent genes in a gene team. However, determining suitable values for this parameter is non-trivial as it depends on the arrangement of genes on the genome. As discussed in He and Goldwasser [2005], a large value of δ may result in many false positives while an overly conservative value may miss many potential conserved clusters. In addition, due to varying rates of rearrangement, different regions of the genome may require different values of δ to discover meaningful gene clusters. The value of δ also depends

C.E. Nelson and S. Vialette (Eds.): RECOMB-CG 2008, LNBI 5267, pp. 100–112, 2008.
© Springer-Verlag Berlin Heidelberg 2008

on the type of conserved gene clusters (pathways, regulons or operons) one is interested in.

Our approach is to find a succinct way to represent all gene teams for every value of δ and compute this representation efficiently. Subsequently, statistical tests [Hoberman et al., 2005] or integration with other information on gene interactions can be used to validate or rank the discovered teams.

The rest of this paper is structured as follows: Section 2 discusses other related efforts in the literature. A formal problem definition is presented in Section 3, followed by a detailed description of our method in Section 4. Section 5 demonstrates the practicality of our methodology using real datasets. We summarize our results and discuss future work in Section 6.

2   Related Work

The notion of a conserved gene cluster may seem intuitive, however, developing a formal definition which captures the essential biological characteristics of such regions is non-trivial [Hoberman and Durand, 2005]. Intuitively, a conserved gene cluster represents a compact region in each genome which contains a large proportion of homologous genes. Due to the effect of rearrangement events, the order of the genes in a conserved gene cluster is usually not conserved.

The gene team model [Béal et al., 2004] is the first formal model to make use of the position of the genes on the genome. The original model assumes that genomes are permutations, i.e. they do not contain duplicated genes. The model was later extended in He and Goldwasser [2005] to handle general sequences, which allows multiple copies of the same gene. Although the term homology team was used in He and Goldwasser [2005] to refer to the extended model, for simplicity we will refer to both of them as the gene team model.

Béal et al. [2004] gave an $O(mn \lg^2 n)$ algorithm for finding the gene teams of m permutations with n genes each. An $O(n_1 n_2)$ time algorithm was proposed in He and Goldwasser [2005] for finding the gene teams for two sequences of length n_1 and n_2 respectively. Both algorithms are similar and are based on the divide-and-conquer paradigm.

In the experimental study presented in He and Goldwasser [2005], the approach used to determine an appropriate value of δ was to select a small number of known operons and pick the minimum value of δ at which the selected operons were reconstructed. There are two drawbacks with this method. Firstly, there may not be any known operons in the genome we are interested in. Furthermore, it is unclear how to select a representative set of known operons.

3   Problem Definition

3.1   Notations and Definitions

The following notations and definitions are adapted from He and Goldwasser [2005].

A homology family is a collection of genes which have descended from a common ancestral gene.

Let Σ denote the set of homology families, then each gene g is a pair (p, f) where $p \in \mathbb{R}$ is the position of the gene and $f \in \Sigma$ is the homology family which the gene belongs to. The distance between two genes $g_i = (p_i, f_i)$ and $g_j = (p_j, f_j)$ is defined as $\Delta(g_i, g_j) = |p_i - p_j|$.

A gene order G is a sequence of genes $\langle g_1, g_2, \ldots, g_n \rangle$, in increasing order of their position. A gene order is a permutation if each homology family appears at most once. We use $\Sigma(G)$ to denote the set of homology families which appears in G.

Definition 1 (Witness and δ-set). *A subsequence G' of a gene order G is a witness to a δ-set of homology families $\Sigma' \subseteq \Sigma$ if $\Sigma(G') = \Sigma'$ and every pair of adjacent genes in G' are separated by a distance of at most δ.*

Definition 2 (δ-team). *A δ-team t, of a set of gene order $\mathcal{G} = \{G_1, G_2, \ldots, G_m\}$, is a δ-set with witnesses in each gene order $G_i, 1 \leq i \leq m$, and it is maximal with respect to subset inclusion of the witnesses. We use $\mathcal{G}(t)$ to denote the set of witnesses of a δ-team t.*

Example 1. Consider two gene orders $G_1 = \langle a_1, b_3, c_5, d_6, e_9, c_{10}, b_{11}, a_{12}, b_{13} \rangle$ and $G_2 = \langle c_1, a_3, d_4, b_6, e_7, b_8, c_9, a_{11}, d_{12} \rangle$ where the letters represent homology families and the numbers in the subscript denote the position of the genes. Then, the 2-teams of G_1 and G_2 are $\{a, c\}$, $\{a, b, c, d\}$, and $\{a, b, c, e\}$.

3.2   Problem Formulation

Given a set of m gene orders $\mathcal{G} = \{G_1, G_2, \ldots, G_m\}$, compute the set of δ-teams for every possible value of δ, denoted by $\mathcal{T}$.

Let GENETEAMS($\mathcal{G}, \delta$) denote the δ-teams of $\mathcal{G}$, then formally,

$$\mathcal{T} = \bigcup_{\delta=0}^{\infty} \text{GENETEAMS}(\mathcal{G}, \delta)$$

4   Our Approach: Gene Team Tree

A naïve approach is to apply a parameterized algorithm for every possible value of δ but this is hopelessly inefficient when the parameter space is large.

There are two gross inefficiencies with the naïve approach. Firstly, not all parameter values lead to new gene teams and secondly, gene teams computed for one value of the parameter can be used to compute the gene teams for other values.

Observe that as the maximum allowed gap length decreases, existing gene teams are split into smaller teams. This allows us to represent the set of gene teams for all possible values of δ compactly in a tree structure.

Definition 3 (Gene Team Tree). *The* gene team tree (GTT) *of a set of gene orders $\mathcal{G}$ has as its nodes the δ-teams for all possible values of δ. The root of the GTT is Σ and a gene team t' is a child of a gene team t if $t' < t$ and there is no other gene team t'' such that $t' < t'' < t$, where $t' < t$ if the witnesses of t' are subsequences of the witnesses for t.*

Remark 1. The leaf nodes of the GTT are singletons and the number of leaves in the GTT for a set gene orders $\mathcal{G}$ is $\sum_{f \in \Sigma} \prod_{G \in \mathcal{G}} \mathrm{occ}(G, f)$, where $\mathrm{occ}(G, f)$ is the number of genes in G from homology family f.

4.1   Basic GeneTeamTree Algorithm

Clearly, the largest possible gene team is Σ. Σ is a gene team when δ is greater than or equal to the largest distance between adjacent genes in each of the m gene orders. For a set of gene orders $\mathcal{G}$, let $\mathrm{MaxGap}(\mathcal{G})$ denote the largest distance between adjacent genes for every gene order in $\mathcal{G}$.

Our algorithm GeneTeamTree takes in a set of gene orders $\mathcal{G}$ and returns the GTT of $\mathcal{G}$. Initially, we start with the gene team $\Sigma(\mathcal{G})$. In each call to GeneTeamTree, we construct a tree node v which stores the gene team. Then, we compute $\mathrm{MaxGap}(\mathcal{G}) - \epsilon$, which is the largest value of δ that will cause the current gene team to be partitioned into smaller sub gene teams. We then make use of the FindTeams algorithm [Béal et al., 2004; He and Goldwasser, 2005] to partition t into a set of smaller gene teams T. For each gene team in T, recursively apply GeneTreamTree to get a tree of gene teams and make it a subtree of v. Finally, the algorithm returns v, which is the root of the GTT. The pseudocode for the algorithm is shown in Algorithm 1.

In practice, if the positions are integers, we set ϵ to be 1. This is the case in our examples and experiments. Otherwise, ϵ can be set to some number that is smaller than the distance between the closest pair of adjacent genes.

Algorithm 1. GeneTeamTree($\mathcal{G}$)

Ensure: Returns the GTT for $\mathcal{G}$

   team(v) := $\Sigma(\mathcal{G})$

   children(v) := $\emptyset$

   if $|\,\text{team}(v)| > 1$ **then**

     δ := $\mathrm{MaxGap}(\mathcal{G}) - \epsilon$

     T := FindTeams($\mathcal{G}, \delta$)

     for each gene team $t \in T$ **do**

       children(v) := children(v) $\cup$ GeneTeamTree($\mathcal{G}(t)$)

     end for

   end if

   return v

Example 2. Running the GeneTeamTree algorithm on the pair of gene order $G_1 = \langle a_1, b_2, a_6, c_8, b_9 \rangle$ and $G_2 = \langle c_1, c_4, b_5, a_6, b_8 \rangle$ produces the tree shown in Figure 1, where a gene team is represented by the pair of its witnesses.

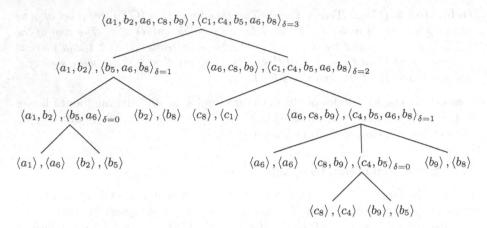

Fig. 1. GTT for $\langle a_1, b_2, a_6, c_8, b_9 \rangle$ and $\langle c_1, c_4, b_5, a_6, b_8 \rangle$ where a gene team is represented by the pair of its witnesses and the value of δ used to split each gene team is shown in subscripts

4.2   Correctness of GeneTeamTree

Let GENETEAMTREE($\mathcal{G}$) denote the result of running algorithm GENETEAMTREE on the set of gene orders $\mathcal{G}$. In order to show that the algorithm GENETEAMTREE is correct, we need to prove that

$$\text{GENETEAMTREE}(\mathcal{G}) = \mathcal{T}$$

First, we show that only certain values δ will lead to new gene teams.

Lemma 1

$$\text{GENETEAMS}(\mathcal{G}, \delta) \begin{cases} = \{\Sigma\} & \text{if } \delta \geq \text{MAXGAP}(\mathcal{G}), \\ \neq \{\Sigma\} & \text{otherwise} \end{cases}$$

Proof Since MAXGAP($\mathcal{G}$) is the largest distance between adjacent genes, when $\delta \geq$ MAXGAP($\mathcal{G}$), Σ clearly satisfies the definition of a gene team.

Otherwise, there exists a pair of adjacent genes g, g' such that $\Delta(g, g') =$ MAXGAP($\mathcal{G}$). g and g' cannot be in the same gene team, thus Σ cannot be a gene team.

The above result can be generalized to an arbitrary gene team, therefore not all values of δ lead to new gene teams. There are certain critical values of δ which lead to the formation of new gene teams, these are the values of δ which are just less than the maximum gap between adjacent genes in a gene team.

The following lemma shows that once a pair of genes is in two different gene teams for a particular value of δ say δ_1, then they will always be in different gene teams when the value of δ decreases to say δ_2. This allows us to apply a divide-and-conquer strategy because gene teams form independent subproblems.

Lemma 2. *If $\delta_1 > \delta_2$ and* GENETEAMS$(\mathcal{G}, \delta_1) = T_1$ *and* GENETEAMS$(\mathcal{G}, \delta_2) =$ T_2 *then for any two genes g and g' if they are in the different gene teams in T_1 they are also in different gene teams in T_2.*

Proof. Suppose on the contrary that g and g' are in different gene teams in T_1 but they are in the same gene team in T_2. Let t be the gene team in T_1 that contains g and t' be the gene team in T_1 that contains g'. Since g and g' are in the same gene team in T_2 it follows that $t \cup t'$ must also be a gene team in T_1. This contradicts the maximality of t and t'.

Theorem 1 (GeneTeamTree is correct)

$$\text{GENETEAMTREE}(\mathcal{G}) = \mathcal{T}$$

Proof GENETEAMTREE finds gene teams for values of δ in decreasing order. Lemma 1 ensures that by using the value of $\delta = \text{MAXGAP}(\mathcal{G}) - \epsilon$ to partition the current gene team, we do not miss out on any gene teams. After partitioning the gene team t, we can consider the sub gene teams in a divide-and-conquer fashion since each sub gene team is independent of the others as a consequence of Lemma 2.

4.3   Time Complexity of GeneTeamTree

Given set of m gene orders, the time complexity of FINDTEAMS is $O(mn \lg^2 n)$ [Béal et al., 2004] when the gene orders are permutations of length n and $O(n_{\max}^m)$ for general sequence [He and Goldwasser, 2005], where $n_{\max}$ is the length of the longest gene order.

A simple implementation of MAXGAP perform a linear scan over each gene order to determine the largest distance between adjacent genes. It has a time complexity of $O(mn_{\max})$.

Let s_i denote the size of the ith gene team generated by FINDTEAMS and let $T(n)$ denote the time complexity of GENETEAMTREE, then

$$T(n) = \begin{cases} c & \text{if } n = 1, \\ \sum_{i=1}^{k} T(s_i) + \text{time for FINDTEAMS} + \text{time for MAXGAP} & \text{otherwise} \end{cases}$$

The running time for FINDTEAMS dominates that of MAXGAP. The worst case occurs when FINDTEAMS splits the gene team into two uneven partitions.

$$T(n) = T(1) + T(n-1) + \text{time for FINDTEAMS} + \text{time for MAXGAP}$$

$$= \begin{cases} O(mn^2 \lg^2 n) & \text{if the input gene orders are permutations,} \\ O(n_{\max}^{m+1}) & \text{otherwise} \end{cases}$$

4.4   Speeding Up GeneTeamTree

The basic algorithm presented in the previous section treats the FINDTEAMS procedure as a black box. It turn out that we can directly compute the GTT within FINDTEAMS to reduce the time complexity.

The key idea is to adjust the value for δ dynamically during the recursion in the FINDTEAMS procedure. When FINDTEAMS reports a gene team t, instead of terminating the recursion, we will dynamically reduce the value of δ to MAXGAP($\mathcal{G}$) $- \epsilon$ and continue recursing. Initially, the value of δ is set to ∞.

The improved algorithm, GENETEAMTREEFAST, is presented in Algorithm 2 and 3.

Algorithm 2. GENETEAMTREEFAST($\mathcal{G}$)

Ensure: Returns the GTT of $\mathcal{G}$
    team(v) := $\emptyset$ {Setup a dummy root node}
    children(v) := $\emptyset$
    return MODIFIEDFINDTEAMS($\mathcal{G}, \infty, v$)

Algorithm 3. MODIFIEDFINDTEAMS($\mathcal{G}, \delta, u$)

    $\delta' := $ MAXGAP($\mathcal{G}$) $- \epsilon$
    if $\delta' \leq \delta$ **then**
        {$\Sigma(\mathcal{G})$ is a gene team}
        team(v) := $\Sigma(\mathcal{G})$
        children(v) := $\emptyset$
        children(u) := children(u) $\cup$ {v}
        MODIFIEDFINDTEAMS($\mathcal{G}, \delta', v$)
        return v
    else
        $\mathcal{G}' := $ EXTRACTRUN($\mathcal{G}, \delta$)
        MODIFIEDFINDTEAMS($\mathcal{G}', \delta, u$)
        MODIFIEDFINDTEAMS($\mathcal{G} - \mathcal{G}', \delta, u$)
        return u
    end if

Recall that EXTRACTRUN [Béal et al., 2004] takes in as input the set of gene orders $\mathcal{G}$ and the parameter δ. It splits one of the gene orders of $\mathcal{G}$ across a gap of length greater than δ into two subsequences, keeping the shorter subsequence, G'. It then returns the set of gene orders obtained by extracting from each gene order in $\mathcal{G}$ the subsequence containing the same homology families as G'.

In order for our GENETEAMTREEFAST algorithm to be efficient, the complexity of MAXGAP should be at most that of EXTRACTRUN. The complexity of EXTRACTRUN in He and Goldwasser [2005] is $O(mn_{\max})$, therefore the simple linear scan implementation of MAXGAP suggested in the previous section suffices. However, the complexity of EXTRACTRUN in Béal et al. [2004] is $O(mp \lg p)$ where p is the size of the smaller sub problem.

We can reduce the time complexity of MAXGAP by storing the length of the gaps between adjacent genes in a priority queue. Then, the complexity of MAXGAP becomes $O(1)$, however we incur overhead maintaining the priority queue.

Creating the priority queue for the sub problem of size p requires mp insertions. The priority queue for the sub problem of size $n - p$ can be obtained by modifying the original priority queue. Extracting a single gene may involve merging two gaps, and is accomplished by deleting the two gaps and inserting the new combined gap into the priority queue. The total number of insertions and deletions needed to update the priority queue is $O(mp)$. Using a binary heap for our priority queue requires $O(\lg n)$ operations for deletions/insertions. Therefore, the total overhead is $O(mp \lg n)$, which is more than $O(mp \lg p)$.

Hence, the running time of our algorithm for m permutation is given by the following recurrence relation,

$$T(n) = T(n - p) + T(p) + cmp \lg n, 1 \leq p \leq n/2$$

Similar to the analysis presented in Béal et al. [2004], in the worst case, $p = n/2$ and $T(n) = O(mn \lg^2 n)$.

We analyzed the expected running time based on the assumption that at each stage , the size of the smaller subproblem, p, is uniformly distributed between 1 and $n/2$. Let $E(n)$ denote the expected running time of the algorithm. Then,

$$E(n) = \frac{1}{n/2} \sum_{p=1}^{n/2} E(n - p) + E(p) + cmp \lg n$$

$$nE(n) = 2 \sum_{p=1}^{n/2} E(n - p) + E(p) + cmp \lg n$$

$$nE(n) - (n - 2)E(n - 2) = 2E(n - 1) + 2E(n - 2) + cmn \lg n$$

$$nE(n) = 2E(n - 1) + nE(n - 2) + cmn \lg n$$

$$E(n) \geq \frac{n + 2}{n} E(n - 2) + cm \lg n$$

$$\geq cm(n + 2) \sum_{x=1}^{(n-3)/2} \frac{\lg(2x + 1)}{2x + 1}$$

$$\geq cm(n + 2) \int_{1}^{(n-1)/2} \frac{\lg(2x + 1)}{2x + 1} \, dx$$

$$= \Omega(mn \lg^2 n)$$

Since $T(n) = O(mn \lg^2 n)$ in the worst case, thus $E(n) = O(mn \lg^2 n)$. Therefore, the expected running time is $\Theta(mn \lg^2 n)$.

Our improved GENETEAMTREEFAST algorithm has a time complexity of $O(mn \lg^2 n)$ for m permutations of length n and $O(n_{max}^m)$ for m sequences of length at most n_{max}. Therefore, our algorithms gracefully extends existing FIND-TEAMS algorithms without increasing their time complexity.

4.5   Handling Multiple Chromosomes

A single chromosome can be directly represented using a gene order, however many genomes are multi-chromosomal. In practice, we would like to compare entire multi-chromosomal genomes and genes in a gene team should not be spread across several chromosomes.

A simple strategy to map multi-chromosomal genomes into a single linear gene order is to order the chromosomes linearly and insert appropriate gaps between the chromosomes. The order does not matter since the chromosomes will be separated immediately.

Let l_{max} be the length of the longest chromosome, then it suffices to insert a gap of length $l_{max} + \epsilon$ between chromosomes. Base on the definition of the GTT, the root consists of all the genes in the genome and the children of the root are $\text{GENETEAMS}(\mathcal{G}, l_{max})$. This will separate the genes in each chromosome but allows teams which consist of an entire chromosome. The subsequent sub problems will only be within a single chromosome.

5   Experiments

5.1   GTT for *E. coli* K-12 and *B. subtilis*

Similar to He and Goldwasser [2005], we use the notion of a *cluster of orthologous groups (COG)* [Tatusov et al., 2001] as an approximation of a homology family.

Using the data from He and Goldwasser [2005], we computed the GTT for *E. coli* K-12 and *B. subtilis* using the starting position of each gene (in base pairs) as its position. There are 1137 homology families in the dataset with 2332 genes in *E. coli* K-12 and 2339 genes in *B. subtilis*. The time taken to compute the GTT is around 4 seconds on a Intel Core 2 Duo E6550 (2.33GHz) processor running Linux. The GTT consists of 19712 nodes, a portion of the GTT is illustrated in Figure 2.

In order to validate our earlier claim that a single value of δ is inadequate, we compared the gene teams in our GTT against a set of known *E. coli* K-12 operons from Gama-Castro et al. [2008]. We use the Jaccard index [Jaccard, 1908] to measure the similarity between a gene team and an operon.

Definition 4. *The* Jaccard index *of two sets A and B is defined as* $\frac{|A \cap B|}{|A \cup B|}$. *It gives a value between zero and one, where a value of one is a perfect match.*

The Jaccard score of an operon is the highest Jaccard index between the operon and some gene team in our GTT. Figure 3 shows a histogram of the Jaccard

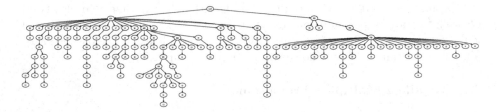

Fig. 2. GTT for *E. coli* K-12 and *B. subtilis* showing gene teams with at least 10 genes

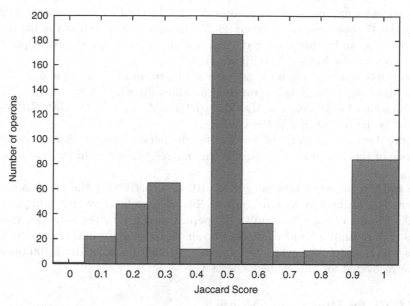

Fig. 3. Histogram of Jaccard scores for 471 operons from Gama-Castro et al. [2008]

scores for the gene teams corresponding to known operons. We only include operons with at least two genes and has at least one gene that is included in our study. There are a total of 471 operons that satisfy the above conditions.

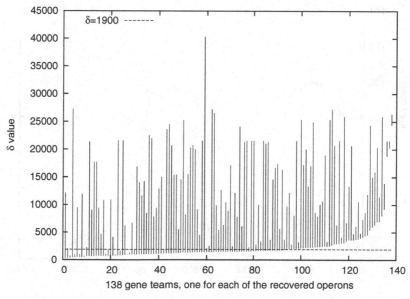

Fig. 4. $[\delta_{\min}, \delta_{\max}]$ for recovered operons, arranged in increasing $\delta_{\min}$

We consider an operon to be recovered if its Jaccard score is more than 2/3. Out of the 471 operons, 138 operons (29.3%) are recovered. Out of the 138 recovered operons, we further analyzed at what range of δ values the corresponding gene teams will be formed (see Figure 4).

Observe that a gene team t is not only a δ-team, but it is really a $[\delta_{min}, \delta_{max}]$-team, where $[\delta_{min}, \delta_{max}]$ is the range of δ values for which t is a δ-team. From the definition of GTT, $\delta_{min} = \text{MAXGAP}(\mathcal{G}(t))$ and $\delta_{max} = \text{MAXGAP}(\mathcal{G}(t')) - \epsilon$, where t' is the parent of t in the GTT.

Figure 4 shows the range of δ values for the gene teams which correspond to each one of the 138 recovered operons, the ranges are sorted in increasing order of δ_{min}.

He and Goldwasser [2005] suggested setting δ to 1900 for the analysis of this dataset. Out of the 138 operons, only 81 (58.7%) can be recovered using a value of 1900. Figure 5 shows the number of operons that can be recovered for various values of δ. It supports our claim that no single value of δ can generate the gene teams which match the entire set of recovered operons as at most 100 operons can be recovered for any single value of δ.

5.2   GTT for Human and Mouse

We downloaded the human and mouse genomes from the MSOAR [Fu et al., 2007] website and formed homology families using the MSOAR hit graph. There are 12662 shared homology families, with 14193 genes in the human genome and 14442 genes in the mouse genome.

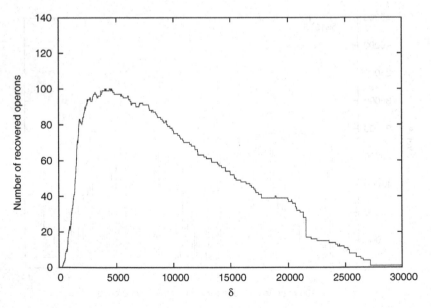

Fig. 5. Number of recovered operons for different values of δ

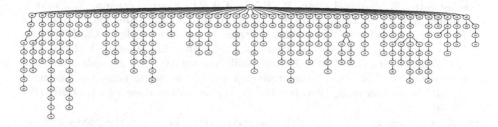

Fig. 6. GTT for human and mouse genomes showing gene teams with more than 100 genes

We computed the GTT for these two genomes using the method for handling multi-chromosomal genomes discussed in Section 4.5. The time taken to compute the GTT is around 5 seconds on a Intel Core 2 Duo E6550 (2.33GHz) processor running Linux. The GTT consists of 35874 nodes, a portion of the GTT is illustrated in Figure 6.

This experiment shows that our method is practical for handling large multi-chromosomal genomes.

6   Conclusion and Future Work

The gene team tree represents the set of all gene teams for every possible value of the parameter δ. Organising the gene teams in the form of a tree allows us to visualize the relationship between the various gene teams and makes explicit the structure of the gene teams. We have developed an efficient algorithm which computes the gene team tree for a given set gene orders. We have also showed that our algorithm is practical by running our algorithm on real datasets. Our analysis of the gene teams which correspond to known operons in *E. coli* showed that no one single value of δ can produce all of these gene teams.

We plan to explore other applications of GTT as a hierarchal structure which represents genomic structures that are conserved during evolution. One such application which we are working on is applying GTT to the problem of ortholog assignment.

References

Béal, M.-P., Bergeron, A., Corteel, S., Raffinot, M.: An algorithmic view of gene teams. Theor. Comput. Sci. 320(2-3), 395–418 (2004)

Ermolaeva, M.D., White, O., Salzberg, S.L.: Prediction of operons in microbial genomes. Nucleic Acids Res. 29(5), 1216–1221 (2001)

Fu, Z., Chen, X., Vacic, V., Nan, P., Zhong, Y., Jiang, T.: Msoar: A high-throughput ortholog assignment system based on genome rearrangement. Journal of Computational Biology 14(9), 1160–1175 (2007)

Gama-Castro, S., Jimnez-Jacinto, V., Peralta-Gil, M., Santos-Zavaleta, A., Pealoza-Spinola, M.I., Contreras-Moreira, B., Segura-Salazar, J., Muiz-Rascado, L., Martnez-Flores, I., Salgado, H., Bonavides-Martnez, C., Abreu-Goodger, C., Rodrguez-Penagos, C., Miranda-Ros, J., Morett, E., Merino, E., Huerta, A.M., Trevio-Quintanilla, L., Collado-Vides, J.: Regulondb (version 6.0): gene regulation model of escherichia coli k-12 beyond transcription, active (experimental) annotated promoters and textpresso navigation. Nucleic Acids Res. 36(Database issue), D120–D124 (2008)

He, X., Goldwasser, M.H.: Identifying conserved gene clusters in the presence of homology families. Journal of Computational Biology 12(6), 638–656 (2005)

Hoberman, R., Durand, D.: The incompatible desiderata of gene cluster properties. In: McLysaght and Huson, pp. 73–87 (2005)ISBN 3-540-28932-1

Hoberman, R., Sankoff, D., Durand, D.: The statistical analysis of spatially clustered genes under the maximum gap criterion. Journal of Computational Biology 12(8), 1083–1102 (2005)

Jaccard, P.: Nouvelles recherches sur la distribution florale. Bulletin de la Société Vaudoise des Sciences Naturelles 44, 223–270 (1908)

Lawrence, J.: Selfish operons: the evolutionary impact of gene clustering in prokaryotes and eukaryotes. Curr. Opin. Genet. Dev. 9(6), 642–648 (1999)

McLysaght, A., Huson, D.H. (eds.): RECOMB 2005. LNCS (LNBI), vol. 3678. Springer, Heidelberg (2005)

Overbeek, R., Fonstein, M., D'Souza, M., Pusch, G.D., Maltsev, N.: The use of gene clusters to infer functional coupling. Proc. Natl. Acad. Sci. USA 96(6), 2896–2901 (1999)

Snel, B., Bork, P., Huynen, M.A.: The identification of functional modules from the genomic association of genes. Proc. Natl. Acad. Sci. USA 99(9), 5890–5895 (2002)

Tatusov, R.L., Natale, D.A., Garkavtsev, I.V., Tatusova, T.A., Shankavaram, U.T., Rao, B.S., Kiryutin, B., Galperin, M.Y., Fedorova, N.D., Koonin, E.V.: The COG database: new developments in phylogenetic classification of proteins from complete genomes. Nucl. Acids Res. 29(1), 22–28 (2001)

Integrating Sequence and Topology for Efficient and Accurate Detection of Horizontal Gene Transfer

Cuong Than, Guohua Jin, and Luay Nakhleh*

Department of Computer Science, Rice University, Houston, TX 77005, USA
nakhleh@cs.rice.edu

Abstract. One phylogeny-based approach to horizontal gene transfer (HGT) detection entails comparing the topology of a gene tree to that of the species tree, and using their differences to locate HGT events. Another approach is based on augmenting a species tree into a phylogenetic network to improve the fitness of the evolution of the gene sequence data under an optimization criterion, such as maximum parsimony (MP). One major problem with the first approach is that gene tree estimates may have wrong branches, which result in false positive estimates of HGT events, and the second approach is accurate, yet suffers from the computational complexity of searching through the space of possible phylogenetic networks.

The contributions of this paper are two-fold. First, we present a measure that computes the support of HGT events inferred from pairs of species and gene trees. The measure uses the bootstrap values of the gene tree branches. Second, we present an integrative method to speed up the approaches for augmenting species trees into phylogenetic networks.

We conducted data analysis and performance study of our methods on a data set of 20 genes from the *Amborella* mitochondrial genome, in which Jeffrey Palmer and his co-workers postulated a massive amount of horizontal gene transfer. As expected, we found that including poorly supported gene tree branches in the analysis results in a high rate of false positive gene transfer events. Further, the bootstrap-based support measure assessed, with high accuracy, the support of the inferred gene transfer events. Further, we obtained very promising results, in terms of both speed and accuracy, when applying our integrative method on these data sets (we are currently studying the performance in extensive simulations). All methods have been implemented in the PhyloNet and NEPAL tools, which are available in the form of executable code from http://bioinfo.cs.rice.edu.

1   Introduction

While the genetic material of an organism is mainly inherited through lineal descent from the ancestral organism, it has been shown that genomic segments in various groups of organisms may be acquired from distantly related organisms through *horizontal DNA, or gene, transfer* (HGT). It is believed that HGT is ubiquitous among prokaryotic organisms (24; 6) and plays a significant role in their genomic diversification (20). Recent studies have also demonstrated evidence of massive HGT in various groups of plants (3; 4).

---

* Corresponding author.

C.E. Nelson and S. Vialette (Eds.): RECOMB-CG 2008, LNBI 5267, pp. 113–127, 2008.
© Springer-Verlag Berlin Heidelberg 2008

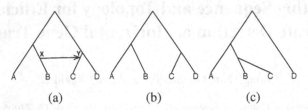

Fig. 1. (a) A phylogenetic network with a single HGT edge from X to Y, ancestors of taxa B and C, respectively. (b) The underlying species tree, which models the evolution of vertically transmitted genomic regions. (c) The tree that models the evolution of horizontally transferred genomic regions. The phylogenetic network *contains* the two trees, in the sense that each of the trees can be obtained from the network by removing all but one incoming edge (branch) for each node in the network.

When HGT occurs among organisms, the evolution of their genomes is best modeled as a *phylogenetic network* (19; 15) which can be viewed as the reconciliation of the evolutionary trees of the various genomic segments, typically referred to as *gene trees*. Figure 1(a) shows an example of a phylogenetic network on four taxa in the presence of an HGT from X, an ancestor of taxon B, to Y, an ancestor of taxon C. In this scenario, genomic regions that are not involved in the horizontal transfer are inherited from the ancestral genome, and their evolution is modeled by the tree in Figure 1(b), while horizontally transferred regions in C's genome are acquired from X, and their evolution is modeled by the tree in Figure 1(c).

As illustrated in Figure 1, the occurrence of HGT may result in gene trees whose topologies are discordant (the trees in (b) and (c)). Detecting discordance among gene trees, particularly with respect to a species tree, and reconciling them into a phylogenetic network, are the fundamental principles on which phylogeny-based HGT detection approaches are built. Several algorithms and software tools based on this approach have been introduced recently (e.g., (8; 22; 2)), all of which infer HGT events by comparing the topologies (shapes) of trees.

A major confounding factor that negatively affects the accuracy of phylogeny-based methods is that reconstructed species/gene trees usually contain error in the form of wrong branches. These errors result in species/gene tree incongruities, thus triggering phylogeny-based methods to make false predictions of HGT events (29). For example, the incongruence between the two tress in Figure 1 could have arisen simply due to poor reconstruction of the gene tree, even though no HGT was involved. In general, the performance of phylogeny-based methods, in terms of both accuracy and speed, is negatively affected by several factors, which include errors in the estimated trees and the exponential space of potential solutions (28; 29). In this paper we address this confounding factor by devising a method that estimates the support of inferred HGT events by using the bootstrap values of the gene tree branches.

Recently, a set of methods were devised to estimate HGT events by augmenting a species tree into a phylogenetic network to improve the fitness of evolution of the gene's sequence data based on the maximum parsimony (MP) criterion (21; 13). While yielding promising results, these methods were very slow, since the problem of inferring a phylogenetic network under the MP criterion is NP-hard and even hard to approximate

(11; 14). In this paper, we present a new integrative method for improving the speed of the MP-based methods by first conducting a topology-based analysis, based on the topology of the tree inferred from the gene's sequences, and then screening the inferred events based on the MP criterion.

We have implemented both approaches and analyzed a data set of 20 genes that exhibited massive HGT in the basal angiosperm *Amborella* according to (4). First, we demonstrated the effects of error in the reconstructed gene trees on the estimates of gene transfer. We found that including poorly supported gene tree branches in the analysis results in a high rate of false positives. Second, the support measure assessed, with high accuracy, the support of the gene transfers inferred by the topology-based analysis. Third, the integrative method, that combines topology comparison with the MP criterion, detected efficiently all but one of HGT edges that were postulated by the authors. Further, our approach detected new candidate HGTs, with high support. The combined accuracy and efficiency of our approach, achieved by integrating topological analysis with sequence information, will enable automated detection of HGT in large data sets.

2 Materials and Methods

2.1 Topology-Based HGT Detection and the RIATA-HGT Tool

When HGT occurs, the evolutionary history of the genomes may be more appropriately represented by a *phylogenetic network*, which is a rooted, directed, acyclic graph (rDAG) that extends phylogenetic trees to allow for nodes with more than a single parent (1; 19; 15; 9). The phylogeny-based HGT reconstruction problem seeks the phylogenetic network with minimum number of HGT edges (equivalently, the minimum number of nodes that have more than a single parent), to reconcile the species and gene trees. The minimization requirement simply reflects a parsimony criterion: in the absence of any additional biological knowledge, the simplest solution is sought. In this case, the simplest solution is one that invokes the minimum number of HGT events to explain tree incongruence.

Definition 1. *(The HGT Detection Problem)*

> **Input:** *Species tree ST and gene tree GT, both of which are rooted.*
> **Output:** *A set with the fewest directed HGT edges, each of which is posited between a pair of branches of ST, such that the resulting phylogenetic network contains the gene tree GT.*

As indicated above, the resulting network is an rDAG, where the network's branches that are also branches in ST are directed away from the root, and the HGT edges are directed as indicated by their description. For example, the phylogenetic network in Figure 1(a) is a solution to the HGT Detection Problem if we consider the trees in 1(b) and 1(c) to be the species and gene trees, respectively, since it has a smallest set of HGT edges (one, in this case) and contains the gene tree. Notice that in this case the gene tree cannot be reconciled with the species tree without HGT edges.

Several heuristics for solving the problem have been recently introduced, e.g., (18; 8; 7; 23; 17; 2; 10; 22). As mentioned above, the performance of methods following this approach, in terms of both accuracy and speed, is negatively affected by several factors, which include errors in the estimated trees and the exponential space of potential solutions (28; 29). We have recently addressed these issues and extended the RIATA-HGT method (22) so that it detects HGT in pairs of trees that are not necessarily binary, and computes multiple minimal solutions very efficiently (27).

2.2   Assessing the Support of HGT Edges

As mentioned above, topology-based HGT detection methods are sensitive to error in the inferred trees, as illustrated on simulated data sets and reported in (28; 29) and on biological data sets, which we report here. In this paper, we propose a measure of the *support* of an inferred HGT edge based on the bootstrap values of the gene tree branches. Roughly speaking, the support value of HGT edge $X \to Y$ in the species tree, where Y' is the sibling of Y, is derived from the bootstrap values of the gene tree branches that separate the clade under Y from the clade under Y'. The rationale behind the idea is that if Y' and Y are well separated in the gene trees (i.e., some branches in the path from Y to Y' have high bootstrap values), an HGT is necessary to move Y away from Y'). For example, the support of HGT edge $X \to Y$ in Figure 2 is calculated based on the bootstrap values of the branches separating B from A in the gene tree.

However, since trees are not necessarily binary in general, and given that multiple HGT edges may involve branches under X or Y, the calculation is more involved (e.g., see Figure 3) and requires a formal treatment, which we now provide.

Given a species tree ST, a gene tree GT, and a set $\varXi$ of HGT edges, create a network N by adding every edge in $\varXi$ to ST. We create two trees ST' and ST'' from N, as follows:

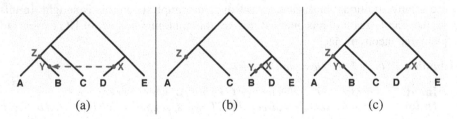

(a)   (b)   (c)

Fig. 2. An illustration of computing the support value of an HGT edge. In this case, the support of HGT edge $X \to Y$ added on the species tree, resulting in the network in (a), is calculated based on the bootstrap of the branches that separate the "moving clade" rooted at Y (which, in this case, is the clade that contains the single leaf B) from its sister clade (which, in this case, is the clade that contains the single leaf A) in the gene tree (b). The species tree is depicted by the solid lines in (a), and the gene tree by the solid lines in (b). The tree in (b), along with all internal nodes, including nodes Y and Z, is the tree ST' used in the procedure for computing the support value, whereas the tree in (c), along with all internal nodes, including nodes X and Y, is the tree ST'' used in the procedure. Notice that nodes Y and Z in (b), as well as nodes X and Y in (c) have in-degree and out-degree 1.

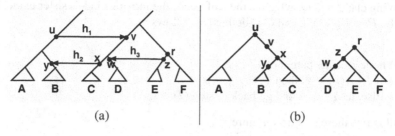

Fig. 3. An illustration of calculating support values of HGT edges when multiple HGT edges are involved. The HGT edges (arrows) h_1, h_2, and h_3, are posited between pairs of branches (lines) in the species tree ST resulting in the phylogenetic network N, part of which is shown in (a). When a genomic segment is transferred across all these three HGT edges, the evolutionary history of that segment in the six taxa A, B, C, D, E, and F is represented by the two clades in (b). Further, these two clades and their internal nodes are part of the tree ST' that is generated in the first step of the support calculation. The tree ST'' is obtained from the tree ST in (a) by adding the endpoints of all HGT edges (the solid circles), but removing the HGT edges themselves. The support of h_1 is derived from the bootstrap support of the gene tree branches that separate the MRCA of set P of leaves (which, in this case, is the set of leaves in clades B and C combined) and the MRCA of set Q of leaves (which, in this case, is the set of leaves in clades D, E, and F combined). The support of h_2 is derived from the bootstrap values of the gene tree branches that separate the MRCA of set P of leaves (in this case, the leaves of clade B) and the MRCA of set Q of leaves (in this case, the leaves of clade A). The support of h_3 is derived from the bootstrap support of the gene tree branches that separate the MRCA of set P of leaves (in this case, the leaves of clade D) and the MRCA of set Q of leaves (in this case, the leaves in clades B and C combined).

- A tree ST' is built from N in such a way that for each edge $X \rightarrow Y$ in Ξ, this edge is kept, and the other edge incident into Y is deleted. The tree in Figure 2(b), including nodes Y and Z is the tree ST' obtained from the network in Figure 2(a).
- A tree ST'' is built from N in a similar fashion, but edge $X \rightarrow Y$ is deleted while the other edge incident into Y is kept. The tree in Figure 2(c), including nodes X and Y is the tree ST'' obtained from the network in Figure 2(a).

Notice that both trees ST' and ST'' have nodes of in-degree and out-degree 1. Retaining these nodes in these two trees ensures the well-definedness of the procedure that we will describe below for computing the support of HGT edge $X \rightarrow Y$. Note that ST' and ST'' can have nodes whose in-degree and out-degree are both 1. One important fact about ST' and ST'' that is necessary for our method for assessing HGT support is that they have the same set of nodes. See Figure 3 for an illustration. We denote by $L_T(v)$ the set of leaves under node v in a tree T; i.e., the set of leaves to which the paths from the root of T must go through node v. We define the support of an HGT edge $h = X \rightarrow Y$, which we denote by $b(h)$, assuming the bootstrap support of the gene tree branches have been computed.

Our task is to find the path of edges between the "moving clade" (the clade of taxa this is "moved" in the HGT event) and its sister clade in the gene tree. In Figure 2, $P = \{B\}$ and $Q = \{A\}$. We define two sets of leaves: P, which is the leaf-set of the moving clade, and Q, which is the leaf-set of the moving clade's sister clade. More formally, $P = L_{ST'}(Y)$, and Q is defined as follows:

1. Let $Y' = Y$.
2. Let node p be the parent of Y' in ST''.
   (a) If $L_{ST'}(p) \neq \emptyset$, then $Q = L_{ST'}(p)$.
   (b) Else, let $Y' = p$, and go back to step (2).

To illustrate these sets from Figure 3:

- For HGT edge h_1: P contains only the leaves in clades B and C, and Q contains only the leaves in clades D, E and F.
- For HGT edge h_2: P contains only the leaves in clade B, and Q contains only the leaves in clade A.
- For HGT edge h_3: P contains only the leaves in clade D, and Q contains only the leaves in clades B and C.

In this example, we get the set Q after only one iteration of the procedure. The need for repeating Step 2 lies in the case where all siblings of Y are moved by HGT events.

Now that we have computed the sets P and Q of leaves, let p and q be the most recent common ancestor nodes of P and Q, respectively, in the gene tree. Let $\mathcal{E}$ be the set of branches in the gene tree between nodes p and q. The support value of the HGT edge h is

$$b(h) = \max_{e \in \mathcal{E}} s(e), \tag{1}$$

where $s(e)$ is the bootstrap value of the branch e in the gene tree. We choose to use the maximum bootstrap value of a branch on the path since that value alone determines whether the donor and recipient involved in an HGT edge truly form a clade in the gene tree. Notice that averaging all values may not work, since, for example, if the path has many branches, only one of which has very high support and the rest have very low support, averaging would reflect low support for the HGT edge.

In case the species tree branches have bootstrap support values associated with them, these can also be incorporated as follows for HGT edge $h : X \to Y$. Let Z be the MRCA of X and Y in the species tree, and let $\mathcal{E}'$ be the set of branches in the species tree on the paths from Z to X and from Z to Y. Then, Formula (1) can be modified to become

$$b(h) = \min\{\max_{e \in \mathcal{E}} s(e), \max_{e' \in \mathcal{E}'} s(e')\}, \tag{2}$$

where $s(e)$ is the bootstrap value of the gene tree branch for $e \in \mathcal{E}$ and the species tree branch for $e' \in \mathcal{E}'$.

2.3 Parsimony-Based HGT Detection and the NEPAL Tool

The relationship between a phylogenetic network and its constituent trees is the basis for the MP extension to phylogenetic networks described in a sequence of papers by Jin, Nakhleh and co-workers (21; 11; 13; 14), which we now review briefly.

The Hamming distance between two equal-length sequences x and y, denoted by $H(x, y)$, is the number of positions j such that $x_j \neq y_j$. Given a fully-labeled tree

T, i.e., a tree in which each node v is labeled by a sequence s_v over some alphabet Σ, we define the Hamming distance of a branch $e \in E(T)$ ($E(T)$ denotes the set of all branches in tree T), denoted by $H(e)$, to be $H(s_u, s_v)$, where u and v are the two endpoints of e. We now define the parsimony length of a tree T.

Definition 2. *The parsimony length of a fully-labeled tree T, is $\sum_{e \in E(T)} H(e)$. Given a set S of sequences, a maximum parsimony tree for S is a tree leaf-labeled by S and assigned labels for the internal nodes, of minimum parsimony length.*

Given a set S of sequences, the MP problem is to find a maximum parsimony phylogenetic tree T for the set S. The evolutionary history of a single (non-recombining) gene is modeled by one of the trees contained inside the phylogenetic network of the species containing that gene. Therefore the evolutionary history of a site s is also modeled by a tree contained inside the phylogenetic network. A natural way to extend the tree-based parsimony length to fit a dataset that evolved on a network is to define the parsimony length for each site as the minimum parsimony length of that site over all trees contained inside the network.

Definition 3. *The parsimony length of a network N leaf-labeled by a set S of sequences, is*

$$NCost(N, S) := \sum_{s_i \in S}(\min_{T \in T(N)} TCost(T, s_i))$$

where $TCost(T, s_i)$ is the parsimony length of site s_i on tree T.

Notice that as usually large segments of DNA, rather than single sites, evolve together, Definition 3 can be extended easily to reflect this fact, by partitioning the sequences S into non-overlapping blocks b_i of sites, rather than sites s_i, and replacing s_i by b_i in Definition 3. This extension may be very significant if, for example, the evolutionary history of a gene includes some recombination events, and hence that evolutionary history is not a single tree. In this case, the recombination breakpoint can be detected by experimenting with different block sizes. Based on this criterion, we would want to reconstruct a phylogenetic network whose parsimony length is minimized. In the case of horizontal gene transfer, a species tree that models vertical inheritance is usually known; e.g., see (16). Hence, the problem of reconstructing phylogenetic networks in this case becomes one of finding a set of edges whose addition to the species tree "best explains" the horizontal gene transfer events, which is defined as the θ-FTMPPN problem in (13). We have implemented heuristics to solve this problem in the NEPAL software tool.

Jin *et al.* demonstrated the high accuracy of HGT detection by solving the θ-FTMPPN problem on a wide array of simulated and biological data sets (13). However, the major drawback of the approach was the high running time (several hours to days for detecting a small number of HGT events on small species trees). In the next section, we propose an approach that integrates topological comparison of trees with the MP criterion to achieve both accuracy and computational efficiency in the detection of HGT events.

2.4   Integrating MP and Topological Comparison

As described above, the extended version of RIATA-HGT performs very well, in terms of speed, but overestimates the number of HGT events (due to inaccuracies in the trees).

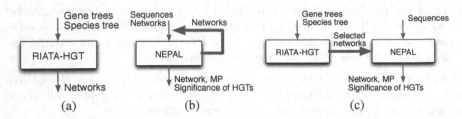

Fig. 4. Diagrams of the three approaches: RIATA-HGT (a), NEPAL (b), and the integrative approach (c). RIATA-HGT takes as input a pair of trees (species and gene trees), and computes a set of minimal phylogenetic networks that contain both trees. NEPAL takes as input a set of sequences and a set of networks (initially, the set of networks includes only the species tree), and solves the θ-FTMPPN problem by finding a set of HGT edges whose addition to the species tree results in an optimal phylogenetic network under the maximum parsimony criterion. In our analysis in this paper, the value of θ was determined based on observing the parsimony improvement, and is data set-dependent. The integrative approach first runs RIATA-HGT on the pair of species/gene trees, and then uses the phylogenetic networks inferred to guide the search of the parsimony search, as implemented in NEPAL.

On the other hand, Jin *et al.* have shown that the MP approach performs very well in terms of accuracy, yet is very slow (13). The computational requirement of the MP criterion stems from the large space of networks that the method has to consider. To achieve good accuracy with high efficiency, we propose to integrate the two approaches.

Figure 4 shows the diagrams of the two separate approaches as well as the proposed integrative approach. In contrast to the MP approach, where a large number of networks is explored due to the exponential number of combinations of candidate HGT edges, our integrative approach focuses on the phylogenetic networks inferred by RIATA-HGT. The rationale behind this approach is that the overestimation of RIATA-HGT, and topology-based HGT detection methods in general, is mainly in the form of *false positives*, whereas the *false negatives* are negligible. In other words, RIATA-HGT in most cases infers all the "true" HGT edges, but also infers additional "false" HGT edges due to errors in the gene trees. Given the accuracy of the MP criterion, we post-process the results of RIATA-HGT by evaluating them based on the MP criterion, with the hope that it would remove the false positive HGT edges. We have developed efficient algorithms for evaluating the parsimony length of phylogenetic networks (14), so both steps of our integrative approach can be carried out efficiently. The algorithm of the integrative approach is outlined in Figure 5.

The time complexity of our integrative algorithm is $O(km2^k)$ using exact network parsimony length computing algorithms and $O(mk^2)$ using the approximation algorithm described in (14).

2.5   Validating the Integrative Approach

To validate the integrative approach, we consider the differences in the HGT edges and parsimony scores computed by each of the two methods. More formally, if N_{int} and N_{mp} are two phylogenetic networks computed by the integrative and MP approaches, respectively, on the same species trees ST and set of sequences S, and $H(N_{\text{int}})$ and

---

TopSeqHGTIdent(ST, GT, S)
INPUT: species tree ST, gene tree GT, and sequence dataset S
OUTPUT: network N with marked significance of each HGT

---

1 Let $\{N_1, \ldots, N_m\}$ be the set of all phylogenetic networks computed by RIATA-HGT, and let $H(N_i)$ be the set of HGT edges in N_i.
2 Let $\mathcal{H} = \cap_{i=1}^{m} H(N_i)$, and $R(N_i) = H(N_i) - \mathcal{H}$. In other words, $\mathcal{H}$ denotes the set of HGT edges that are shared by all networks, and $R(N_i)$, for $1 \leq i \leq m$, the set of HGT edges that are in N_i but not shared by all other networks.
3 Apply NEPAL to $N' = ST + \mathcal{H}$.
4 For each network N_i, $1 \leq i \leq m$, apply NEPAL by incrementally adding (in no particular order) the HGT edges in $R(N_i)$ to N', and compute the minimum parsimony length of the phylogenetic network.
5 Let $N = ST$, N_{opt} be the best network according to maximum parsimony criterion, that is $NCost(N_{opt}, S) = min_{i=1}^{m}(NCost(N_i, S))$.
   Apply NEPAL by adding to ST each time one of the HGT events $h \in H(N_{opt})$ that results in the most significant drop in the parsimony score and let $N = N \cup h$. Stop this process when the drop is smaller than a specified threshold.

Fig. 5. An outline of the integrative approach. Steps 1 through 4 seek the most parsimonious network among the several networks computed by RIATA-HGT. Step 5 conducts one last pass of parsimony length calculation to identify the HGT edges in the optimal network whose contribution to lowering the parsimony length is significant. This last step is necessary since, even though the most parsimonious network is optimal among all networks computed by RIATA-HGT, it may still have " parsimoniously unnecessary" HGT edges whose addition by RIATA-HGT was necessitated by the incongruence between the two trees.

$H(N_{\mathrm{mp}})$ are the sets of HGT edges in N_{int} and N_{mp} respectively, then we have two measures of quality:

- $m^{\mathrm{HGT}}(N_{\mathrm{int}}, N_{\mathrm{mp}}) = (H(N_{\mathrm{int}}) - H(N_{\mathrm{mp}})) \cup (H(N_{\mathrm{mp}}) - H(N_{\mathrm{int}}))$. This measure reflects the difference in the locations of the HGT edges between the two networks.
- $m^{\mathrm{pars}}(N_{\mathrm{int}}, N_{\mathrm{mp}}) = |NCost(N_{\mathrm{int}}, S) - NCost(N_{\mathrm{mp}}, S)|$. This measure quantifies the difference in the MP scores of the two networks.

2.6   Data

We studied 20 out of the 31 mitochondrial gene data sets, which were collected from 29 diverse land plants and analyzed in (4). These are *cox2, nad2, nad3, nad4(ex4), nad4(exons), nad5, nad6, nad7, atp1, atp8, ccmB, ccmC, ccmFN1, cox3, nad1, rpl16, rps19, sdh4,* and three introns *nad2intron, nad5intron* and *nad7intron*. We used a species tree for the dataset based on information at NCBI (http://www.ncbi.nih.gov) and analyzed the entire dataset with both seed and nonseed plants together. For each gene data set, we restricted the species tree to those species for which the gene sequence is available. For the parsimony analysis, we analyzed each gene data set separately, by solving the θ-FTMPPN problem, as implemented in NEPAL, on the gene DNA sequence with respect to the species tree. The θ-FTMPPN problem seeks a set of HGT edges, each of whose addition to the species tree improves the parsimony score beyond the specified threshold θ. In this paper, we determined the value of θ based on observing the parsimony improvement, as HGT edges were added, and the value of θ was dependent on the data set; i.e., data sets did not necessarily have the same value of θ. Our current criterion for determining the value of θ is very simple, yet shows very good accuracy: we observe the slope of the decrease in the parsimony score as the HGT edges are added, and stop adding new edges when the slope changes (slows down) significantly. In all data sets we have analyzed so far, the change in slope has been very sharp, which makes it very straightforward to determine the number of HGT edges to add. For the analysis by RIATA-HGT (which takes as input a pair of trees), for each gene, we used the species tree as the reference tree and the gene tree reported in (4) as the second tree. Bergthorsson *et al.* also calculated and reported the bootstrap support of the gene tree branches, which we used in calculating the support values of HGT edges inferred by RIATA-HGT. For the integrative approach, we used the phylogenetic networks produced by RIATA-HGT and the gene sequences, and applied the algorithm described above. It is important to note that in their analyses, Bergthorsson *et al.* focused only on genes that were horizontally transferred to the mitochondrial genome of *Amborella*.

3   Results and Discussion

Table 1 summarizes information about the HGT edges reported by Bergthorsson *et al.* for the 20 genes (4), the HGT edges inferred by the MP criterion as implemented in the NEPAL tool (13; 14), the HGT edges inferred by the RIATA-HGT method for topology-based HGT detection (22; 27), and the HGT edges inferred by the integrative approach of topology- and MP-based analysis. Figure 6 shows the HGT edges inferred for two data sets; we omit the phylogenetic networks for the other data sets.

Bergthorsson *et al.* reported the groups of species to which the donor(s) of horizontally transferred genes belong, rather than the specific donor. In particular, they focused on four groups: Bryophytes, Moss, Eudicots, and Angiosperms. For the recipient, the authors only focused on *Amborella*. Further, for each HGT event (edge), they computed and reported its significance based on the SH test (25). Of the 25 HGT events that Bergthorsson *et al.* postulated, 13 were supported, 9 unsupported, and 3 (the 3 intron data sets) had no reported support.

Of the 13 HGTs reported in (4) with high support according to the SH test, the MP analysis identified 12, missing only the HGT involving gene *nad5* from the Angiosperms. RIATA-HGT also identified 12 out of the 13 HGT edges, missing only the HGT involving gene *nad6*. Further, all these 12 HGT edges identified by RIATA-HGT had support (out of 100) higher than 95, based on Formula (1) above. The integrative approach identified 11 of the 13 well-supported HGT edges, missing only the HGTs involving genes *nad5* and *nad6*. In the former case, the HGT was identified by RIATA-HGT, but deemed insignificant in the parsimony analysis phase, and in the latter case, RIATA-HGT missed the HGT edge, and since the parsimony analysis in the integrative approach uses those edges that RIATA-HGT identifies, this HGT edge was not identified.

Table 1. Mitochondrial gene data sets and HGTs postulated in (4) and those computed by the MP analysis (NEPAL), RIATA-HGT, and the integrative approach (RIATA-HGT+MP). The '#HGTs' value is the number of HGTs found; 'donor' denotes the group from which the gene was transferred (in all cases, the recipient is *Amborella*; 'SH' denotes support of the HGT events as computed by the SH test (25) and reported in (4) (values lower than 0.05 indicate high support, and NS indicates support is not significant). The 'F?' column indicates which of the HGTs postulated by the authors were found by the MP analysis, and the '#Nets' shows the number of networks analyzed by the parsimony method to identify the correct HGTs. For the RIATA-HGT column, each row shows the minimum number of HGTs computed, the number of minimal solutions, and the number of distinct HGTs in all minimal solutions, respectively. B=Bryophyte, M=Moss, E=Eudicot, and A=Angiosperm.

Gene	Bergthorsson et al.			MP			RIATA-HGT			RIATA-HGT+MP		
	#HGTs	donor	SH	#HGTs	F?	#Nets	#HGTs	#Nets	#events	#HGTs	F?	#Nets
cox2	3	M	<0.001	1	Y	8482	9	4	12	1	Y	23
		E	NS		N	8482	9	4	12		N	23
		E	NS		N	8482	9	4	12		N	23
nad2	2	M	<0.001	1	Y	3500	7	6	11	1	Y	21
		E	NS		N	3500	7	6	11		N	21
nad4(exons)	1	M	<0.001	1	Y	1620	4	2	5	1	Y	9
nad4(ex4)	1	E	NS	2	Y	1832	6	3	8	2	Y	21
nad5	2	M	<0.001	1	Y	3292	6	6	9	1	Y	17
		A	0.025		Y	3292	6	6	9		N	17
nad6	1	B	<0.001	1	Y	2484	6	3	8	1	N	15
nad7	2	M	<0.001	1	Y	2948	7	1	7	1	Y	13
		E	NS		N	2948	7	1	7		N	13
atp1	1	E	0.001	1	Y	2817	6	18	14	1	Y	27
atp8	1	E	0.008	2	Y	9059	5	6	11	1	Y	21
ccmB	1	E	NS	2	Y	66015	6	3	14	2	Y	101
ccmC	1	E	0.03	1	Y	2786	7	21	15	1	Y	29
ccmFN1	1	E	0.004	2	Y	4412	7	18	13	2	Y	46
cox3	1	A	NS	1	N	3466	8	15	18	1	N	52
nad1	1	E	<0.001	1	Y	2812	9	12	14	1	Y	27
rpl16	1	E	NS	3	Y	21632	10	27	23	1	Y	67
rps19	1	E	0.003	1	Y	1476	5	4	7	1	Y	13
sdh4	1	E	NS	3	Y	18670	9	18	18	3	Y	540
nad2intron	1	M	—	2	Y	5904	8	2	10	2	Y	44
nad5intron	1	M	—	2	Y	10280	9	5	18	2	Y	51
nad7intron	1	M	—	1	Y	3284	12	48	26	1	Y	51

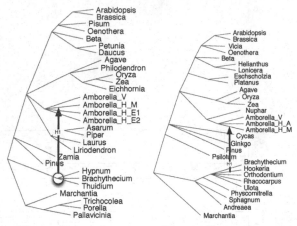

The *cox2* gene data set   The *nad5* gene data set

Fig. 6. HGT edges identified computationally. In the *cox2* data set, the MP analysis identified four equally "good" HGT edges, all of which represent HGT to Amborella_H_M, and each of which denotes a different donor (the four branches inside the circle): Hypnum, Brachythecium, Thuidium, and the clade containing all three. Under the MP criterion, those four donors contribute equally to the improvement in the parsimony length. RIATA-HGT identified the HGT edge with the clade of all three species as the donor, and Amborella_H_M as the recipient, along with many other "false positive" HGT edges, which are not shown here. The integrative approach finds the single HGT edge from the clade of all three species to Amborella_H_M. In the case of the *nad5* data set, the MP analysis identified the only shown HGT edge, which was identified as well by RIATA-HGT along with other HGT edges (not shown). The integrative approach identified only the shown HGT edge.

The only HGT that the MP analysis missed in the *nad5* data set involves an HGT from the Angiosperms. This transfer had support of 0.025 based on the SH test as reported by Bergthorsson *et al.*. Further, this edge was identified by RIATA-HGT with support of 94. In the MP analysis, the only significant *Amborella* transfer comes from the Moss group. The grouping of Eudicots and Monocots results in much less significant improvement in the parsimony length. The transfer from the Angiosperms to *Amborella* has even less impact on the parsimony length.

The three HGT edges postulated by Bergthorsson *et al.* for the intron data sets, and which had no support values based on the SH test reported, were identified by the MP analysis of NEPAL, RIATA-HGT, and the integrative approach. Further, those three edges had support around 50, based on Formula (1) above.

Of the other 9 HGT events reported by the authors with no significant support based on the SH test, the MP analysis, RIATA-HGT and the integrative approach did not identify five of them. However, all three approaches identified four HGT edges that had no significant support based on the SH test. All these four HGTs were from the Eudictots to *Amborella*, and they were in the *nad4(ex4)*, *ccmB*, *rpl16*, and *sdh4* data sets. In all four cases, the support ranged from low ($\approx$ 50) to high ($\approx$ 95), based on Formula (1) above.

In eight data sets, the MP analysis identified HGT edges in addition to those reported in (4). However, none of these edges involved Amborella. One possible explanation for why these edges were not reported in (4) is probably because the authors focused only on HGT events involving *Amborella*. Another explanation may be the inaccuracy of the parsimony criterion for evaluating HGT edges in these cases. RIATA-HGT identified the same HGT edges in these eight cases, support ranging from low ($\approx$ 50) to high ($\approx$ 90), based on Formula (1) above. The integrative approach in these cases also identified the same HGT edges.

It is important to note that all other HGT edges identified by RIATA-HGT had support ranging from very low to very high ($\approx$ 100 in some cases), based on Formula (1) above, but were rendered insignificant based on the parsimony phase in the integrative approach as well. Further, we analyzed the parsimony scores of the networks computed by NEPAL and by the integrative approach. In both cases, the scores were very similar.

Finally, the parsimony approach analyzed several thousand networks (using the best available heuristic (11)) to identify the HGTs, and that took several hours on each gene, and up to two days on some. Table 1 shows the exact number of networks that our parsimony analysis checks even with the branch-and-bound heuristics. The integrative approach, on the other hand, took a few minutes on each of the data sets.

4 Conclusions and Future Work

In this paper, we presented a measure to assess the support of HGT edges inferred by phylogeny-based methods that compare species and gene tree topologies. Further, we presented a method for speeding up approaches the infer HGT events by augmenting species trees into phylogenetic networks to improves the fitness of evolution of gene sequence data. We obtained promising results on 20 mitochondrial gene data sets, previously analyzed in (4).

While we used the maximum parsimony criterion in the second phase of our integrative approach, it is also possible to use stochastic models of HGT (e.g., (26; 12; 5)) to probabilistically screen the phylogenetic networks produced in the first phase. This is one of the future directions we intend to pursue. The immediate task for us in our future work is to study the performance of these measures in extensive simulations. The advantage that simulations provide over real data is that the true HGT events are known, which allows us to make absolute quantification of the performance of the methods.

Acknowledgments

This work is supported in part by the Department of Energy grant DE-FG02-06ER25734, the National Science Foundation grant CCF-0622037, and the George R. Brown School of Engineering Roy E. Campbell Faculty Development Award. Further, the work was supported in part by the Rice Computational Research Cluster funded by NSF under Grant CNS-0421109, and a partnership between Rice University, AMD and Cray.

The authors would like to thank Aaron O. Richardson for providing the *Amborella* gene data sets, and the anonymous reviewers for comments on the technical details as well as the readability of the manuscript.

References

[1] Baroni, M., Semple, C., Steel, M.: A framework for representing reticulate evolution. Annals of Combinatorics 8(4), 391–408 (2004)

[2] Beiko, R.G., Hamilton, N.: Phylogenetic identification of lateral genetic transfer events. BMC Evolutionary Biology 6 (2006)

[3] Bergthorsson, U., Adams, K.L., Thomason, B., Palmer, J.D.: Widespread horizontal transfer of mitochondrial genes in flowering plants. Nature 424, 197–201 (2003)

[4] Bergthorsson, U., Richardson, A., Young, G.J., Goertzen, L., Palmer, J.D.: Massive horizontal transfer of mitochondrial genes from diverse land plant donors to basal angiosperm Amborella. Proc. Nat'l Acad. Sci., USA 101, 17747–17752 (2004)

[5] Galtier, N.: A model of horizontal gene transfer and the bacterial phylogeny problem. Systematic Biology 56(4), 633–642 (2007)

[6] Gogarten, J.P., Doolittle, W.F., Lawrence, J.G.: Prokaryotic evolution in light of gene transfer. Mol. Biol. Evol. 19(12), 2226–2238 (2002)

[7] Gorecki, P.: Reconciliation problems for duplication, loss and horizontal gene transfer. In: Proc. 8th Ann. Int'l Conf. Comput. Mol. Biol. (RECOMB 2004), pp. 316–325 (2004)

[8] Hallett, M.T., Lagergren, J.: Efficient algorithms for lateral gene transfer problems. In: Proc. 5th Ann. Int'l Conf. Comput. Mol. Biol. (RECOMB 2001), pp. 149–156. ACM Press, New York (2001)

[9] Huson, D.H., Bryant, D.: Application of phylogenetic networks in evolutionary studies. Molecular Biology and Evolution 23(2), 254–267 (2006)

[10] Huson, D.H., Kloepper, T., Lockhart, P.J., Steel, M.A.: Reconstruction of reticulate networks from gene trees. In: Miyano, S., Mesirov, J., Kasif, S., Istrail, S., Pevzner, P.A., Waterman, M. (eds.) RECOMB 2005. LNCS (LNBI), vol. 3500, pp. 233–249. Springer, Heidelberg (2005)

[11] Jin, G., Nakhleh, L., Snir, S., Tuller, T.: Efficient parsimony-based methods for phylogenetic network reconstruction. Bioinformatics 23, e123–e128 (2006); Proceedings of the European Conference on Computational Biology (ECCB 2006)

[12] Jin, G., Nakhleh, L., Snir, S., Tuller, T.: Maximum likelihood of phylogenetic networks. Bioinformatics 22(21), 2604–2611 (2006)

[13] Jin, G., Nakhleh, L., Snir, S., Tuller, T.: Inferring phylogenetic networks by the maximum parsimony criterion: a case study. Molecular Biology and Evolution 24(1), 324–337 (2007)

[14] Jin, G., Nakhleh, L., Snir, S., Tuller, T.: A new linear-time heuristic algorithm for computing the the parsimony score of phylogenetic networks: theoretical bounds and empirical performance. In: Măndoiu, I.I., Zelikovsky, A. (eds.) ISBRA 2007. LNCS (LNBI), vol. 4463, pp. 61–72. Springer, Heidelberg (2007)

[15] Kunin, V., Goldovsky, L., Darzentas, N., Ouzounis, C.A.: The net of life: reconstructing the microbial phylogenetic network. Genome Research 15, 954–959 (2005)

[16] Lerat, E., Daubin, V., Moran, N.A.: From gene trees to organismal phylogeny in prokaryotes: The case of the γ-proteobacteria. PLoS Biology 1(1), 1–9 (2003)

[17] MacLeod, D., Charlebois, R.L., Doolittle, F., Bapteste, E.: Deduction of probable events of lateral gene transfer through comparison of phylogenetic trees by recursive consolidation and rearrangement. BMC Evolutionary Biology 5 (2005)

[18] Makarenkov, V.: T-REX: Reconstructing and visualizing phylogenetic trees and reticulation networks. econstructing and visualizing phylogenetic trees and reticulation networks 17(7), 664–668 (2001)

[19] Moret, B.M.E., Nakhleh, L., Warnow, T., Linder, C.R., Tholse, A., Padolina, A., Sun, J., Timme, R.: Phylogenetic networks: modeling, reconstructibility, and accuracy. IEEE/ACM Transactions on Computational Biology and Bioinformatics 1(1), 13–23 (2004)

[20] Nakamura, Y., Itoh, T., Matsuda, H., Gojobori, T.: Biased biological functions of horizon-tally transferred genes in prokaryotic genomes. Nature Genetics 36(7), 760–766 (2004)

[21] Nakhleh, L., Jin, G., Zhao, F., Mellor-Crummey, J.: Reconstructing phylogenetic networks using maximum parsimony. In: Proceedings of the 2005 IEEE Computational Systems Bioinformatics Conference (CSB 2005), pp. 93–102 (2005)

[22] Nakhleh, L., Ruths, D., Wang, L.S.: RIATA-HGT: A fast and accurate heuristic for recon-strucing horizontal gene transfer. In: Wang, L. (ed.) COCOON 2005. LNCS, vol. 3595, pp. 84–93. Springer, Heidelberg (2005)

[23] Nakhleh, L., Warnow, T., Linder, C.R.: Reconstructing reticulate evolution in species–theory and practice. In: Proc. 8th Ann. Int'l Conf. Comput. Mol. Biol. (RECOMB 2004), pp. 337–346 (2004)

[24] Ochman, H., Lawrence, J.G., Groisman, E.A.: Lateral gene transfer and the nature of bac-terial innovation. Nature 405(6784), 299–304 (2000)

[25] Shimodaira, H., Hasegawa, M.: Multiple comparisons of log-likelihoods with applications to phylogenetic inference. Molecular Biology and Evolution 16, 1114–1116 (1999)

[26] Suchard, M.A.: Stochastic models for horizontal gene transfer: taking a random walk through tree space. Genetics 170, 419–431 (2005)

[27] Than, C., Nakhleh, L.: SPR-based tree reconciliation: Non-binary trees and multiple solu-tions. In: Proceedings of the Sixth Asia Pacific Bioinformatics Conference, pp. 251–260 (2008)

[28] Than, C., Ruths, D., Innan, H., Nakhleh, L.: Identifiability issues in phylogeny-based de-tection of horizontal gene transfer. In: Bourque, G., El-Mabrouk, N. (eds.) RECOMB-CG 2006. LNCS (LNBI), vol. 4205, pp. 215–219. Springer, Heidelberg (2006)

[29] Than, C., Ruths, D., Innan, H., Nakhleh, L.: Confounding factors in HGT detection: Sta-tistical error, coalescent effects, and multiple solutions. Journal of Computational Biol-ogy 14(4), 517–535 (2007)

An Evolutionary Study of the Human Papillomavirus Genomes

Dunarel Badescu[1,*], Abdoulaye Baniré Diallo[1,2,*], Mathieu Blanchette[2], and Vladimir Makarenkov[1]

[1] Département d'informatique, Université du Québec à Montréal, C.P. 8888, Succursale Centre-Ville, Montréal (Québec), H3C 3P8, Canada
[2] McGill Centre for Bioinformatics and School of Computer Science, McGill University, 3775 University Street, Montréal, Québec, H3A 2B4, Canada

Abstract. In this article, we undertake a study of the evolution of Human Papillomaviruses (HPV), whose potential to cause cervical cancer is well known. First, we found that the existing HPV groups are monophyletic and that the high-risk carcinogenicity taxa are usually clustered together. Then, we present a new algorithm for analyzing the information content of multiple sequence alignments in relation to epidemiologic carcinogenicity data to identify regions that would warrant additional experimental analyses. The new algorithm is based on a sliding window procedure looking for genomic regions being responsible for disease. Examination of the genomes of 83 HPVs allowed us to identify specific regions that might be influenced by insertions, deletions, or simply by mutations, and that may be of interest for further analyses.

1 Introduction

Human papillomaviruses (HPV) have a causal role in cervical cancer with almost half a million new cases identified each year [1,3,18]. The HPV genomic diversity is well known [2]. About a hundred of HPV types are identified, and the whole genomes of more than eighty of them are sequenced (see the latest Universal Virus Database report by International Committee on Taxonomy of Viruses (ICTV)). A typical HPV genome is a double-stranded, circular DNA genome of size close to 8 Kbp, with complex evolutionary relationships and a small set of genes. In general, the E5, E6, and E7 genes modulate the transformation process, the two regulatory proteins, E1 and E2, modulate transcription and replication, and the two structural proteins L1 and L2 compose the viral capsid. Protein E4 has an unclear function in the HPV life cycle, however, several studies indicate that it could facilitate the viral genome replication and the activation of viral late functions [32], and it could also be responsible for virus assembly [22]. A HPV is considered to belong to a new HPV type if its complete genome has been cloned and the DNA sequence of the gene L1 differs by more

---

* The two first authors contributed equally to the work and should be considered as joint first authors.

C.E. Nelson and S. Vialette (Eds.): RECOMB-CG 2008, LNBI 5267, pp. 128–142, 2008.

than 10% from the closest known HPV type. The comparison of HPV genomes, conducted by ICTV, is based on nucleotide substitutions only [19,8]. Older HPV classifications were built according to their higher or lower risk of cutaneous or mucosal diseases. Most of the HPV studies were based on single gene (usually E6 or E7) analyses. The latter genes are predominantly linked to cancer due to the binding of their products to the p53 tumor suppressor protein and the retinoblastoma gene product pRb [29]. To define carcinogenic types, we used epidemiologic data recruited in 25 countries from a large international survey on HPVs in cervical cancer and from a multicenter case-control study conducted on 3,607 women with incident, histologically confirmed cervical cancer recruited in 25 countries [19,20]. HPV DNA detection and typing in cervical cells or biopsies were centrally done using PCR assays which attest for the quality of the study [19]. More than 89% of them had squamous cell carcinoma (i.e. Squam cancer) and about 5% had adenosquamous carcinoma (i.e. Adeno cancer) see Table 1 adapted from [19]. More than half of the infection cases are due to the types 16 and 18 of HPV, which are referred to as high-risk HPVs [5].

Table 1. Distribution of carcinogenic HPVs for the Squam and Adeno types of cancer. Complete genomic sequence data are not available yet for HPVs-35, HR, 68, and X.

HPV types	Squamous cell carcinoma		Adenocarcinoma and adenosquamous carcinoma	
	Number	% positive	Number	% positive
HPV-16	1,452	54.38	77	41.62
HPV-18	301	11.27	69	37.30
HPV-45	139	5.21	11	5.95
HPV-31	102	3.82	2	1.08
HPV-52	60	2.25		
HPV-33	55	2.06	1	0.54
HPV-58	46	1.72	1	0.54
HPV-56	29	1.09		
HPV-59	28	1.05	4	2.16
HPV-39	22	0.82	1	0.54
HPV-51	20	0.75	1	0.54
HPV-73	13	0.49		
HPV-82	7	0.26		
HPV-26	6	0.22		
HPV-66	5	0.19		
HPV-6	2	0.07		
HPV-11	2	0.07		
HPV-53	1	0.04		
HPV-81	1	0.04		
HPV-55	1	0.04		
HPV-83	1	0.04		
Total	**2,293**	**85.89**	**168**	**90.37**

In this paper, we first studied a whole genome phylogenetic classification of the HPV and the insertion and deletion (indel) distribution among HPV lineages leading to the different types of cancer. First, we inferred a phylogenetic tree of 83 HPVs based on whole HPV genomes. We found that the evolution of the L1 gene, used by ICTV to establish the HPV classification, generally reflects the whole genome evolution. Second, we compared the differences between gene trees built for the 8 most important HPV genes (E1, E2, E4, E5, E6, E7, L1 and L2) using the normalized Robinson and Foulds topological distance [23]. Then, we described a new algorithm for analyzing the information content of multiple sequence alignments in order to identify regions that may be responsible for the carcinogenicity. This algorithm is based on a new formula taking into account the sequence similarity among carcinogenic taxa and the sequence dissimilarity between the carcinogenic and non-carcinogenic taxa computed for a genomic region bounded by the position of the sliding window. Using the new technique, we examined all available genes in 83 HPV genomes and identified the specific genomic regions that would warrant interest for future biological studies.

2   Indel Analysis of HPV Genomes and Reconciliation of HPV Gene Trees

The 83 completely sequenced HPV genomes (all identified by ICTV) were downloaded and aligned using ClustalW [28], producing an alignment with 10426 columns. The phylogenetic tree of 83 HPVs (Figure 2) was inferred using the PHYML program [12] with the HKY substitution model. Bootstrap scores were computed to assess the robustness of the branches using 100 replicates. Most branches obtain support above 80%, but for a better readability, they are not represented in Figure 2. However, they are given in the supplemental materials[1]. As suggested in [29], the bovine PV of type 1 was used as outgroup to root this phylogeny. To the best of our knowledge, the constructed phylogenetic tree is the first whole genome phylogenetic tree of HPVs.

Our analysis revealed the presence of 12 known monophyletic HPV groups that are denoted by numerated nodes, labeled according to the ICTV annotation, in Figure 2. The other monophyletic groups obtained were not depicted by numbers. The obtained whole-genome phylogeny obtained usually corresponds to the HPV classification provided by ICTV on the basis of the L1 gene. Most of the dangerous HPVs (see Table 1) can be found in the sister subtrees rooted by the nodes 16 and 18.

As carcinogenicity may be introduced into a HPV by an insertion or deletion (indel) of a group of nucleotides, we first addressed the problem of indel distribution in the evolution of HPV. Thus, the most likely indel scenario was inferred using a heuristic method described in [9,10]. Such a scenario includes the distribution of the predicted indel and base conservation events for all HPV genes. Table 2

---

[1] Supplemental materials are available at:
http://www.labunix.uqam.ca/~makarenv/RecombCG2008.zip

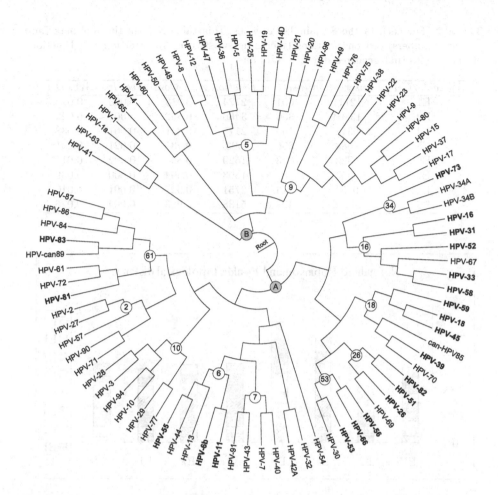

Fig. 1. Phylogenetic tree of 83 HPVs obtained with PHYML. The 21 carcinogenic HPV are shown in bold. The white nodes identify the existing HPV groups according to the ICTV and NCBI taxonomic classifications; the shaded nodes (A and B) distinguish between the non-carcinogenic and carcinogenic families. Bootstrap scores are above 80% for most of the branches; for a better readability, they are not represented. The HPVs 1 and 34 are present in two copies, (1 and 1a) and (34A and 34B), respectively.

reports, for each of the 8 main genes of HPV, the total number of conservations, insertions and deletions of nucleotides that occurred during their evolution. Genes E1, L1 and L2 show more than 90% conservation at the nucleotide level, E2, E4 and E6 between 80 and 90%, and E5 and E7 respectively 73% and 59%.

The highest indel frequencies are in the subtrees rooted by the node 61 where there are only low risks of carcinogenicity (Figure 2). The groups included in the subtree A have low percentage of indels on in each branch. It is likely that the organisms of this subtree inherited their carcinogenicity from their closest common ancestor.

Table 2. For each of the 8 main HPV genes, this table reports the numbers (and average numbers) of Conservations (including substitutions), Insertion and Deletions of nucleotides that occurred during evolution

Variable/Gene	Conservation	Insertion	Deletion	Avg. Cons.	Avg. Ins.	Avg. Del.
E1	12111	601	2774	0.918	0.003	0.010
E2	13304	306	3460	0.852	0.001	0.022
E4	6318	195	2117	0.851	0.001	0.038
E5	1688	356	503	0.731	0.021	0.031
E6	7323	613	1529	0.890	0.002	0.011
E7	3457	0	1393	0.594	0.000	0.039
L1	9664	314	2751	0.927	0.001	0.010
L2	21716	494	5138	0.923	0.004	0.026

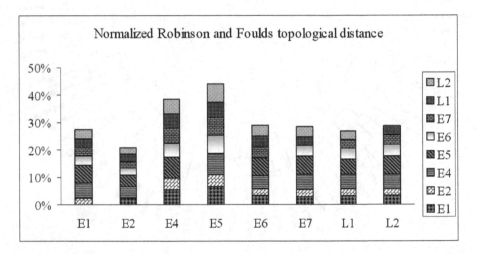

Fig. 2. Average normalized Robinson and Foulds topological distance for each of the 8 main HPV genes. Each column of the diagram represents a gene and consists of the stacked rectangles whose heights are proportional to the values of the normalized Robinson and Foulds topological distances between the phylogeny of this gene and those represented by the stacked rectangles. The column heights depicts the total average distance. For the sake of presentation the percentage values on the ordinate axis were divided by 7 (which is the number of pairwise comparisons made for each gene tree).

We also carried out an analysis intended to compare the topologies of the gene phylogenies built for the 8 main HPV genes. Thus, we first aligned, using ClustalW [28], the HPV gene sequences, separately for each gene, and inferred 8 gene phylogenies using the PHYML program [12] with the HKY model. In order to measure their degree of difference, we computed the Robinson and Foulds (RF) topological distances between each pair of gene trees [23]. As the number of tree leaves varied from 70 to 83 (due to the non-availability of some gene sequences for a few HPVs), we reduced the size of some trees prior to this pairwise topological comparison and normalized all distances by the largest possible value of the RF distance, which is $2n - 6$ for two binary trees with n leaves. Figure 2 shows the results obtained, with RF distances are depicted as stacked rectangles. The results suggest that the trees representing the evolution of the E4 and E5 genes differ the most, on average, from the other gene phylogenies, whereas the phylogeny of E2 reconciles the most the topological differences of this group of gene trees. Two HPV gene phylogenies differ from each other by about 32%, on average. In the future, it might also be interesting to compare the gene trees we obtained using Maximum Likelihood tests such as Shimodaira-Hasegawa [24] or Kishino-Hasegawa [17] and to assess the confidence of phylogenetic tree selection using a program such as CONSEL [25].

These results confirm the hypothesis made in a number of HPV studies (see for instance [21,30]), that most HPV genes undergo frequent recombination events. Uncritical phylogenetic analyses performed on recombinant sequences could lead to the impression of novel, relatively isolated branches. Recently, Angulo and Carvajal-Rodriguez (2007) have provided new support to the recent evidence of recombination in HPV. They found that the gene with recombination in most of the groups is L2 but the highest recombination rates were detected in L1 and E6. Gene E7 was recombinant only within the HPV16 type. The authors concluded that this topic deserves further study because recombination is an important evolutionary mechanism that could have a high impact both in pharmacogenomics and for vaccine development.

3 Algorithm for the Identification of Putatively Carcinogenic Regions

This section describes a new algorithm intended for finding genomic regions that may be responsible for HPV carcinogenicity. The algorithm is based on the hypothesis that sequence regions responsible for cancer are very similar among the carcinogenic HPVs while they differ a lot from the homolog regions in the non-carcinogenic HPVs. The following procedure was adopted. First, 83 available HPV genomes were downloaded and inserted into a relational database along with the clinical information regarding identified HPV types and histological type of cancer occurrences [19,20]. We constructed three HPV Types Datasets: "High-Risk", containing HPVs16 and 18, "Squamous", containing HPV types responsible for Squamous Cell Carcinoma (HPV-6, 11, 16, 18, 26, 31, 33, 39, 45, 51, 52, 53, 55, 56, 58, 59, 66, 73, 81, 82, 83) and Adeno with types responsible for Adenocarcinoma (HPV-16, 18, 31, 33, 35, 39, 45, 51, 58, 59). See Table 1 for more detail. HPV types with incomplete genome information or without annotations were excluded from the dataset. As previously, we used the gene sequences aligned separately for each gene.

Then, we scanned all gene sequence alignments using a sliding window of a fixed width (in our experiments the window width ranged from 3 to 20 nucleotides, see Figure 3). First, a detailed scan of each gene with increments of 1 nucleotide was performed to identifying the regions with a potential for causing

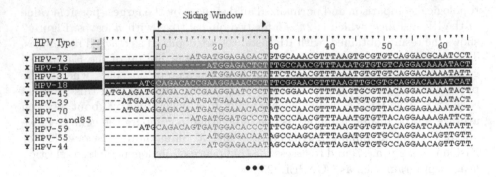

Fig. 3. A sliding window of a fixed width was used to scan all sequences of each HPV gene separately. The sequences in black belong to the set X (carcinogenic HPVs; in this example HPVs 16 and 18), all other sequences belong to the set Y (non-carcinogenic HPVs). The organism is indicated in the column on the extreme left.

carcinogenicity (the main results are reported in Table 3), and called here *hit regions*. Second, a non-overlapping windows of width 20 nucleotides was carried out for plotting Figures 4, 5, and 6[2]. Three separate analyses were made for the three above-described carcinogenic families: High-Risk, Squamous and Adeno HPVs.

Once the window position is fixed and the taxa are assigned to the sets X (carcinogenic HPVs) and Y (non-carcinogenic HPVs), the hit region identification function, denoted here as Q, can be computed. This function is defined as a difference between the means of the squared distances computed among the sequence fragments (bounded by the sliding window position) of the taxa from the set X and those computed only between the sequence fragments from the distinct sets X and Y. The mean of the squared distances computed among the sequence fragments of the carcinogenic taxa from the set X, and denoted here as $V(X)$, is computed as follows:

$$V(X) = \frac{1}{(N(X)(N(X) - 1)/2)} \sum_{\{x_1, x_2 \in X \mid x_1 \neq x_2\}} dist_h^2(x_1, x_2), \tag{1}$$

and the mean of the squared distances computed only between the sequence fragments from the distinct sets X and Y, and denoted here as $D(X, Y)$, is computed as follows:

$$D(X, Y) = \frac{1}{N(X)N(Y)} \sum_{\{x \in X, y \in Y\}} dist_h^2(x, y), \tag{2}$$

---

[2] Figures 5 and 6 are in Appendix available at:
http://www.labunix.uqam.ca/~makarenv/Appendix_RECOMB_2008_Paris.pdf

Table 3. Selected high-scoring regions with respect to the values of the hit region identification function Q. The best results for the contiguous regions of size 13 to 20 are reported. The best entry by HPV type (High-Risk, Squam, Adeno) and by gene is presented. The largest values of Q are in bold.

Dataset	Gene	Q	Index	Window width	$D(X,Y)$	$V(X)$.
High-Risk	E1	0.417	695	16	0.74	0.22
Squam	E1	0.345	575	14	0.50	0.08
Adeno	E1	0.353	307	20	0.52	0.09
High-Risk	**E2**	**0.553**	**1289**	**13**	**0.76**	**0.02**
Squam	E2	0.385	613	16	0.47	0.00
Adeno	E2	0.415	1265	20	0.66	0.14
High-Risk	E4	0.480	606	17	0.62	0.00
Squam	E4	0.373	1035	15	0.46	0.01
Adeno	E4	0.395	549	15	0.49	0.00
High-Risk	E5	0.339	88	13	0.41	0.01
Squam	E5	0.401	72	16	0.50	0.00
Adeno	E5	0.363	72	16	0.44	0.00
High-Risk	E6	0.496	725	17	0.69	0.05
Squam	**E6**	**0.531**	**725**	**17**	**0.76**	**0.06**
Adeno	**E6**	**0.521**	**725**	**17**	**0.75**	**0.06**
High-Risk	E7	0.258	206	13	0.34	0.05
Squam	E7	0.263	445	16	0.38	0.08
Adeno	E7	0.262	110	16	0.40	0.10
High-Risk	**L1**	**0.574**	**241**	**14**	**0.79**	**0.02**
Squam	L1	0.294	1159	15	0.34	0.00
Adeno	L1	0.302	1181	17	0.56	0.20
High-Risk	L2	0.310	1751	14	0.65	0.28
Squam	L2	0.320	1916	15	0.38	0.00
Adeno	L2	0.313	1914	17	0.37	0.00

where $N(X)$ and $N(Y)$ are the cardinalities of the sets X and Y, respectively, and $dist_h(x_1, x_2)$ is the Hamming distance between the sequence fragments corresponding to the taxa x_1 to x_2.

Then, the hit region identification function Q is defined as follows:

$$Q = ln(1 + D(X,Y) - V(X)). \tag{3}$$

The larger the value of this function for a certain genomic region, the more distinct are the carcinogenic taxa from the non-carcinogenic ones. The use of the Hamming distance instead of the well-adapted sequence to distance transformations such as the Jukes-Cantor (1969), Kimura 2-parameter (1980) or Tamura-Nei (1993) corrections, is justified by the two following facts: first, often the latter transformation formulae are not applicable to short sequences (remember that in our experiments the sequence lengths, equal to the sliding window width, varied from 3 to 20 nucleotides), and second, most of the well-known transformation models either ignore gaps or assign a certain penalty to them. As the carcinogenicity of HPVs can be related to an insertion or deletion of a group

---

Algorithm 1. Algorithmic scheme(MSA, *MSA_L*,X, *N(X)*, *Y*, *N(Y)*, *WIN_MIN*, *WIN_MAX*, *S*, *TH*)

Require: MSA: Multiple sequence alignment (considered as a matrix),
   MSA_L :   Length of MSA,
   X: Set of carcinogenic taxa,
   N(X):   Cardinality of the set *X*,
   Y: Set of non-carcinogenic taxa,
   N(Y):   Cardinality of the set *Y*,
   WIN_MIN: Minimum sliding window width,
   WIN_MAX: Maximum sliding window width,
   S: Sliding window step,
   TH: Minimum Q value for Hit (i.e., hit threshold).

Ensure: Set of Hit Regions: (win_width, idx, Q), where
   win_width :   Current sliding window width,
   idx :   Hit Index (i.e., its genomic position),
   Q :   Value of the hit region identification function.

1: **for** *win_width* **from** *WIN_MIN* **to** *WIN_MAX* **do**
2:      **for** *idx* **from** 0 **to** $MSA\_L - win\_width$ **with step** S **do**
3:          $MSA\_X \leftarrow MSA[X][idx..idx + win\_width]$
4:          $MSA\_Y \leftarrow MSA[Y][idx..idx + win\_width]$
5:          $V(X) \leftarrow D(X,Y) \leftarrow 0$
6:          **for all distinct** $i, j \in X$ **do**
7:              $V(X) \leftarrow V(X) + dist_h^2(MSA\_X[i], MSA\_X[j])$
8:          **end for**
9:          $V(X) \leftarrow 2 \times V(X)/(N(X) \times (N(X) - 1))$
10:         **for each** $i \in X$ **and** $j \in Y$ **do**
11:             $D(X,Y) \leftarrow D(X,Y) + dist_h^2(MSA\_X[i], MSA\_Y[j])$
12:         **end for**
13:         $D(X,Y) \leftarrow D(X,Y)/(N(X) \times N(Y))$
14:         $Q \leftarrow ln(1 + D(X,Y) - V(X))$
15:         **if** $Q > TH$ **then**
16:             *identify the current region* (win_width, idx, Q) *as a hit region*
17:         **end if**
18:     **end for**
19: **end for**

---

of nucleotides, the gaps should not be ignored but rather considered as valid characters, with the same weight as the other nucleotides, when computing the pairwise distances between the genomic regions.

The time complexity of this algorithm executed with overlapping sliding windows of a fixed width, and advancing one alignment site by step, is $O(l \times n^2 \times w)$, where l is the length of the multiple sequence alignment, n the number of taxa, and w the window width. However, this complexity can be reduce to $O(n^2 \times l)$ if we avoid recomputing the Hamming distance for neighbouring overlapping windows. This can be done by only removing the value of the left column of the sliding window while taking into account the value of added column in the

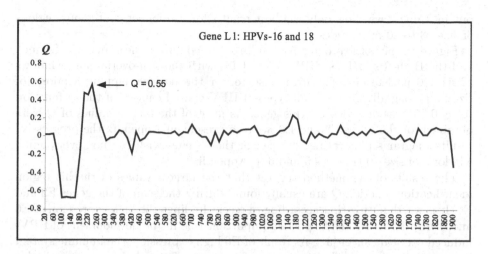

Fig. 4. The variation of the hit identification function Q for the High-Risk HPVs (HPVs-16 and 18) obtained with the non-overlapping sliding widows of width 20 during the scan of the gene L1. The abscissa axis represents the window position.

Hamming distance of the sliding window. For a non-overlapping sliding window, the time complexity is $O(n^2 \times l)$. If the width of the sliding window varies, as it was the case in our experiments, the time complexity should be obviously multiplied by the difference between the maximum and minimum window widths. The detailed algorithmic scheme is presented on the previous page.

4   Results, Discussion and Conclusion

The procedure for identifying hit regions in the 83 available HPV genomes was carried out twice: first, with overlapping windows of width w ($w = 3..20$), advancing one alignment site by step, and second, with non-overlapping windows of width 20. The 8 most important HPV genes (see Table 3) were scanned in such a way. The scan based on the overlapping windows provided over 35,000 values of Q bigger than 0.25. From the best 100 results obtained for each gene, we manually selected (see Table 3) the longest contiguous regions (up to 20 nucleotides) corresponding to the largest values of the hit region identification function Q. The values of Q were dependent on the window width, with better results usually associated with small windows. For instance (see Table 3), for larger window sizes, the largest values of Q were found during the scans of genes E2 and E6 for all types of HPVs, with the exception of the overall best score obtained during the scan of the gene L1 for the High-Risk HPV types (the value of 0.574 for a 14-nucleotide region starting with the index 241, see Table 3). For windows of small width, the largest values of Q were observed during the scan of the gene E4 for the High-Risk HPV category but in Table 3 we show only the best results for the longer contiguous regions of size 13 to 20 nucleotides.

Figure 3 depicts the progressive results obtained during the scan of the L1 gene and the High-Risk HPVs (HPVs-16 and 18) with the non-overlapping windows of size 20 nucleotides. The highest score, for the non-overlapping windows of size 20 among all genes and all types of HPV-caused cancer, of the Q function ($Q = 0.55$) was obtained for this gene. As most of the largest values of Q were obtained for the genes E2 and E6, we also present in Appendix the progressive results diagrams illustrating the scan of these genes with the non-overlapping windows of size 20 (Figures 5 and 6 in Appendix[2]).

The results of our method suggest that the largest values of the hit region identification function Q are usually found during the scan of the genes E2 and E6. It is worth noting that according to recent findings the high expression of E6 and disruption of E2 might play an important role in the development of HPV-induced cervical cancer [31]. As result of E6 high expression, the immune system is potentially evaded [7]. Disruption of the gene E2 was observed in invasive carcinomas [4] and in high-grade lesions [11]. Surprisingly, the overall largest value of Q was obtained for a specific region of the L1 gene. This underlines the possible use of our method for investigating particular regions of capsidal proteins in relation with vaccine design. It has been shown that linear epitopes within the protein L1 that induce neutralizing antibodies exist [6].

We noticed that the obtained results usually depend on the window width. As substitutions affect individual sites whereas indels often involve several consecutive nucleotides, small window sizes will tend to favor the former. However, the use of the Hamming distance, which does not ignore gaps in calculation, and variable window width allow us to account for both substitution and indel events. In the future, it would be interesting to study in more detail, in collaboration with virologists, all genomic regions providing the highest scores of the hit region identification function Q (the particular attention here should be paid to the genes E2, E6 and L1), determine, for each selected region, the evolutionary events (substitutions or indels) responsible for the observed differences in the carcinogenic and non-carcinogenic HPVs, and then establish at which level (i.e. on which branch) of the associated gene phylogeny this event has occurred. It may also be interesting to consider merging our results to those given by methods for detecting sequences under lineage-specific selection such as DLESS [26]. Next, we plan to compare this work with other approaches of the computational virology, which used some simpler methods, such as signatures, to analyze other viruses. Another interesting development would be to design a statistical test allowing one to measure the significance of the obtained results.

Acknowledgement. Dunarel Badescu is an NSERC fellow. We thank Alix Boc and Emmanuel Mongin for their useful comments.

References

1. Angulo, M., Carvajal Rodriguez, A.: Evidence of recombination within human alpha-papillomavirus. Virology Journal 4, 33 (2007)
2. Antonsson, A., Forslund, O., Ekberg, H., Sterner, G., Hansson, B.G.: The Ubiquity and Impressive Genomic Diversity of Human Skin Papillomaviruses Suggest a Commensalic Nature of These Viruses. Journal of Virology 74(24), 11636–11641 (2000)
3. Bosch, F.X., Manos, M.M., Muoz, N., Sherman, M., Jansen, A.M., Peto, J., Schiffman, M.H., Moreno, V., Kurman, R., Shan, K.V.: Prevalence of Human Papillomavirus in Cervical Cancer: a Worldwide Perspective. International Biological Study on Cervical Cancer (IBSCC) Study Group. Journal of the National Cancer Institute 87(11), 796–802 (1995)
4. Chan, P.K., Cheung, J.L., Cheung, T.H., Lo, K.W., Yim, S.F., Siu, S.S., Tang, J.W.: Profile of viral load, integration, and E2 gene disruption of HPV58 in normal cervix and cervical neoplasia. Journal of Infectious Diseases 196(6), 868–875 (2007)
5. Chan, S.Y., Delius, H., Halpern, A.L., Bernard, H.U.: Analysis of genomic sequences of 95 papillomavirus types: uniting typing, phylogeny, and taxonomy. Journal of Virology 69(5), 3074–3083 (1995)
6. Combita, A.-L., Touz, A., Bousarghin, L., Christensen, N.D., Coursaget, P.: Identification of Two Cross-Neutralizing Linear Epitopes within the L1 Major Capsid Protein of Human Papillomaviruses. Journal of Virology 76(13), 6480–6486 (2002)
7. Cordano, P., Gillan, V., Bratlie, S., Bouvard, V., Banks, L., Tommasino, M., Campo, M.S.: The E6E7 oncoproteins of cutaneous human papillomavirus type 38 interfere with the interferon pathway. Virology 377(2), 408–418 (2008)
8. de Villiers, E.M., Fauquet, C., Broker, T.R., Bernard, H.U., Zur Hausen, H.: Classification of papillomaviruses. Virology 324(1), 17–27 (2004)
9. Diallo, A.B., Makarenkov, V., Blanchette, M.: Exact and Heuristic Algorithms for the Indel Maximum Likelihood Problem. Journal of Computational Biology 14(4), 446–461 (2007)
10. Diallo, A.B., Makarenkov, V., Blanchette, M.: Finding maximum likelihood indel scenarios. In: Proceeding of the fourth Recomb satellite conference on Comparative Genomics, pp. 171–185 (2006)
11. Graham, D.A., Herrington, C.S.: HPV-16 E2 gene disruption and sequence variation in CIN 3 lesions and invasive squamous cell carcinomas of the cervix: relation to numerical chromosome abnormalities. Molecular Pathology 53, 201–206 (2000)
12. Guindon, S., Gascuel, O.: A simple, fast, and accurate algorithm to estimate large phylogenies by maximum likelihood. Systematic Biology 52, 696–704 (2003)
13. Goldman, N., Anderson, J.P., Rodrigo, A.G.: Likelihood-based tests of topologies in phylogenetics. Systematic Biology 49, 652–670 (2000)
14. Bchen-Osmond: ICTVdB - The Universal Virus Database C (ed). Columbia University, New York, USA
15. Jukes, T.H., Cantor, C.R.: Evolution of protein molecules. In: Munro, H.N. (ed.) Mammalian protein metabolism, pp. 21–123. Academic Press, London (1969)
16. Kimura, M.: A simple method for estimating evolutionary rate of base substitution through comparative studies of nucleotide sequences. Journal of Molecular Evolution 16, 111–120 (1980)
17. Kishino, H., Hasegawa, M.: Evaluation of the maximum likelihood estimate of the evolutionary tree topologies from DNA sequence data, and the branching order in Hominoidea. Journal of Molecular Evolution 29, 170–179 (1989)

18. Muñoz, N.: Human papillomavirus and cancer: the epidemiological evidence. Journal of Clinical Virology 19(1-2), 1–5 (2000)
19. Muñoz, N., Bosch, F.X., de Sanjos, S., Herrero, R., Castellsagu, X., Shah, K.V., Snijders, P.J.F., Meijer, C.J.L.M.: Epidemiologic classification of human papillomavirus types associated with cervical cancer. New England Journal of Medecine 384, 518–527 (2003)
20. Muñoz, N., Bosch, F.X., Castellsagu, X., Daz, M., de Sanjose, S., Hammouda, D., Shah, K.V., Meijer, C.J.: Against which human papillomavirus types shall we vaccinate and screen? The international perspective. International Journal of Cancer 111, 278–285 (2004)
21. Narechania, A., Chen, Z., DeSalle, R., Burk, R.D.: Phylogenetic incongruence among oncogenic genital alpha human papillomaviruses. Journal of Virology 79, 15503–15510 (2005)
22. Prétet, J.L., Charlot, J.F., Mougin, C.: Virological and carcinogenic aspects of HPV. Bulletin Academic National de Medecine 191(3), 611–613 (2007)
23. Robinson, D.R., Foulds, L.R.: Comparison of phylogenetic trees. Mathematical Biosciences 53, 131–147 (1981)
24. Shimodaira, H., Hasegawa, M.: Multiple comparisons of log-likelihoods with applications to phylogenetic inference. Molecular Biology and Evolution 16, 1114–1116 (1999)
25. Shimodaira, H., Hasegawa, M.: CONSEL: for assessing the confidence of phylogenetic tree selection. Bioinformatics 17, 1246–1247 (2001)
26. Siepel, A., Pollard, K.S., Haussler, D.: New methods for detecting lineage-specific selection. In: Apostolico, A., Guerra, C., Istrail, S., Pevzner, P.A., Waterman, M. (eds.) RECOMB 2006. LNCS (LNBI), vol. 3909, pp. 190–205. Springer, Heidelberg (2006)
27. Tamura, K., Nei, M.: Estimation of the number of nucleotide substitutions in the control region of mitochondrial DNA in humans and chimpanzees. Molecular Biology and Evolution 10, 512–526 (1993)
28. Thompson, J.D., Higgins, D.G., Gibson, T.J.: CLUSTAL W: improving the sensitivity of progressive multiple sequence alignment through sequence weighting, positions-specific gap penalties and weight matrix choice. Nucleic Acids Research 22, 4673–4680 (1994)
29. Van Ranst, M., Kaplanlt, J.B., Burk, R.D.: Phylogenetic Classification of Human Papillomaviruses: Correlation with clinical manifestations. Journal of General Virology 73, 2653–2660 (1992)
30. Varsani, A., Van der Walt, E., Heath, L., Rybicki, E.P., Williamson, A.L., Martin, D.P.: Evidence of ancient papillomavirus recombination. Journal of General Virology 87, 2527–2531 (2006)
31. Wang, J.T., Ding, L., Gao, E.S., Cheng, Y.Y.: Analysis on the expression of human papillomavirus type 16 E2 and E6 oncogenes and disruption of E2 in cervical cancer. Zhonghua Liu Xing Bing Xue Za Zhi 28(10), 968–971 (2007)
32. Wilson, R., Ryan, G.B., Knight, G.L., Laimins, L.A., Roberts, S.: The full-length E1^E4 protein of human papillomavirus type 18 modulates differentiation-dependent viral DNA amplification and late gene expression. Virology 362(2), 453–460 (2007)

Appendix

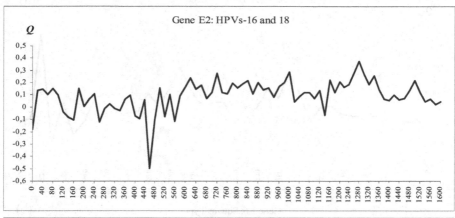

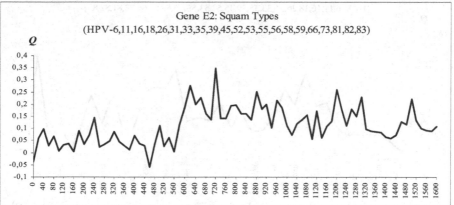

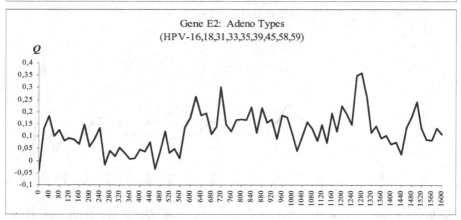

Fig. 5. The variation of the hit identification function Q for: (a) High-Risk HPVs (HPV-16 and 18), (b) Squam cancer causing HPVs, and (c) Adeno cancer causing HPVs obtained with the non-overlapping sliding widows of width 20 during the gene E2 scan

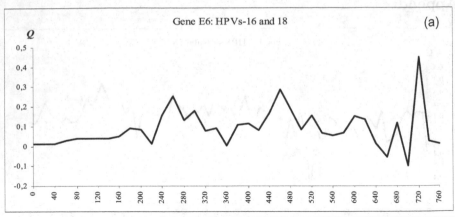

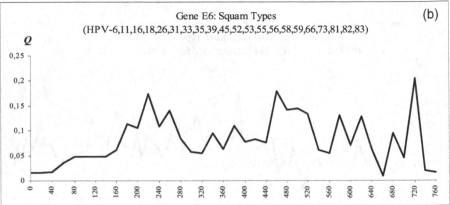

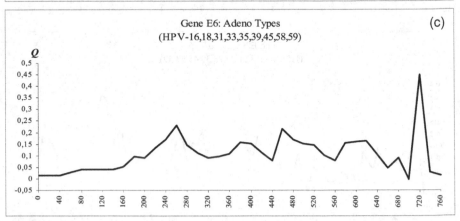

Fig. 6. The variation of the hit identification function Q for: (a) High-Risk HPVs (HPV-16 and 18), (b) Squam cancer causing HPVs, and (c) Adeno cancer causing HPVs obtained with the non-overlapping sliding widows of width 20 during the gene E6 scan

An Algorithm for Inferring Mitogenome Rearrangements in a Phylogenetic Tree

Matthias Bernt[1], Daniel Merkle[2], and Martin Middendorf[1],*

[1] Parallel Computing and Complex Systems Group, Department of Computer
Science, University of Leipzig, Germany
{bernt,middendorf}@informatik.uni-leipzig.de
[2] Department of Mathematics and Computer Science,
University of Southern Denmark
daniel@imada.sdu.dk

Abstract. Given the mitochondrial gene orders and the phylogenetic
relationship of a set of unichromosomal taxa, we study the problem of
finding a plausible and parsimonious assignment of genomic rearrangement events to the edges of the given phylogenetic tree. An algorithm
called algorithm TreeREx (tree rearrangement explorer) is proposed for
solving this problem heuristically. TreeREx is based on an extended version of algorithm CREx (common interval rearrangement explorer, [4])
that heuristically computes pairwise rearrangement scenarios for gene
order data. As phylogenetic events in such scenarios reversals, transpositions, reverse transpositions, and tandem duplication random loss
(TDRL) operations are considered. CREx can detect such events as patterns in the signed strong interval tree, a data structure representing gene
groups that appear consecutively in a set of two gene orders. TreeREx
then tries to assign events to the edges of the phylogenetic tree, such
that the pairwise scenarios are reflected on the paths of the tree. It is
shown that TreeREx can automatically infer the events and the ancestral
gene orders for realistic biological examples of mitochondrial gene orders. In an analysis of gene order data for teleosts, algorithm TreeREx is
able to identify a yet undocumented TDRL towards species *Bregmaceros
nectabanus*.

1 Introduction

Phylogenetic hypothesis are often supported by the computation of parsimonious scenarios based on genome-wide rearrangement operations. Especially mitochondrial gene orders became a very fruitful source for such investigations as
the number of genes is not too large and for more than 1000 species the mitochondrial gene order is known. In literature inversions and transpositions are the
most often considered genomic rearrangement operation for phylogenetic reconstruction [5, 18]. Even when only inversions and a small number of gene orders

---

* This work was supported by the German Research Foundation (DFG) through the
project "Deep Metazoan Phylogeny" within SPP 1174.

C.E. Nelson and S. Vialette (Eds.): RECOMB-CG 2008, LNBI 5267, pp. 143–157, 2008.

are considered, recovering a most parsimonious scenario is usually NP-complete (e.g. [8]). Considering combinations of rearrangement operations in event-based reconstruction methods is done very rarely.

In recent biological studies it was shown that the so called tandem duplication random loss (TDRL) operation is a genomic rearrangement operation that can be found several times in the mitochondrial gene order evolution, e.g., in millipedes [12] and deep-sea gulper eels [11]. A TDRL duplicates a contiguous segment of genes, followed by the loss of one copy of each of the duplicated genes. The biological fact that gene groups are often preserved during evolution lead to the utilization of so-called common intervals [10, 20] and strong interval trees (SIT) [7, 16] (also called PQ-trees), which are data structures that reflect properties of contiguous gene groups. The recently proposed algorithm CREx infers heuristically a rearrangement scenario between two gene orders [4]. This algorithm computes a SIT for a pair of genomes and identifies patterns in the SIT that indicate certain genomic rearrangement operations. Reverse transpositions, transpositions, reversals, and TDRLs can be identified by this strategy. With this set of operations CREx is well suited for studying mitochondrial gene order evolution. In [17] the algorithm CREx was used to manually infer the evolution of mitochondrial gene orders in Echinoderms based on pairwise inspection of the gene orders.

In this paper we introduce an extension of algorithm CREx so that it can better handle TDRL/reversal combinations. Moreover, we propose an algorithm called TreeREx (tree rearrangement explorer) which takes as input a binary rooted phylogenetic tree and the gene orders of a set of taxa and heuristically infers the corresponding rearrangement operations on the edges of the tree. Algorithm TreeREx utilizes algorithm CREx. We show the applicability of TreeREx on several biological examples.

The paper is structured as follows. In Section 2 basic definitions are given. Algorithm CREx is presented in Section 3. In Section 4 algorithm TreeREx is introduced, and in Section 5 algorithm TreeREx is applied to a small biological example. Results are given in Section 6. Section 7 concludes the paper.

2   Basic Definitions

2.1   Rearrangement Operations

A *permutation of size n* is a permutation of the elements in $\{1, 2, \ldots, n\}$. A *signed permutation* of size n is a permutation of size n where every element has an additional sign ("+" or "−") that defines its orientation ("+" is usually omitted). In the following we call a signed permutation $\pi = (\pi^1, \ldots, \pi^n)$ just permutation. A *reversal* $\rho_R(i, j)$, $1 \le i \le j \le n$ applied to a signed permutation π of size n transforms it into $\pi \circ \rho_R(i, j) = (\pi^1, \ldots, \pi^{i-1}, -\pi^j, \ldots, -\pi^i, \pi^{j+1}, \ldots, \pi^n)$. A *transposition* $\rho_T(i, j, k)$, $1 \le i \le j < k \le n$ applied to π transforms it into $\pi \circ \rho_T(i, j, k) = (\pi^1, \ldots, \pi^{i-1}, \pi^{j+1}, \ldots, \pi^k, \pi^i, \ldots, \pi^j, \pi^{k+1}, \ldots, \pi^n)$. A *reverse transposition* $\rho_{rT}(i, j, k)$, with $1 \le i \le j \le n$ and $(1 \le k < i)$ or $(j < k \le n)$, applied to π transforms it (here shown for $j < k$) into

$\pi \circ \rho_{\mathrm{rT}}(i,j,k) = (\pi^1, \ldots, \pi^{i-1}, -\pi^k, \ldots, -\pi^{j+1}, \pi^i, \ldots, \pi^j, \pi^{k+1}, \ldots \pi^n)$. A *tandem duplication random loss* ρ_{TDRL} duplicates a contiguous segment of genes in tandem, followed by the loss of one copy of each of the duplicated genes. Note that TDRLs which have no effect on the gene order are excluded. Furthermore TDRLs which have the same effect as a transposition are handled as a transposition. A *scenario* for two signed permutations π and σ is a sequence of rearrangement operations that transforms π into σ. A sequence with a minimal (weighted) number of operations is called *parsimonious*.

2.2   Common Intervals and Strong Interval Trees

An interval of a permutation π is a set of consecutive elements of the permutation π. Let Π be a set of signed permutations of size n. A *common interval* [10, 20] of Π is a subset of $\{1, 2, \ldots, n\}$ that is an interval in each $\pi \in \Pi$. The singletons $\{i\}, i \in \{1, 2, \ldots, n\}$ and the set $\{1, 2, \ldots, n\}$ of all elements are called *trivial common intervals*. Let $C(\Pi)$ be the set of all common intervals of Π. Two intervals c and c' *overlap* if $c \cap c' \neq \emptyset$, $c \not\subset c'$, and $c' \not\subset c$. If two intervals do not overlap they *commute*. A common interval is called a *strong common interval*, if it does not overlap with any other common interval. The set of all strong common intervals can be computed in time $O(kn)$ for k signed permutations of size n [3]. The *strong interval tree* of two permutations $\Pi = \{\pi_1, \pi_2\}$ is a tree $T(\Pi)$ where the nodes are exactly the strong common intervals of Π such that the root node is the interval containing all elements, the leaves are the singletons, and the edges are defined by the minimal inclusion relation of the intervals (i.e. there is an edge between node c and c' iff $c' \subset c$ and there is no node c'' with $c' \subset c'' \subset c$). Each node is given a sign ($+$ or $-$). If the children of a node appear in the same order in both input gene orders, the node is called linear increasing ($+$); if the children of a node appear in opposite order in the two gene orders, it is linear decreasing ($-$); otherwise the node is called prime. For a more comprehensive introduction of SITs see [2]. The importance of the SIT is that it greatly facilitates the identification of the genome rearrangement operations in algorithms CREx and TreeREx.

A genomic rearrangement operation ρ applied to one of the permutations $\pi \in \Pi$ is said to be *preserving for* Π if it does not destroy any common interval $c \in C(\Pi)$ (i.e., $C(\Pi) = C(\Pi \cup \{\pi \circ \rho\})$). An operation is not preserving, if there exists a common interval, such that it does not exists after applying the rearrangement operation.

3   Algorithm CREx

CREx [4] is an algorithm to heuristically determine preserving rearrangement scenarios for pairs of unichromosomal genomes. The algorithm uses the fact that each of the four rearrangement operations that are considered here leads to a pattern in the SIT. To illustrate this each of the four rearrangement operations is applied to the identity permutation and the resulting SIT is computed. Figure 1

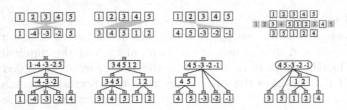

Fig. 1. Top: Genomic rearrangement events considered in CREx and TreeREx; Bottom: Strong interval tree of the identity permutation and the resulting permutation after applying the corresponding genomic rearrangement event; From left to right: reversal, transposition, reverse transposition, and tandem duplication random loss

shows the applied rearrangement operations and the resulting SITs. More formally, the following patterns appear for the different operations when applied to a permutation π.

- If a reversal $\rho_{\mathrm{R}}(i,j)$ is applied, a linear node with a linear parent node of opposite sign occurs in the corresponding SIT (see also [2]). The linear node reflects the common interval of all elements that are inverted.
- If a transposition $\rho_{\mathrm{T}}(i,j,k)$ is applied, the corresponding SIT has a linear node with elements $\{\pi^i,\dots,\pi^k\}$ that has two linear children reflecting the common intervals $\{\pi^i,\dots,\pi^j\}$ and $\{\pi^{j+1},\dots,\pi^k\}$. The sign of the node must be different from the signs of the child nodes.
- If a reverse transposition $\rho_{\mathrm{rT}}(i,j,k)$ is applied, the corresponding SIT has a linear node with elements $\{\pi^i,\dots,\pi^k\}$. One child is a linear node reflecting the common interval of elements $\{\pi^i,\dots,\pi^j\}$ that are not inverted due to the reverse transposition. This child must have the same sign as its parent. The other involved elements are singletons as child nodes of node $\{\pi^i,\dots,\pi^k\}$ which must have a different sign.
- A tandem duplication loss operation ρ_{TDRL} leads to a prime node reflecting all the elements involved in the rearrangement operation.

Algorithm CREx computes for two input permutations π_1 and π_2 the strong interval tree for these permutations. Then CREx searches for patterns corresponding to rearrangement operations. If a pattern is identified, the corresponding rearrangement operation ρ is included in the scenario to be computed and the next pattern is searched in the strong interval tree of $\pi_1 \circ \rho$ and π_2 (the pattern for ρ will not occur in this strong interval tree). This process is repeated until a complete scenario is inferred. Note that the described patterns may be obfuscated if overlapping rearrangements have taken place, e.g. in the extreme case of rearrangement hot spots prime nodes will emerge. Operations inferred from different nodes are commutative as the corresponding strong intervals commute. Obviously, the search order for patterns is very important. If reversals are identified before reverse transpositions and transpositions, then all reverse transpositions and transpositions would be inferred as being reversal operations in the scenario. Therefore, the search order for the patterns of the genomic rearrangement operations is i.) transpositions, ii.) reverse transpositions, iii.) reversals, and iv.)

TDRL operations. Note that this order introduces a bias which is resolved in an extension described in Section 4.3.

Special care has to be taken when prime nodes occur in the SIT. In algorithm CREx a prime node is an indicator for one or several TDRLs. As a TDRL operation will not change the sign of the elements involved, reversals are utilized to equalize the sign of all the elements in a prime node. Algorithm CREx uses a heuristic approach to identify a parsimonious number of reversals and TDRLs for the corresponding prime node. Let π_1 and π_2 be the two permutations of the elements in the prime node. Two variants are now included in the latest version of CREx to infer the reversals that are needed: i.) (reversals first) a set of reversals is applied to the origin permutation π_1 to equalize the signs (with respect to π_2), and then, starting from the resulting permutation, the minimum of number of TDRLs is computed [9]; or ii.) (reversals last) first a set of reversals is applied to π_2 (resulting in π_2', such that all the signs a equalized with respect to π_1). Then a minimal number of TDRLs is inferred to transform permutation π_1 to permutation π_2'. Note that the number of different possible parsimonious scenarios to equalize the signs grows exponentially in the number of blocks of elements that have different signs in both permutations. Algorithm CREx uses a brute-force approach and each possible minimal set of reversals is tested, resulting in a potentially different number of TDRLs per reversal set. Scenarios, for which the sum of the number of reversals and TDRLs is minimal, are considered as possible scenarios. Furthermore, note that the resulting scenarios for a prime node is not guaranteed to be parsimonious, as a mixed sequence of reversals and TDRLs may result in a smaller scenario. Algorithm CREx, a tutorial, and several detailed examples are available online at http://pacosy.informatik.uni-leipzig.de/CREx.

4   Algorithm TreeREx

Although algorithm CREx supports a user to find parsimonious rearrangement scenarios for a given phylogenetic hypothesis with more than two input genomes, this process has to be done by manual inspection of pairwise scenarios. The overlap of pairwise scenarios for different pairs of taxa can be utilized to infer events on the edges of a given phylogenetic tree. Algorithm TreeREx (tree rearrangement explorer) automates this procedure. A given phylogentic tree is analyzed in a bottom-up manner by iteratively considering triples and quadruples of gene orders. TreeREx utilizes the pairwise comparisons suggested by CREx, assigns genomic rearrangement events to edges of the phylogenetic tree, and computes the permutations assigned to ancestral nodes.

4.1   Consistency

Let $\Pi := \{\pi_1, \ldots, \pi_m\}$ be a set of m input permutations and $T = (V, E)$ be a binary phylogenetic tree with the permutations $\pi_1, \ldots, \pi_m$ assigned to the leaf nodes $v_1, \ldots, v_m$. Let $r(\pi_i, \pi_j)$ be the set of rearrangement events for the pairwise scenarios between π_i and π_j, $1 \leq i, j \leq m$ (potentially inferred by algorithm CREx). Let π be the permutation to be assigned to the the parent node v of v_i and v_j. Let r_ϵ, $\epsilon \in \{i, j\}$ be the inferred rearrangement events on edge $(v, v_\epsilon) \in E$ by intersecting all pairwise scenarios from any permutation towards π_ϵ. Formally r_ϵ is computed by

$$r_\epsilon := \bigcap_{\pi_l \in \Pi \setminus \pi_\epsilon} r(\pi_l, \pi_\epsilon), \epsilon \in \{i, j\} \tag{1}$$

Note that this intersection is well defined if the scenarios are commutative because the rearrangements in the scenarios can be handled as a set. As pointed out above scenarios inferred by CREx are commutative as long as the corresponding SITs do not imply scenarios for prime nodes consisting of more than one operation. The treatment of ordered scenarios is explained in more detail in Section 4.3.

If the genomic rearrangement operations r_i applied inversely to π_i lead to the same permutation as applying the operations r_j inversely to π_j, then the inferred ancestral permutation $\pi = \pi_i \circ r_i^{-1} = \pi_j \circ r_j^{-1}$ of node v and the events r_i and r_j are said to be *consistent* with tree T. An example with $m = 4$ is illustrated in Figure 2.

The definition of consistency can be relaxed by restricting the set of intersected scenarios in Equation 1. Suppose the number of intersected pairwise scenarios is reduced by k, such that at least two sets are intersected for inferring the events on an edge. Due to this relaxation, an ancestral permutation π may be inferred by inversely applying potentially different intersections r_i and r_j to π_i and π_j (nevertheless, also discarding scenarios may still lead to inconsistency). Note that usually there exist several possible reductions by k scenarios, such that still an ancestral permutation π can be inferred by inversely applying the result of the intersection. The inferred ancestral permutations may be different depending on the set of discarded scenarios. If a majority of all reductions by k scenarios lead to the same permutation π, then π is said to be *k-consistent* (with tree T). Obviously, a node is 0-consistent, iff it is consistent. If no k can be found, such that a node is k-consistent, the node is *inconsistent*.

Example: Suppose in Figure 2 permutation π can not be inferred consistently. Suppose when $r(\pi_3, \pi_1)$ not considered for computing r_1, an ancestor π is

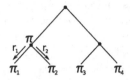

Fig. 2. Consistency: For permutations $\Pi = \{\pi_1, \ldots, \pi_4\}$ the inferred events on the edges in the tree T towards π_1 (respectively π_2) are $r_1 := \bigcap_{\pi_l \in \Pi \setminus \pi_1} r(\pi_l, \pi_1)$ (respectively $r_2 := \bigcap_{\pi_l \in \Pi \setminus \pi_2} r(\pi_l, \pi_2)$); if operations r_1 applied inversely to π_1 lead to the same permutation as applying the operations of r_2 inversely to π_2, then π is consistent with the tree T

inferred by inversely applying r_1 to π_1. Furthermore suppose, when $r(\pi_4, \pi_2)$ is not considered for computing r_2, the same ancestor can be inferred by inversely applying r_2 to π_2. That is, two scenarios have been discarded to infer an ancestral permutation. If most of the possible reductions by two scenarios lead to π, π is 2-consistent (with T).

4.2   Method

Algorithm TreeREx traverses subtrees in a given (binary) phylogenetic tree in a bottom-up manner beginning with subtrees induced by the permutations assigned to leaf nodes. More precisely, TreeREx iteratively selects a subtree with three or four leaf nodes, which has a height of 2, and for which the permutations assigned to the leaf nodes of the subtree are known. Then for the parents of the leaf nodes of the subtree the ancestral permutations are computed and the next subtree is selected. This continues until all (but the root node) have an assigned permutation. More formally, TreeREx proceeds as follows. Let π_1 and π_2 be permutations assigned to two nodes in the phylogenetic tree, and let these two nodes be siblings. Let π be an unknown permutation to be assigned to the parent node v of these siblings. Let π' be the permutation assigned the sibling v' of v, and let π'_1 and π'_2 be the permutations assigned to the child nodes of v'. If the permutation of π' is known, TreeREx infers the permutation for π by utilizing the subtree induced by π_1, π_2, and π'. If the permutation of v' is not known, TreeREx infers the permutation for π by utilizing the subtree induced by π_1, π_2, π'_1, and π'_2. Note that due to the bottom-up traversal such an induced subtree with three or four permutations assigned to leaf nodes can always be found, as long as at least two inner nodes (the root and one child node of the root) of the complete phylogenetic tree have no permutation assigned, yet.

   Assigning permutations to inner nodes and events to the edges of an induced subtree T_S is done as follows. In a first step TreeREx checks if consistent genomic rearrangement operations and a consistent permutation can be found by utilizing the necessary pairwise CREx scenarios of the leaves of T_S. If this fails, TreeREx tries to infer genomic rearrangement events by iteratively checking, if a k-consistent permutation can be assigned to an inner node. Therefore k is increased from 1 to its maximal possible value or until a k-consistent permutation is found. In the case that no k-consistent permutation can be found for an inner node, a fall-back strategy in applied as follows. For the scenario of π_1 and π_2 — computed by CREx — each possible ancestral permutation for π, based on each possible partition of the events in the scenario of π_1 and π_2, is computed. Let $\Gamma(\pi_1, \pi_2)$ be the set of these permutations. To chose the best ancestral permutation the scenarios of the assumed ancestral permutation $\pi \in \Gamma(\pi_1, \pi_2)$ and a permutation $\pi' \in \Gamma'(\pi'_1, \pi'_2)$ assigned to its sibling v' are taken into account, where $\Gamma'(\pi'_1, \pi'_2)$ is the set of possible ancestral permutations for π'. (In the case that π' and its children π'_1 and π'_2 are consistent, it holds $|\Gamma'(\pi'_1, \pi'_2)| = 1$). Then for each combination of ancestral permutations $\pi \in \Gamma(\pi_1, \pi_2)$ and $\pi' \in \Gamma(\pi'_1, \pi'_2)$ the set of rearrangement events is computed. π and π' are then chosen, such that the sum of the (weighted) number of rearrangement events is minimal. Hence a weighting function (denoted by q in the pseudo-code of TreeREx) has to be defined. The pseudo-code of algorithm TreeREx is given in the Appendix.

Algorithm TreeREx is designed to support biologist when analyzing real biological data and aims at inferring biologically evident events. Therefore we followed the presented heuristic approach that includes the four most known phylogenetic rearrangement operation in mitogenomes. Note that the outcome of TreeREx includes the consistency of internal nodes, which are a good indicator for the support of the inferred events. Having only consistent internal nodes in a subtree strongly supports the infered rearrangements and ancestral gene orders. Obviously, as even simplifications of the underlying problems are NP-complete and as the number of possible scenarios for two species only can be immense, this trade-off between usability and optimality is needed. If the outcome includes inconsistent and a large sequence of rearrangement operations on one edge only, then the support for this is very small. But, in contrast, if the outcome includes mainly k-consistent nodes for small values of k, then the support for the inferred events is very strong.

4.3   Extensions

So far algorithm TreeREx has been explained only in its basic version. Several extensions are used in order to improve the reliability of the inferred phylogenies. This includes the handling of non-commutative rearrangement operations, using the direction information of TDRL events, and including shared adjacency scenarios as alternative scenarios.

Ordered and Alternative Scenarios. The pairwise scenarios as computed by CREx are inferred by heuristically identifying patterns in the SIT. So far we assumed that there is only one pairwise scenario. In the data structure that handles the scenario between two permutations, alternative scenarios can be stored. A transposition can be replaced by three reversals or by a reversal and a reverse transposition. A reverse transposition can be replaced by two reversals or by a transposition and a reversal. For handling prime nodes CREx computes different alternatives of reversals and subsequent TDRLs, or TDRLs with subsequent reversals (comp. Section 3). Furthermore, we assumed so far, that events of a scenario can be applied in a commutative manner, which is only true if the SIT has linear nodes only. The individual events (combinations of reversals and TDRLs) inferred by a prime node can not be applied commutatively but have to be in a certain order.

Both alternative and ordered scenarios have to be handled properly when the intersection of CREx scenarios are computed according to Equation 1. Each alternative is handled seperately, e.g. the intersection of two alternative scenarios is an alternative scenario consisting of the intersections of all-vs-all alternatives. The intersection of an ordered sequence of events is the largest common suffix shared with the other scenario. As a formal description is very technical, this is illustrated with a small example.

Example: Let $r_1 = \rho_{\text{TDRL}} \rightarrow \rho_{R_2} \rightarrow \rho_{R_1}$ denote an ordered sequence of two reversals and one TDRL. Let $r_2 = \rho_{\text{T}} || \{\rho_{R_4}, \rho_{R_5}, \rho_{R_6}\}$ denote an alternative of either a transposition ρ_{T} or three commutative unordered reversals. Let $r_3 = \rho_{R_2} \rightarrow \rho_{R_1}$ and $r_4 = \rho_{\text{T}} || \{\rho_{R_4}, \rho_{R_6}\}$ denote two other sequences. Let $r_1 || r_2$ and $r_3 || r_4$ represent two pairs of alternative scenarios inferred by CREx. The intersection of these scenarios leads to the scenario represented by $\rho_{R_2} \rightarrow \rho_{R_1} || (\rho_{\text{T}} || \{\rho_{R_4}, \rho_{R_6}\})$.

Handling of Tandem Duplication Random Loss Events. TDRLs are in general irreversible [9] and hence imply a direction of the corresponding edge in the phylogenetic tree. TreeREx discards scenarios with TDRLs leading towards the root node. Furthermore, applying a TDRL inversely for checking the consistency of a permutation (comp. Section 4.1), needs special attention in TreeREx, as only the TDRL is not symmetric. That is, if a TDRL is applied to a permutation π_1 leading to π_2, applying the same TDRL to π_2 usually does not lead to π_1.

Inclusion of Shared Adjacency Scenarios. As presented so far CREx infers only TDRLs and (if needed) reversals for prime nodes. In [21] a method was presented to heuristically infer (ordered) scenarios of transpositions and reversals. The basic idea is to identify reversals and transpositions by a proper analysis of shared adjacencies of the two input permutations. We included this method and use the inferred reversal/transposition scenario as an alternative scenario for handling prime nodes with algorithm TreeREx.

5 A Detailed Small Biological Example

The echinoderm phylogeny has been investigated intensely (e.g. [13]), but is still heavily discussed [19]. To show the practical usefulness of TreeREx a small biological example of mitochondrial gene orders of four echinoderms is used. The gene orders are from the four taxonomic gropus *Asteroidea* (A), *Echinodea* (E), *Holothuridea* (H), and *Crinoidea* (C). The gene orders derived from the corresponding GenBank entries are given in the Appendix. The used topology is given by $((A, E), H), C)$ [17] (see Figure 3). TreeREx traverses the tree in a bottom up manner as follows.

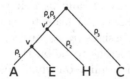

Fig. 3. Phylogenetic tree as used in the illustrative small biological example; genomic rearrangement operation inferred by algorithm TreeREx are denoted by $\rho_1, \ldots, \rho_5$; inner nodes for which TreeREx inferred ancestral permutations are denoted by v and v'

Fig. 4. TDRL inferred by algorithm CREx for the scenario of *Strongylocentrotus purpuratus* (E) towards *Cucumaria miniata* (H); (operation ρ_2 in Figure 3)

Fig. 5. Operations inferred by algorithm CREx for the scenario of *Florometra serratissima* (C) towards *Strongylocentrotus purpuratus* (E); left: reverse transposition (operation ρ_3 in Figure 3); right: TDRL (operation ρ_5 in Figure 3)

i.) The first subtree to be analyzed is given by $((A, E), H)$. Let v denote the parent of A and E. The sibling of v is a node which has already an assigned permutation (permutation H). Two events are predicted by CREx for the evolution between H and A, i.e., CREx(A,H) $= \{\rho_1, \rho_2\}$, with ρ_1 being a reversal of (-rrnL ... -P), and ρ_2 being the TDRL (towards H) as shown in Figure 4. Furthermore, CREx(E,A) $= \{\rho_1\}$ and CREx(E,H) $= \{\rho_2\}$. The intersections define the predicted events on the edges (v, A) and (v, E), i.e., $r(v, A) =$ CREx(E,A) $\cap$ CREx(H,A) $= \{\rho_1\}$ and $r((v, E)) =$ CREx(A,E) $\cap$ CREx(H,E) $= \{\}$. Applying ρ_1 inversely to A leads to the same permutation as applying no event to E. This is a consistent case and therefore the ancestor of A and E is E, which is assigned to node v. Furthermore, event ρ_1 is assigned to edge e_1.

ii.) The second subtree analyzed by TreeREx is $((v, H), C)$, with E assigned to v. Let v' denote the parent of v and H (see Figure 3). The sibling of v' is the leaf node C and has therefore an assigned permutation. As stated above, CREx(E,H) $= \{\rho_2\}$. Furthermore, CREx(C,H) also includes ρ_2 and therefore the intersection is $r(v', H) = \{\rho_2\}$. The intersection of CREx(H,E) (3 TDRLs) and CREx(C,E) is empty. As applying ρ_2 inversely to H gives the same permutation as applying no operation (inversely) to E, v' is also consistent and permutation E is also assigned to node v'. To infer the operations on the two edges incident to the root node, we compute the pairwise scenario of C and E. CREx(C,E) $= \{\rho_3, \rho_4, \rho_5\}$ with ρ_3 being a reverse transposition as shown in the left part of Figure 5, ρ_4 being a reversal of (-L2 ... -nad1), and ρ_5 being a TDRL event as shown in the right part of Figure 5. CREx(E,C) leads to more than 3 events (not shown here), and therefore TreeREx assigns ρ_3, ρ_4, and ρ_5 to the edges incident to the root node. As ρ_5 is a TDRL it has to be on the edge towards v'. Without an outgroup it is impossible to determine on which edge ρ_3 and ρ_4 occurred. Note, that a manual analysis for inferring the rearrangement scenarios for this small example has been presented in [17]. Algorithm TreeREx is capable of inferring all these events automatically.

6   Results

In this section the mitogenomes of teleosts and echinoderms are analyzed with algorithm TreeREx. Note that to the best of our knowledge there is no other algorithm for inferring mitogenome rearrangements based on the four different events as used in algorithm TreeREx. We utilized the mitochondrial gene orders from [6] which were marked as complete. All gene orders were removed which did not have the standard set of 37 mitochondrial genes (13 protein coding-, 2 rRNA-, and 22 tRNA- genes). For the tree topologies we utilized published phylogenies.

6.1   Teleosts

For the analysis of the teleost mitogenomes we merged the phylogenies suggested in [14] and [11]. Most of the teleosts have the typical vertebrate gene order ('TV') which can be found e.g. in human mitochondrial genomes. Therefore subtrees with identical gene order are collapsed. The result of the TreeREx analysis is given in Figure 6. In can be seen, that all but one of the ancestral gene orders were inferred consistent or k-consistent, $k > 1$. The only inconsistency occurred for the ancestral permutation of *Bregmaceros nectabanus*. In [11] a mechanism of mitochondrial gene rearrangement in gulper eels was proposed, which exactly corresponds to the TDRL as found by TreeREx leading towards *Eurypharynx pelecanoides*. The involved nodes in the phylogenetic tree are all inferred consistently, hence there is a very strong support for this TDRL. Interestingly, another TDRL was found in the mitogenomes of teleosts, namely the TDRL leading towards *Bregmaceros nectabanus*). To the best of our knowledge, this TDRL has not been documented in literature, yet. The TDRL overlaps with the transposition towards *Caelorinchus kishinouyei*, which leads to an inconsistent parent of *Bregmaceros nectabanus*. Nevertheless, the subtree of the corresponding three

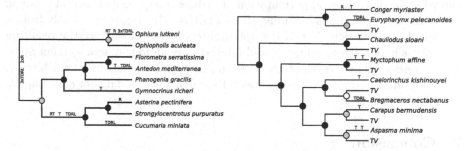

Fig. 6. Rearrangement events inferred by algorithm TreeREx; left: echinoderms; right: teleosts; abbreviations used: R=reversal, T=transposition, rT=reverse transposition, TDRL=tandem duplication random loss; nodes indicate consistency: black node=consistent, grey node=k-consistent with $k > 0$, white node=inconsistent; TV stands for species that have the typical the vertebrate gene order

Fig. 7. TDRL inferred by algorithm `TreeREx` for the scenario of a typical vertebrate gene order towards the gene order of *Bregmaceros nectabanus*

species has a consistent root node, and can be explained with only one transposition and one TDRL. Therefore we conclude, that the genomic rearrangement operations found by `TreeREx` are very likely. The TDRL and transposition are depicted in Figure 7. The mitogenomes of teleosts can be seen as a relatively easy data set, as many of the leaf nodes have the typical vertebrate gene order as the assigned permutation. Nevertheless, besides the reverse transposition all considered types of rearrangement operations occur. The computation time for `TreeREx` was 0.2 seconds on a Laptop with 2Ghz processor.

6.2 Echinoderms

In Section 5 the gene order of four echinoderms was used to describe algorithm `TreeREx`. In this subsection we utilize all known mitochondrial echinoderm gene orders for the analysis of `TreeREx`. The phylogenetic tree for this echinoderm data set has been obtained by a careful analysis of the mitochondrial protein sequences in [17]. The operations inferred by `TreeREx` are depicted in Figure 6. In [17] the same results were found for this tree by manual inspection of pairwise scenarios. None of the ancestral permutations was inferred inconsistently, and only two permutations were k-consistent with $k > 0$. The TDRL, that separates *Cucumaria miniata* from the ancestral gene order of echinoids was discussed in [1]. The ancestral state of ophiuroids is very difficult to infer, as the corresponding permutations are known to be heavily rearranged (in [19] the ancestral state of ophiuroids remains also unresolved). Although algorithm `TreeREx` infers a permutation for this ancestral permutation, it utilizes a sequence of several genomic rearrangement operations including three TDRLs. This is not very likely from a biological point of view. Yet, if the ophiuroids are not considered, the resulting operations have a strong support, and interestingly all four rearrangement operations considered in this paper are necessary to explain the evolutionary history. The computation time of `TreeREx` was 0.1 seconds for this data set on a Laptop with 2Ghz processor.

7   Conclusion

In this paper algorithm `CREx` has been extended to better handle alternative scenarios, ordered scenarios, and combinations of reversals and tandem duplication loss events. `CREx` is an an algorithm to heuristically infer pairwise scenarios of two given unichromosomal gene orders ([4]). Four biologically evident operations, namely reversals, transpositions, reverse transpositions and tandem duplication loss events are considered. Furthermore CREx tries to preserve conserved gene groups in rearrangement scenarios. The main contribution of this paper is algorithm TreeREx, that utilizes the pairwise scenarios as computed by CREx, to automatically infer ancestral permutations and genomic rearrangement operations in a given binary phylogenetic tree. TreeREx was applied to biological data sets of mitochondrial gene orders of echinoderms and of teleosts. In both data sets we could identify genome rearrangement operations that are in strong correspondence with published results. Furthermore, algorithm TreeREx was able to identify a new strongly supported TDRL operation towards species *Bregmaceros nectabanus*.

References

1. Arndt, A., Smith, M.J.: Mitochondrial gene rearrangement in the sea cucumber genus cucumaria. Mol. Biol. Evol. 15(8), 9–16 (1998)
2. Bérard, S., Bergeron, A., Chauve, C., Paul, C.: Perfect sorting by reversals is not always difficult. IEEE/ACM Transaction on Computational Biology and Bioinformatics 4(1), 4–16 (2007)
3. Bergeron, A., Chauve, C., de Montgolfier, F., Raffinot, M.: Computing common intervals of k permutations, with applications to modular decomposition of graphs. In: Brodal, G.S., Leonardi, S. (eds.) ESA 2005. LNCS, vol. 3669, pp. 779–790. Springer, Heidelberg (2005)
4. Bernt, M., Merkle, D., Ramsch, K., Fritzsch, G., Perseke, M., Bernhard, D., Schlegel, M., Stadler, P.F., Middendorf, M.: Crex: inferring genomic rearrangements based on common intervals. Bioinformatics 23, 2957–2958 (2007)
5. Blanchette, M., Kunisawa, T., Sankoff, D.: Gene order breakpoint evidence in animal mitochondrial phylogeny. J. Mol. Evol. 49, 193–203 (1999)
6. Boore, J.L.: Mitochondrial gene arrangement database (2006), http://evogen.jgi.doe.gov/
7. Booth, K.S., Lueker, G.S.: Testing for the consecutive ones property, interval graphs, and graph planarity using pq-tree algorithms. Journal of Computer and System Sciences 13, 335–379 (1976)
8. Caprara, A.: The reversal median problem. INFORMS Journal on Computing 15, 93–113 (2003)
9. Chaudhuri, K., Chen, K., Mihaescu, R., Rao, S.: On the tandem duplication-random loss model of genome rearrangement. In: SODA, pp. 564–570 (2006)
10. Heber, S., Stoye, J.: Algorithms for finding gene clusters. In: Gascuel, O., Moret, B.M.E. (eds.) WABI 2001. LNCS, vol. 2149, pp. 252–263. Springer, Heidelberg (2001)
11. Inoue, J.G., Miya, M., Tsukamoto, K., Nishida, M.: Evolution of the deep-sea gulper eel mitochondrial genomes: large-scale gene rearrangements originated within the eels. Mol. Biol. Evol. 20, 1917–1924 (2003)
12. Lavrov, D.V., Boore, J.L., Brown, W.M.: Complete mtdna sequences of two millipedes suggest a new model for mitochondrial gene rearrangements: duplication and nonrandom loss. Mol. Biol. Evol. 19, 163–169 (2002)
13. Littlewood, D.T.J., Smith, A.B., Cloug, h.K.A., Emson, R.H.: The interrelationships of the echinoderm classes: morphological and molecular evidence. Biol. J. Linn. Soc. 61, 409–438 (1997)

14. Miya, M., Takeshima, H., Endo, H., Ishiguro, N.B., Inous, J.G., Mukai, T., Satoh, T.P., Yamagucki, M., Kawaguchi, A., Mabuchi, K., Shirai, S.M., Nishida, M.: Major patterns of higher teleost phylogenies: a new perspective based on 100 complete mitochondrial dna sequences. Mol. Phyl. Evol. 26, 121–138 (2003)
15. Mooi, R., David, B.: Skeletal homologies of echinoderms. Paleont. Soc. Papers 3, 305–335 (1997)
16. Parida, L.: Using pq structures for genomic rearrangement phylogeny. Journal of Computational Biology 13(10), 1685–1700 (2006)
17. Perseke, M., Fritzsch, G., Ramsch, K., Bernt, M., Merkle, D., Middendorf, M., Bernhard, D., Stadler, P.F., Schlegel, M.: Evolution of mitochondrial gene orders in echinoderms. Mol. Phyl. Evol. (in press, 2008)
18. Sankoff, D.: Analytical approaches to genomic evolution. Biochimie 75, 409–413 (1993)
19. Scouras, A., Beckenbach, K., Arndt, A., Smith, M.J.: Complete mitochondrial genome dna sequence for two ophiuroids and a holothuroid: the utility of protein gene sequence and gene maps in the analyses of deep deuterostome phylogeny. Mol. Phyl. Evol. 31(1), 50–65 (2004)
20. Uno, T., Yagiura, M.: Fast algorithms to enumerate all common intervals of two permutations. Algorithmica 26(2), 290–309 (2000)
21. Zhao, H., Bourque, G.: Recovering true rearrangement events on phylogenetic trees. In: Tesler, G., Durand, D. (eds.) RECMOB-CG 2007. LNCS (LNBI), vol. 4751, pp. 149–161. Springer, Heidelberg (2007)

Appendix

A1. Mitogenomes Used in the Biological Example

- *Acanthaster brevispinus (A):*
  cox1 R nad4L cox2 K atp8 atp6 cox3 -S2 nad3 nad4 H S1 nad5 -nad6 cob F rrnS E
  T CR -rrnL -nad2 -I -nad1 -L2 -G -Y D -M V -C -W A -L1 -N Q -P
- *Strongylocentrotus purpuratus (E):*
  cox1 R nad4L cox2 K atp8 atp6 cox3 -S2 nad3 nad4 H S1 nad5 -nad6 cob
  F rrnS E T CR P -Q N L1 -A W C -V M -D Y G L2 nad1 I nad2 rrnL
- *Cucumaria miniata (H):*
  cox1 R E CR P N L1 W -V nad4L cox2 K atp8 atp6 cox3 -S2 nad3 nad4 H
  S1 nad5 -nad6 cob F rrnS T -Q -A C M -D Y G L2 nad1 I nad2 rrnL
- *Florometra serratissima (C):*
  cox1 R nad4L cox2 K atp8 atp6 cox3 -S2 nad3 nad4 H S1 nad5 -nad6 cob P -Q N L1
  -A W C -V M -D -CR -T -E -rrnS -F -L2 -G -rrnL -Y -nad2 -I -nad1

A2. Pseudo-Code of Algorithm TreeREx

1: **INPUT:** A phylogenetic binary tree $T = (V, E)$ with leaf nodes $\{\pi_1, \ldots, \pi_n\}$
   OUTPUT: A mapping $E \rightarrow \mathcal{R}^*$ (phylogenetic events on edges)
   and   a mapping $V \rightarrow \Pi$ (permutations for internal nodes)
2: **while** ($\exists$ induced subtree T_S of height 2 with 3 or 4 leaf nodes, for which permutations are assigned to leaf nodes only) **do**
3:    Let $\Pi = \{\pi_1, \ldots, \pi_m\}$ be the set of permutations assigned to the leaf nodes of T_S ($m = 3$ or $m = 4$ holds)

4:    Let π_i, π_j be permutations assigned to sibling leaf nodes of T_S, for which no permutation is assigned to their parent node v

5:    Let v' be the sibling of v

6:

7:    // Check if a consistent permutation can be inferred:

8:    $r_i := \bigcap_{\pi \in \Pi \setminus \pi_i} \text{CREx}(\pi, \pi_i)$

9:    $r_j := \bigcap_{\pi \in \Pi \setminus \pi_j} \text{CREx}(\pi, \pi_j)$

10:    Let $\varpi_i = \pi_i \circ r_i^{-1}$ be the permutation computed by inversely applying r_i to π_i

11:    Let $\varpi_j = \pi_j \circ r_j^{-1}$ be the permutation computed by inversely applying r_j to π_j

12:    **if** $(\varpi_i == \varpi_j)$ **then**

13:      Assign all events r_i and r_j to their corresponding edges.

14:      Assign the permutation $\pi := \varpi_i$ to v.

15:    **else**

16:      // Check if a k-consistent permutation can be inferred:

17:      k=0

18:      **while** (no k-consistent permutation was found) $\wedge$ (k is not maximal) **do**

19:        k=k+1

20:        Similarly as in the consistent case, try to infer i.) a k-consistent permutation to be assigned to v and ii.) k-consistent events to the corresponding edges

21:      **end while**

22:

23:      **if** no permutation was assigned to node v **then**

24:        // Inconsistent case:

25:        Compute all possible ancestral permutations $\Gamma(\pi_i, \pi_j)$ for node v

26:        Let π_k, π_l be the permutations assigned to the children of v'

27:        **if** v' has an assigned permutation π' **then**

28:          $\Gamma(\pi_k, \pi_l) := \{\pi'\}$

29:        **else**

30:          Compute all possible ancestral permutations $\Gamma(\pi_k, \pi_l)$ of node v'

31:        **end if**

32:        // Assign permutation to v (and v' if necessary) in a parsimonious manner:

33:        **for all** $(\pi \in \Gamma(\pi_i, \pi_j)$ and $\pi' \in \Gamma(\pi_k, \pi_l))$ **do**

34:          Compute the weighted number of events $q(\text{CREx}(\pi, \pi'))$

35:        **end for**

36:        Assign π to v (and π' to v' if necessary), such that $q(\text{CREx}(\pi, \pi'))$ is minimal

37:        Assign the infered events to the edges

38:      **end if**

39:    **end if**

40: **end while**

Perfect DCJ Rearrangement

Sèverine Bérard[1,2], Annie Chateau[2], Cedric Chauve[3], Christophe Paul[2], and Eric Tannier[4]

[1] Université Montpellier 2, UMR AMAP, Montpellier, F-34000 France
[2] CNRS, LIRMM, CNRS UMR55076, Université Montpellier 2, Montpellier, France
[3] Department of Mathematics, Simon Fraser University, Burnaby (BC), Canada
[4] INRIA, LBBE, CNRS UMR5558, Université de Lyon 1; Villeurbanne, France

Abstract. We study the problem of transforming a multichromosomal genome into another using Double-Cut-and-Join (DCJ) operations. We introduce the notion of DCJ scenario that does not break families of common intervals (groups of genes co-localized in both genomes). Such scenarios are called perfect, and generalize the notion of perfect reversal scenarios. While perfect sorting by reversals is NP-hard if the family of common intervals is nested, we show that finding a shortest perfect DCJ scenario can be answered in polynomial time in this case. Moreover, while perfect sorting by reversals is easy when the family of common intervals is weakly separable, we show that the corresponding problem is NP-hard in the DCJ case. These contrast with previous comparisons between the reversal and DCJ models, that showed that most problems have similar complexity in both models.

1 Introduction

A generic formulation of genome rearrangement problems is, given two genomes and some allowed edit operations, to transform one genome into the other using a minimum number of operations. The solutions are used to estimate an evolutionary distance between species, and to propose possible scenarios that could explain the differences in terms of gene order between the considered genomes (see [10,23,11] for example). Probably the most used algorithmic results related to genome rearrangements concern the problem of sorting signed permutations by reversals. This problem aims at computing a shortest sequence of reversals that transforms one signed permutation into another, and can be solved in polynomial time [16,7,25]. It was later generalized to handle multichromosomal genomes with linear chromosomes, using rearrangements such as translocations and chromosomes fusions and fissions [17]. Here, we study a more general rearrangement model on multichromosomal genomes, the Double-Cut-and-Join model (DCJ), that was considered in several recent works [28,8,2,20,21]. In this model, temporary circular chromosomes can be created, which allows to simulate rearrangements such as transpositions and block-interchanges using two consecutive DCJs [28].

Another way (than pure parsimony) of handling gene order data is to consider groups of genes that are co-localized with the homologous genes (genes having

C.E. Nelson and S. Vialette (Eds.): RECOMB-CG 2008, LNBI 5267, pp. 158–169, 2008.
© Springer-Verlag Berlin Heidelberg 2008

a single common ancestor) in the genomes of different species.These groups are likely together in the common ancestral genome and not disrupted during evolution. For two permutations, such groups of co-localized genes can be modeled by *common intervals*. Following the assumption that such common intervals are preserved during evolution leads naturally to the study of rearrangement scenarios that preserve common intervals. Such scenarios, which may not be shortest among all scenarios, are called *perfect* [14]. Computing a reversal scenario of minimum length that preserves a given subset of the common intervals of two signed permutations is NP-hard [14] and several papers have explored this problem, describing families of instances that can be solved in polynomial time [3,4,24,13] and fixed-parameter tractable algorithms [4,5].

When comparing algorithmic properties of the reversal and DCJ models, the classical problems seem to have similar behaviors: the distance and scenario computations can be solved in polynomial time, yet the best complexity varies for the latter by an $O(\sqrt{n})$ factor [25,8]; the median problems are both NP-hard[1]. In this paper we extend the notion of perfect scenario to the DCJ model. We define a notion of scenario preserving common intervals that also allows to use the property of the DCJ model to create temporary circular chromosomes. While the general problem of computing a shortest DCJ scenario that preserves a family $\mathcal{F}$ of common intervals (the $\mathcal{F}$-perfect rearrangement problem) is still NP-hard, our results point to interesting differences between the reversal and DCJ models. If the family of common intervals is *nested*, we show that finding a perfect DCJ scenario of minimum length is solvable in polynomial time, while it is NP-hard for reversals [14]; if the family is *weakly separable*, we show that the DCJ problem is NP-hard, while this case was solved in polynomial time for reversals [4].

The paper is organized as follows: in Section 2, we introduce genomes, DCJ operations and common intervals. Then in Section 3, we define perfect DCJ scenarios for multichromosomal genomes and describe some basic properties of such scenarios. In Section 4, we define the different properties of families of common intervals, that result in different complexity status for the perfect rearrangement problems. In Section 5, we describe a polynomial algorithm for the $\mathcal{F}$-perfect rearrangement problem if $\mathcal{F}$ is nested, and in Section 6, we prove NP-hardness of the general problem.

2   Genomes, Intervals and Rearrangements

We follow the modeling of a genome used in [8]. A *gene* a is an oriented sequence of DNA, identified by its *tail* a_t and its *head* a_h. Tails and heads are the *extremities* of the genes. An *adjacency* is an unordered pair of gene extremities. A *genome* is a set of adjacencies on a set of genes. Each adjacency in a genome means that two gene extremities are consecutive on the DNA molecule.

---

[1] The median problem consists in, given three genomes as an input, find a fourth one (the median) that minimizes the sum of the distance from the median to the three input genomes. This problem has been proved to be NP-hard for permutations in [12], and for multichromosomal genomes in [26].

In a genome, each gene extremity is adjacent to zero or one other extremity. An extremity x that is not adjacent to any other extremity is called a *telomere*, and can be written as a *telomeric adjacency* xT with a symbol T (we use the same notation for all telomeres).

For a genome Π on a set of genes, we define the graph G_Π: its vertex set is the set of all gene extremities, and its edge set is composed of $a_t a_h$ for every gene a, plus the adjacencies of Π, except telomeric adjacencies. An example of such a graph is drawn on Figure 1.

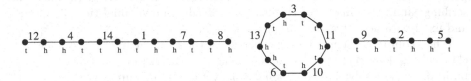

Fig. 1. The graph $G_{\Pi^{ex}}$, where Π^{ex} is given by the union of $C_1 = \{T12_t, 12_h4_h, 4_t14_t, 14_h1_t, 1_h7_h, 7_t8_t, 8_hT\}$, $C_2 = \{3_t11_t, 11_h10_t, 10_h6_t, 6_h13_h, 13_t3_h\}$ and $C_3 = \{T9_t, 9_h2_t, 2_h5_h, 5_tT\}$

The graph G_Π is composed of disjoint paths and cycles. Each connected component of G_Π is called a *chromosome* of Π. A chromosome is said to be *linear* if it is a path, and *circular* if it is a cycle.

An *interval* of Π is a set of genes I, such that the subgraph of G_Π induced by the extremities of genes in I is connected. For example, $\{12, 4, 14, 1, 7, 8\}$ and $\{14, 1, 7, 8\}$ are intervals of genome Π^{ex}, which is represented on Figure 1. An interval I is said to be a *common interval* of two genomes Π and Γ if it is an interval of both.

Given a genome Π, a *Double-Cut-and-Join* is an operation ρ acting on two adjacencies pq and rs of Π (p, q, r, s are gene extremities, some being possibly T symbols; in particular, we consider valid the adjacency TT). The DCJ operation *cuts* both pq and rs and *joins* either pr and qs, or ps and qr, creating two new adjacencies. Examples of DCJ operations are shown in Figure 2.

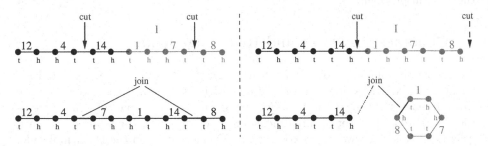

Fig. 2. Two examples of DCJ operations. Left: the DCJ cuts 4_t14_t and 7_t8_t and joins 4_t7_t and 14_t8_t (it is a reversal). Right: the DCJ cuts 14_h1_t and 8_hT and joins 14_hT and 8_h1_t. This operation produces a circular chromosome. The first operation breaks the interval $I = \{1, 7, 8\}$ whereas the second preserves it.

A DCJ operation can reverse an interval in a genome, fuse two chromosomes into one, fisse one chromosome into two, or exchange two intervals from two different chromosomes, both containing a telomere (reciprocal translocation). Two consecutive DCJs may result in a block interchange (two intervals exchange their positions), or a transposition (if these two intervals are consecutive): the first DCJ extracts a set of genes and creates a circular chromosome, while the second DCJ reinserts these genes elsewhere in a chromosome. The DCJ operation is thus a very general framework, where temporary circular chromosomes allow to simulate a wide range of genome rearrangements, introduced by Yancopoulos *et al.* [28] and since adopted by many others [8,20,1], sometimes under the name "2-break rearrangements" [2].

A sequence S of k DCJ operations transforming one genome Π into another genome Γ is called a *DCJ scenario of length* k for the two genomes. The minimum number of DCJ operations needed to transform Π into Γ is the *DCJ-distance* and denoted by $d(\Pi, \Gamma)$.

3   Perfect DCJ Scenarios

The adjacencies of a genome Π can be partitioned into three classes with respect to a subset I of its genes: an adjacency pq (p and q possibly being T symbols) is said to be *inside* I if the two genes of which p and q are extremities belong to I; it is called *outside* if the two genes of which p and q are extremities do not belong to I; it is a *border* adjacency if one of the genes of which p and q are extremities belongs to I but not the other. In these three definitions, a T symbol is considered to be outside I.

Note that an interval of Π has zero or two border adjacencies. Let I be any set of genes of a genome Π, which has at most two border adjacencies. A DCJ acting on Π *preserves* I if, in the resulting genome, I still has at most two border adjacencies. For example, on Fig. 2, the DCJ operation on the left does not preserve the interval $\{1, 7, 8\}$ but the operation on the right does preserve this interval. A DCJ that does not preserve I is said to *break* I.

Given a family $\mathcal{F}$ of common intervals of two genomes Π and Γ, a DCJ scenario transforming Π into Γ is said to be $\mathcal{F}$-perfect if every DCJ preserves all intervals in $\mathcal{F}$. The $\mathcal{F}$-**Perfect DCJ problem** consists in, given Π, Γ and $\mathcal{F}$, computing a $\mathcal{F}$-perfect DCJ scenario of minimum length transforming Π into Γ. When genomes are restricted to signed permutations (they have only one chromosome) and temporary circular chromosomes are not allowed, this definition coincides with the one of perfect scenarios of reversals [14,3,4,5,24,13].

With this definition of $\mathcal{F}$-perfect DCJ scenarios the elements of an interval I of $\mathcal{F}$ can be not consecutive at some point of such a scenario, provided that the elements of I are split into at most one linear segment and possibly several circular segments. This allows to use the property of the DCJ model to create temporary circular chromosomes.

4   Families of Common Intervals

Given two genomes Π and Γ, two common intervals are said to *overlap* if their intersection is not empty and none is contained in the other. A common interval I of Π and Γ is *strong* if I does not overlap any other common interval. It is *maximal* if it is strong and not contained in another common interval.

A family $\mathcal{F}$ of common intervals is *weakly partitive* if for every two overlapping intervals I and J of $\mathcal{F}$, $I \cup J$, $I \cap J$, $I - J$ and $J - I$ belong to $\mathcal{F}$. We denote by $\mathcal{F}^*$ the unique smallest weakly partitive family that contains $\mathcal{F}$; $\mathcal{F}^*$ can be computed in polynomial time. It follows immediately from [4] that a DCJ scenario is $\mathcal{F}$-perfect if and only if it is $\mathcal{F}^*$-perfect. A family $\mathcal{F}$ is called *nested* if every element of $\mathcal{F}$ is strong (note this implies that $\mathcal{F} = \mathcal{F}^*$). $\mathcal{F}$ is called *weakly separable* if every *strong* interval of $\mathcal{F}$ with at least three elements is the union of two overlapping intervals of $\mathcal{F}$. Of course, as soon as there are intervals of $\mathcal{F}$ with at least three elements, the nested property and the weakly separable property are mutually exclusive.

By definition, the sub-family of strong intervals of $\mathcal{F}^*$ for a family $\mathcal{F}$ is nested. It follows that we can represent the strong common intervals of Π and Γ by a forest, in which each node is a strong common interval of $\mathcal{F}^*$, and its children are the maximal strong common intervals of $\mathcal{F}^*$ it properly contains (see [4,18]). Each component of this forest is a rooted tree, in which the root is a maximal common interval of Π and Γ. An example of such tree is given in Fig. 3. Given a maximum common interval, the tree can be computed in linear time and space [6,18]. A node of the forest of strong intervals is called *prime* if it has at least four children and it properly contains no common interval including more than one of its children. It is *linear* if it has two elements or it is the union of two overlapping common intervals, both containing a subset of its children. Any strong interval of $\mathcal{F}^*$ is either prime or linear (see for instance [4]).

It is known [14] that given a nested family of common intervals $\mathcal{F}$ of two permutations, it is NP-hard to compute a perfect scenario of reversals of minimum

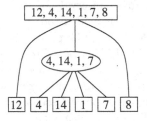

Fig. 3. The tree that represents the strong common intervals of the maximal common interval $I = \{12, 4, 14, 1, 7, 8\}$ of Π^{ex} and Γ^{ex}, given by the union of $C_1 = \{T12_t, 12_h14_h, 14_t7_h, 7_t4_t, 4_h1_h, 1_t8_t, 8_h2_t, 2_h6_t, 6_hT\}$ and $C_2 = \{T9_t, 9_h3_t, 3_h10_t, 10_h5_t, 5_h11_h, 11_t13_h, 13_tT\}$. Prime nodes are surrounded by an ellipse, while linear nodes are framed by a rectangle.

length. Conversely, if $\mathcal{F}$ is weakly separable[2], the algorithm described in [4] computes an $\mathcal{F}$-perfect reversal scenario in polynomial time. We prove here the exact opposite results for multichromosomal genomes with DCJ operations.

5 $\mathcal{F}$ Nested: A Polynomial-Time Solvable Case

We give here an algorithm to solve the $\mathcal{F}$-perfect DCJ problem if $\mathcal{F}$ is a nested family.

Sorting an interval. We say that a common interval I is *sorted* in Π with respect to Γ if the set of adjacencies inside I in Π contains the set of adjacencies inside I in Γ. If a DCJ scenario results in a genome where I is sorted, we say that this scenario *sorts* I.

We can distinguish different kinds of DCJ with respect to a common interval I. A DCJ ρ cuts *inside* I if it cuts either two inside adjacencies or one inside and one border adjacency. On the contrary, a DCJ ρ cuts *outside* I if it cuts either two outside adjacencies, one outside and one border adjacency, two border adjacencies, or one inside and one outside adjacency in the case I does not have any border adjacency. Note that a DCJ does not break I if and only if it cuts inside or outside I.

Lemma 1. *If a DCJ scenario S_0 between two genomes Π and Γ does not break a common interval I, then there exists a DCJ scenario $S = S_1 S_2$ of same length as S_0 for which all operations in S_1 cut inside I and all operations in S_2 cut outside I.*

This lemma, that is an equivalent to a lemma stated for reversals in [14], implies that the DCJs that sort I may always be applied before the ones that rearrange the remaining of the genome.

Outline of the algorithm. The algorithm can be decomposed into three main steps:

1. Compute the maximal common intervals of Π and Γ. This can be done by computing the maximal common connected components of G_Π and G_Γ, with techniques presented for example in [15].
2. For each maximal common interval I, compute the tree of (strong) intervals that lie in I. By a preorder traversal of this tree, sort each node assuming its children have been sorted, by a technique we describe further.

---

[2] The terminology *weakly separable* is inspired by the notion of separable permutations, that are the permutations whose common intervals with the identity define a strong interval tree with no prime node. For a weakly separable family of common intervals, the strong intervals forest can have prime nodes, but no edge can be incident to two prime nodes, and these prime nodes belong to $\mathcal{F}^*$ but not $\mathcal{F}$ and are then only implicitly defined by $\mathcal{F}$.

3. Finally, after all maximal common intervals have been sorted, compute a parsimonious series of DCJ that creates all the remaining adjacencies of Γ, that are not inside any maximal common interval, for example with the technique described in [8].

The first and last step use known techniques that are described in the literature. The core of our method is the second step, which we now describe into details. Lemma 1 implies that a perfect scenario between two genomes can be computed during a preorder traversal of each tree of common intervals in such a way that, when processing a node I, all its children are already sorted with respect to Γ.

The sorting direction of an interval. We now consider a strong common interval I of Π and Γ. If I has border adjacencies in Π let x_Π and y_Π be the extremities of genes that *are not in* I and belong to the two border adjacencies of I in Π (they may be T symbols). If I has no border adjacencies in Π, let $x_\Pi = y_\Pi = T$. If I has border adjacencies in Γ, let m_Γ and M_Γ be the extremities of genes that *are in* I and belong to the two border adjacencies of I in Γ.

We wish to sort the interval I with respect to Γ, that is we want to obtain a genome Π' from Π which contains every adjacency inside I in Γ. We will use only DCJs that cut inside I, so in Π', there is a limited number of possibilities regarding the border adjacencies of I in Π'. If I has no border adjacencies in Γ, the set of adjacencies in Π' is unambiguous: it is the set of adjacencies inside I in Γ, plus the adjacency $x_\Pi y_\Pi$. But if I has border adjacencies in Γ, there are three possibilities for the border adjacencies:

1. $x_\Pi m_\Gamma$ and $M_\Gamma y_\Pi$ are adjacencies in Π', and in this case we say that I is sorted *positively*;
2. $x_\Pi M_\Gamma$ and $m_\Gamma y_\Pi$ are adjacencies in Π', and in this case Π is sorted *negatively*;
3. $m_\Gamma M_\Gamma$ and $x_\Pi y_\Pi$ are adjacencies in Π', and in this case Π is sorted *neutrally*

The way I is sorted is called its *sorting direction* [3] in Π'.

We denote by $\Pi \setminus I^+$ (resp. $\Pi \setminus I^-$ and $\Pi \setminus I^N$) the genome obtained from Π, in which I is sorted positively (resp. negatively and neutrally) with respect to Γ. Note that $\Pi \setminus I^N$ contains a circular chromosome. It is clear that $d(\Pi \setminus I^-, \Pi \setminus I^+) = d(\Pi \setminus I^-, \Pi \setminus I^N) = d(\Pi \setminus I^+, \Pi \setminus I^N) = 1$.

As it was shown in [5] for the reversal model, the main difficulty for sorting a genome while preserving common intervals is to choose among the sorting directions of these intervals. The following lemma greatly simplifies this choice in the DCJ model.

---

[3] Note that the notions of positive or negative sorting direction of a common interval are strongly related to the choice of gene extremities x_Π, y_Π, m_Γ and M_Γ (two different choices are possible, and it will swap the positive and negative sorts). We choose arbitrarily and independently for all strong intervals.

Lemma 2. *Let Π and Γ be two genomes and let I be a set of genes that has two border adjacencies in Γ and at most two in Π. Then one and only one of the three following possibilities holds:*

- *$d(\Pi, \Pi \setminus I^+) = d(\Pi, \Pi \setminus I^-) - 1 = d(\Pi, \Pi \setminus I^N) - 1$ and $d(\Pi, \Gamma) = d(\Pi, \Pi \setminus I^+) + d(\Pi \setminus I^+, \Gamma)$;*
- *$d(\Pi, \Pi \setminus I^-) = d(\Pi, \Pi \setminus I^+) - 1 = d(\Pi, \Pi \setminus I^N) - 1$ and $d(\Pi, \Gamma) = d(\Pi, \Pi \setminus I^-) + d(\Pi \setminus I^-, \Gamma)$;*
- *$d(\Pi, \Pi \setminus I^N) = d(\Pi, \Pi \setminus I^+) - 1 = d(\Pi, \Pi \setminus I^-) - 1$ and $d(\Pi, \Gamma) = d(\Pi, \Pi \setminus I^N) + d(\Pi \setminus I^N, \Gamma)$.*

The algorithm. This yields a method for sorting a maximal common interval, where the DCJ operations to apply can be computed using the algorithm presented in [8]:

Algorithm 1. $\mathcal{F}$-Perfect sorting of a maximal common interval I of genomes Π and Γ, given the strong interval tree T of a nested family $\mathcal{F}$ of common intervals in I

LET $\Pi' = \Pi$
FOR each interval $I' \subseteq I$ of Π and Γ in a post-traversal order of T
    {Note: all children of I' are sorted}
    IF I' has no border adjacencies in Γ *{Note: possible only if $I' = I$}*
        Sort I' with a minimum number of DCJs inside I' and outside its children
    Else
        Compute $k = \min(d(\Pi', \Pi' \setminus I'^+), d(\Pi', \Pi' \setminus I'^-), d(\Pi', \Pi' \setminus I'^N))$
        Sort I' with k DCJs inside I' and outside its children
    LET Π' denote the resulting genome.

Lemma 3. *Given two genomes Π and Γ, a nested family $\mathcal{F}$ of common intervals and a maximal element I of $\mathcal{F}$, Algorithm 1 computes a DCJ scenario that sorts I with respect to Γ and preserves all the intervals of $\mathcal{F}$ contained in I. The scenario is minimum, and no scenario achieves the same number of operations and sorts I with another direction.*

Lemma 2, together with Lemma 3, provides the general result:

Theorem 1. *Given two genomes Π and Γ on n genes, and a nested family $\mathcal{F}$ of common intervals, a minimum $\mathcal{F}$-perfect scenario of length $d(\Pi, \Gamma)$ can be computed in time $O(n^2)$.*

Note that this result defines a class of instances where a perfect scenario is also parsimonious. These instances are defined only in terms of the structure of the considered common intervals and not in terms of their breakpoint graph, which differs from similar results in the reversal model [3,24,13].

6 $\mathcal{F}$ General (Even Weakly Separable): A Hardness Result

In general, the problem of $\mathcal{F}$-perfect DCJ rearrangement is hard, and even with weakly separable families of common intervals. This is the DCJ version of NP-hardness for reversals [14], but contrasts with the linear time solution when $\mathcal{F}$ is supposed to be weakly separable [4].

Theorem 2. *The $\mathcal{F}$-perfect DCJ problem is NP-hard, even if $\mathcal{F}$ is weakly separable.*

The NP-hardness proof relies on a very simple pattern: it uses the fact that it is possible to sort an interval of shape (3 2 1) in Π and (1 2 3) in Γ either neutrally or negatively in three operations, and it is impossible to choose between the two directions. No DCJ scenario sorts this pattern positively in less than 4 operations, while preserving the intervals $\{1, 2\}$ and $\{2, 3\}$. From this, we can deduce two interesting remarks that will be developed in an extended version of this paper:

 - The behavior of this perfect DCJ problem is different from the perfect rearrangement problem where temporary circular chromosomes have to be reinserted immediately to simulate block-interchanges. Indeed, for the latter, a block interchange would have sorted (3 2 1) into (1 2 3) while preserving all intervals. It is not the case when the block-interchange has to be simulated by two consecutive DCJs. This points an interesting difference between DCJ problems and block-interchange problems, and calls for further thoughts on the relationship between the DCJ model and the reversals and block-interchange model.
 - The pattern that causes NP-hardness is limited to linear strong intervals with three elements. It is then possible to devise FPT algorithms based on the number of such patterns in the genomes. Like the FPT algorithms for reversals [4,5], this should lead to efficient algorithms to solve the perfect DCJ problem.

7   Conclusion

We proved in this paper that $\mathcal{F}$-perfect sorting by DCJ is NP-hard in general, and even if $\mathcal{F}$ is a weakly separable family of common intervals. On the other hand, it has a polynomial time solution when $\mathcal{F}$ is nested. This contrasts with perfect sorting by reversals that is hard if $\mathcal{F}$ is nested, and easy if $\mathcal{F}$ is a weakly separable. The key to these results is the ability of DCJ to create temporary circular chromosomes, that was already the important factor in the fact that sorting with DCJ is simpler than with reversals [8]. This illustrates that the DCJ model, both by its combinatorial simplicity and its pertinence for modeling genome rearrangements, offers an interesting way to attack several genome rearrangement problems [22,27].

In an extended version of this paper, we will describe a fixed parameter polynomial algorithm for the problem of perfect DCJ rearrangement, using the number of patterns used in the NP-hardness proof as a parameter. A natural problem that could benefit from such an algorithm is the perfect reversal median [9], or perfect DCJ-median [1,19]. We also plan to investigate the relationships between the general DCJ model and the reversal/translocation/block-interchange model, as the problem of computing a perfect scenario seems to be the first one where these two models differ. This seems to be surprising, as those two models have always been considered to be equivalent, since two DCJs simulate block-interchanges. We will also address the case of genomes with circular chromosomes using the notion of PC-trees [18]. Eventually, the algorithm we describe for nested families of common intervals runs in quadratic time, but we think there is a linear time solution, with a smart treatment of prime nodes.

Acknowledgments

C. Chauve is supported by grants from NSERC and SFU. C. Paul is supported by the ANR grant ANR-O6-BLAN-0148-01 "GRAAL". E. Tannier is funded by ANR JC05_49162 "REGLIS" and NT05-3_45205 "GENOMICRO". A. Chateau is supported by the ANR BLAN07-1_185484 "CoCoGen".

References

1. Adam, Z., Sankoff, D.: The ABC of MGR with DCJ. Evol. Bioinformatics 4, 69–74 (2008)
2. Alekseyev, M., Pevzner, P.: Multi-break rearrangements and chromosomal evolution. Theor. Comput. Sci. (in press, 2008)
3. Bérard, S., Bergeron, A., Chauve, C.: Conservation of combinatorial structures in evolution scenarios. In: Lagergren, J. (ed.) RECOMB-WS 2004. LNCS (LNBI), vol. 3388, pp. 1–14. Springer, Heidelberg (2005)
4. Bérard, S., Bergeron, A., Chauve, C., Paul, C.: Perfect sorting by reversals is not always difficult. IEEE/ACM Trans. Comput. Biol. Bioinform. 4, 4–16 (2007)
5. Bérard, S., Chauve, C., Paul, C.: A more efficient algorithm for perfect sorting by reversals. Inform. Proc. Letters 106, 90–95 (2008)
6. Bergeron, A., Chauve, C., de Montgolfier, F., Raffinot, M.: Computing common intervals of k permutations, with applications to modular decomposition of graphs. In: Brodal, G.S., Leonardi, S. (eds.) ESA 2005. LNCS, vol. 3669, pp. 779–790. Springer, Heidelberg (2005)
7. Bergeron, A., Mixtacki, J., Stoye, J.: The inversion distance problem. In: Mathematics of Evolution and Phylogeny. Oxford University Press, Oxford (2005)
8. Bergeron, A., Mixtacki, J., Stoye, J.: A unifying view of genome rearrangements. In: Bücher, P., Moret, B.M.E. (eds.) WABI 2006. LNCS (LNBI), vol. 4175, pp. 163–173. Springer, Heidelberg (2006)
9. Bernt, M., Merkle, D., Middendorf, M.: A fast and exact algorithm for the perfect reversal median. In: Măndoiu, I.I., Zelikovsky, A. (eds.) ISBRA 2007. LNCS (LNBI), vol. 4463, pp. 305–316. Springer, Heidelberg (2007)

10. Bourque, G., Pevzner, P.: Genome-scale evolution: reconstructing gene orders in the ancestral species. Genome Res. 12, 26–36 (2002)
11. Braga, M., Sagot, M.-F., Scornavacca, C., Tannier, E.: Exploring the solution space of sorting by reversals with experiments and an application to evolution. IEEE/ACM Trans. Comput. Biol. Bioinform (2008)
12. Caprara, A.: The reversal median problem. INFORMS J. Comp. 15, 93–113 (2003)
13. Diekmann, Y., Sagot, M.-F., Tannier, E.: Evolution under reversals: Parsimony and conservation of common intervals. IEEE/ACM Trans. Comput. Biol. Bioinform. 4, 301–309 (2007)
14. Figeac, M., Varré, J.-S.: Sorting by reversals with common intervals. In: Jonassen, I., Kim, J. (eds.) WABI 2004. LNCS (LNBI), vol. 3240, pp. 26–37. Springer, Heidelberg (2004)
15. Habib, M., Paul, C., Raffinot, M.: Common connected components of interval graphs. In: Sahinalp, S.C., Muthukrishnan, S.M., Dogrusoz, U. (eds.) CPM 2004. LNCS, vol. 3109, pp. 347–358. Springer, Heidelberg (2004)
16. Hannenhalli, S., Pevzner, P.: Transforming cabbage into turnip: Polynomial algorithm for sorting signed permutations by reversals. J. ACM 46, 1–27 (1999)
17. Hannenhalli, S., Pevzner, P.A.: Transforming men into mice: polynomial algorithm for genomic distance problem. In: FOCS 1995, pp. 581–592 (1995)
18. Hsu, W.-L., McConnell, R.M.: PC trees and circular-ones arrangements. Theor. Comput. Sci. 296, 99–116 (2003)
19. Lenne, R., Solnon, C., Stutzle, T., Tannier, E., Birattari, M.: Reactive stochastic local search algorithms for the genomic median problem. In: van Hemert, J., Cotta, C. (eds.) EvoCOP 2008. LNCS, vol. 4972, pp. 266–276. Springer, Heidelberg (2008)
20. Lin, Y., et al.: An efficient algorithm for sorting by block-interchange and its application to the evolution of vibrio species. J. Comput. Biol. 12, 102–112 (2005)
21. Lu, L., Huang, Y., Wang, T., Chiu, H.-T.: Analysis of circular genome rearrangement by fusions, fissions and block-interchanges. BMC Bioinformatics 7, 295 (2006)
22. Mixtacki, J.: Genome halving under DCJ revisited. In: Hu, X., Wang, J. (eds.) COCOON 2008. LNCS, vol. 5092. Springer, Heidelberg (2008)
23. Murphy, W., et al.: Dynamics of mammalian chromosome evolution inferred from multispecies comparative maps. Science 309, 613–617 (2005)
24. Sagot, M.-F., Tannier, E.: Perfect sorting by reversals. In: Wang, L. (ed.) COCOON 2005. LNCS, vol. 3595, pp. 42–51. Springer, Heidelberg (2005)
25. Tannier, E., Bergeron, A., Sagot, M.-F.: Advances on sorting by reversals. Discrete Appl. Math. 155, 881–888 (2007)
26. Tannier, E., Zheng, C., Sankoff, D.: Multichromosomal genome median and halving problems. In: Proceedings of WABI 2008 (2008)
27. Warren, R., Sankoff, D.: Genome halving with double cut and join. In: APBC 2008, pp. 231–240 (2008)
28. Yancopoulos, S., Attie, O., Friedberg, R.: Efficient sorting of genomic permutations by translocation, inversion and block interchange. Bioinformatics 21, 3340–3346 (2005)

A   The DCJ Distance and The Breakpoint Graph

The formula for the DCJ distance based on the breakpoint graph is much used in the proofs of our results. The *breakpoint graph* of two genomes Π and Γ on the same set of genes, denoted by $BP(\Pi, \Gamma)$, is the bipartite graph which vertex set is the set of extremities of the genes, and in which there is an edge between two vertices x and y if xy is an adjacency in either Π (these are Π-edges) or Γ (Γ-edges). Note that T symbols do not participate. Vertices in this graph have degree zero, one or two; so the graph is a set of paths and cycles, where some paths may have no edge (see Fig. 4).

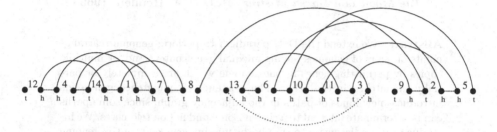

Fig. 4. The breakpoint graph of the genomes Π^{ex} (see Figure 1) and Γ^{ex}, given by the union of $C_1 = \{T12_t, 12_h14_h, 14_t7_h, 7_t4_t, 4_h1_h, 1_t8_t, 8_h2_t, 2_h6_t, 6_hT\}$ and $C_2 = \{T9_t, 9_h3_t, 3_h10_t, 10_h5_t, 5_h11_h, 11_t13_h, 13_tT\}$. Π^{ex}-edges are dotted lines, and Γ^{ex}-edges are plain lines.

The *DCJ-distance* is immediately readable from the breakpoint graph, as stated by Theorem 3, that restates the main result of [8] in terms of the breakpoint graph in place of the *adjacency graph*[4].

Theorem 3. [8] *For two genomes Π and Γ, let $c(\Pi, \Gamma)$ be the number of cycles of the breakpoint graph $BP(\Pi, \Gamma)$, and $p(\Pi, \Gamma)$ be the number of paths with an even number of edges. The DCJ distance is*

$$d(\Pi, \Gamma) = n - \left(c(\Pi, \Gamma) + \frac{p(\Pi, \Gamma)}{2} \right).$$

The basis for this result is Lemma 4, that is implicit in [8], and states that any–greedy– DCJ that creates an adjacency that is present in Γ but not in Π is optimal.

Lemma 4. *For two genomes Π and Γ, if a DCJ operation on Π results in a genome Π' containing an adjacency that is present in Γ but not in Π, then $d(\Pi, \Gamma) = d(\Pi', \Gamma) + 1$.*

---

[4] The breakpoint graph $BP(\Pi, \Gamma)$, introduced for permutations in [16], is the line-graph of the *adjacency graph* introduced in [8].

Sorting Genomes with Insertions, Deletions and Duplications by DCJ

Sophia Yancopoulos[1] and Richard Friedberg[2]

[1] The Feinstein Institute for Medical Research, Manhasset NY 11030, USA
[2] Department of Physics, Columbia University, NY, NY 10027, USA

"There Ain't No Such Thing As A Free Lunch."
The Moon Is a Harsh Mistress **Robert A. Heinlein 1966**

Abstract. We extend the DCJ paradigm to perform genome rearrangements on pairs of genomes having unequal gene content and/or multiple copies by permitting genes in one genome which are completely or partially unmatched in the other. The existence of unmatched gene ends introduces new kinds of paths in the adjacency graph, since some paths can now terminate internal to a chromosome and not on telomeres. We introduce ghost adjacencies to supply the missing gene ends in the genome not containing them. Ghosts enable us to close paths that were due to incomplete matching, just as null points enable us to close even paths terminating in telomeres. We define generalalized DCJ operations on the generalized adjacency graph, and give a prescription for calculating the DCJ distance for the expanded repertoire of operations which includes insertions, deletions and duplications.

1 Introduction

Insertions, deletions, and whole genome as well as tandem and segmental duplications are an important component of genome evolution [1] and should be part of the arsenal of genome rearrangement algorithms. El-Mabrouk [2] included insertions/deletions in the problem of sorting signed permutations by reversals, and Marron *et al* [3] examined genomic distances under deletions and insertions. We wished to extend the DCJ paradigm [4] to allow the computation of genomic distance by DCJ in these contexts so as to have greater applicability in examples arising in evolution, cancer, genetic disruptions leading to disease, analyses of structural and copy number variation in humans [5] and between species [6].

The double cut and join operation (DCJ) [4] allows for the efficient sorting of genomes having equal synteny block content (hereafter called "genes") by a series of rearrangement operations. These include translocations, inversions, fissions, fusions, and the creation and absorption of circular intermediates. Genomes to be sorted can be comprised of linear chromosomes, circular chromosomes, or a combination of both. We propose a generalization of this treatment for genomes having unequal gene content and/or duplications by a relatively straightforward extension of the model.

C.E. Nelson and S. Vialette (Eds.): RECOMB-CG 2008, LNBI 5267, pp. 170–183, 2008.
© Springer-Verlag Berlin Heidelberg 2008

We commence (Section 2) with a review of the *adjacency graph* introduced by Bergeron *et al* [7] for comparing two genomes with equal gene content and no duplications. In Section 3 we present our generalization of the adjacency graph for treating insertions and deletions. The central feature is the introduction of new vertices ("ghosts") to stand for gene ends that are absent in one of the genomes. Ghost vertices are reminiscent of, but not the same as, *null vertices* [4] used to furnish the missing caps in a genome with fewer linear chromosomes than the other. In Section 4 we discuss issues that arise in constructing a rule for the distance associated with a given sorting route between two genomes, and propose a "surcharge" rule to deal with these issues. We also suggest the possibility of allowing elements that are half ghost and half null, which in some cases shorten the distance; we leave open the question whether these elements should be allowed. In Section 5 we extend the treatment to genome pairs with multiple copies, ending in Section 6 with some concluding remarks.

2   The Adjacency Graph

We show genome graphs of the initial and target genomes of a genomic transformation in Figure 1. An arc joining the first and last vertices of the initial genome indicates genes 3 and 1 are connected to each other in a circular chromosome. The genes themselves can be thought of as existing in the white spaces between the points and have been labeled by number in the figure. A negative number indicates a gene in a reversed or inverted orientation such as in the final genome which has a reversal of gene 2.

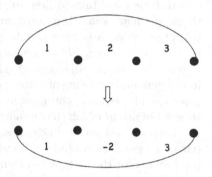

Fig. 1. Genome transformation of a gene inversion within a circular.

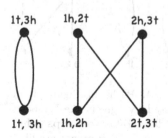

Fig. 2. Adjacency Graph for transformation in Fig. 1.

The transformation can be elegantly represented by the *adjacency graph* (Figure 2) introduced by Bergeron *et al* [7]. In this graph, the initial genome is displayed as a series of *points* at the top representing the adjacencies between gene ends. The target genome is similarly portrayed by its respective adjacencies at the bottom of the graph. In the adjacency graph, vertices with corresponding gene ends are connected between initial and target genome representations. We call these connecting lines *green lines*. The gene ends comprising the vertices can be individually labeled by *head* "h" and *tail* "t" as shown in Figure 2. If the gene content is the same in both genomes and there are no duplications, each vertex in the graph has at most two green lines emanating from it. A **DCJ** swaps two gene ends in two different vertices of the same genome in the adjacency graph.

The adjacency graph is *dual* [8] to the usual representation of the transformation, known as the *breakpoint* or *edge graph* [9], in that all points in the adjacency graph correspond to lines in the edge graph, and conversely all lines in the adjacency graph (representing the gene ends) correspond to points in the edge graph. For circular initial and target genomes, the resulting genomic distance by DCJ operations [4] is simply

$$D = N - C \qquad (1)$$

where N is the number of genes in the initial (and target) genome and C the number of *cycles* in the adjacency graph [10]. Cycles follow green lines along continuous closed paths starting (and ending) at any particular adjacency. They correspond one for one with cycles in the edge graph.

When linear chromosomes are present, some vertices in the adjacency graph have only one line connecting them to the other genome. Gene ends representing ends of chromosomes are called *telomeres*. In addition to closed cycles, there are now *paths*, beginning and ending at telomeres. A path is *even* if it begins and ends in the same genome, *odd* if in opposite genomes. Figure 3 shows the adjacency graph between an initial genome consisting of a linear chromosome, gene 1, and a circular chromosome, gene 2, and target genome in which the circular chromosome

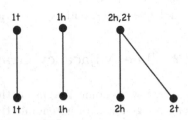

Fig. 3. An Adjacency Graph containing **paths**

has been opened into a linear one. This graph contains two odd and one even path. For genomes containing linear chromosomes, Bergeron *et al* [7] give a simple formula for the distance in terms of the numbers of cycles and of odd paths.

It is possible to recast [10] this formulation by completing each telomere vertex with a *cap*; this fictitious gene end can be the origin of a second green line which terminates on a cap in the other genome, making it possible to close all paths in the adjacency graph. The result of acting with this procedure on Figure 3 is shown in Figure 4. To carry it out, it may be necessary (such as for the even path from 2h to 2t) to add "null" vertices consisting of two caps in one of the genomes so as to equalize the number of caps in the two genomes. With this addition, the total number of adjacencies as well as the number of chromosomes become equalized for initial and target genomes since a null point is counted as

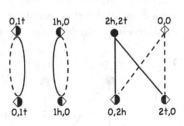

Fig. 4. Paths in Figure 3 are closed by "green lines" (shown as dashed) between **caps** (white triangles). **Telomeres** are filled semicircles.

an adjacency as well as a chromosome. Since all paths are closed into cycles, the formula for genomic distance resembles that for genomes containing only circular chromosomes:

$$D = N' - C'. \qquad (2)$$

Here N' includes all vertices in either genome, including telomere-cap vertices and null (double-cap) vertices, and C' includes the closed cycles in the original graph as well as those formed by closing paths. The two formulations are equivalent and can be shown to result in the same genomic distance [10].

We propose a straightforward extension of this model to genomes having unequal gene content as well as duplications by the addition of "ghost" adjacencies. This device, similar to the addition of null points or chromosomes, allows us to perform the Bergeron *et al* [7] analysis for genomes of unequal gene content.

3 Insertions and Deletions

Consider two genomes having unequal gene content, but without duplications. We show it is possible to represent both genomes via an adjacency graph. For equal gene content all gene ends in adjacencies in the initial genome have partner gene ends in the target genome; with unequal gene content there exist gene ends in one genome without corresponding partners in the other.

As an example of a transformation with unequal gene content but no duplications, we show a pair of genomes in Figure 5. The initial genome consists of a single linear chromosome with two genes. The target genome also consists of a single linear chromosome but with an extra gene (#2) inserted between the outer two.

Figure 6 shows the adjacency graph for this transformation without caps. The vertices containing two gene ends carry two labels, while those which are telomeres contain only one label. There are two paths terminating in telomeres at each end, and one path terminating in the unpaired gene ends 2t and 2h in the lower genome.

In Figure 7 we have added caps to the figure as well as (dashed) green lines connecting caps in the upper genome to those in the lower. This could have been done in

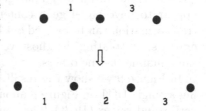

Fig. 5. Insertion of gene 2

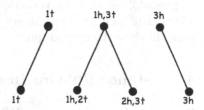

Fig. 6. Fig. 5 Adjacency Graph

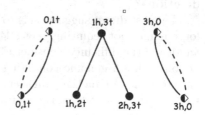

Fig. 7. Addition of caps

either of two ways since the different "0" labels are equivalent; we have chosen the way that yields two cycles rather than one since maximizing the number of cycles yields the smaller distance (also applicable to Figure 4). Now all vertices carry two labels. There are no null vertices (double "0") in this example because both genomes have the same number of chromosomes. "True" adjacency vertices

such as (1h,3t) contain gene ends that exist in both genomes and correspond to
a pair of green lines. For "telomere vertices" one label represents the telomere
and the other ("0") is a cap. The telomere
paths are now closed. A single path terminating in unpaired gene ends, belongs to the unmatched gene.

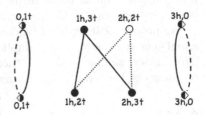

In Figure 8 we have added a ghost vertex
in the upper genome which bears the missing labels 2t, 2h. The path of Figure 7 can
now be closed, as shown by dotted lines, and
the number of cycles counted in the usual
manner. The initial and final genomes now
appear to have equal gene content, and the
transformation can be carried out by DCJ operations, by treating the ghost vertex in the
same manner as the others.

Fig. 8. Addition of the **"ghost"**
vertex (white circle) along with
caps allows all paths to be "closed"

In Figure 9 we show the result of performing a single DCJ on Figure 8 about the ghost
vertex and vertex (1h,3t). The graph becomes
sorted into 1-cycles containing only one vertex
in each genome.

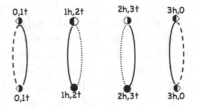

The addition of ghost vertices, as with the
addition of null vertices, allows us to close
paths in the adjacency graph.

Fig. 9. Adjacency Graph post DCJ

4   Distance Rule for Insertions/Deletions

We now address the question: **What distance should be assigned between
genomes of unequal content, for transformations with insertions/
deletions?**

The most direct suggestion is to count one unit for each DCJ, just as we have
for genomes not requiring ghosts [4]. But this leads to certain paradoxes. To get
an idea of the difficulty, consider Figure 10a, which presents us with the problem
of converting a genome consisting of a single circular chromosome (gene 1) to
one consisting of another single circular chromosome (gene 2). The problem may
seem bizarre since the two genomes have no content in common. Let us, however,

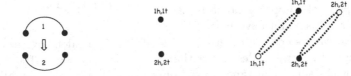

(a) Genome Graph   (b) Adjacency Graph   (c) After adding ghosts

Fig. 10. Conversion of a circular containing gene 1 to one containing gene 2

follow the suggestion. The adjacency graph in Figure 10b for this transformation has no green lines because no gene ends can be matched.

Bravely, we forge ahead, adding ghost vertices to complete the matching (Figure 10c). With the addition of ghosts, the new prescription becomes:

$$D = N'' - C'' \tag{3}$$

where N'' now includes all vertices in either genome including telomere-caps, nulls and ghosts, and C'' includes all the closed cycles including those formed by closing paths using caps and ghosts. We find two vertices in each genome, and two cycles, giving a distance $N'' - C'' = 2 - 2 = 0$. So it appears that the initial genome of Figure 10 can be converted to the target genome at no cost at all!

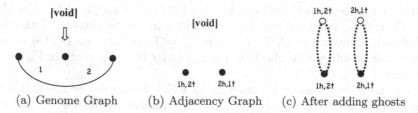

(a) Genome Graph   (b) Adjacency Graph   (c) After adding ghosts

Fig. 11. Creation of a circular chromosome containing genes 1 and 2

A similar situation is depicted in Figure 11 which appears to show (using equation 3 for genomes expanded to include ghosts) that a "void" genome (consisting of nothing at all) can be transformed into one containing a single circular chromosome of two genes without any cost.

4.1   The Triangle Inequality

Such phenomena can easily be found to take place in more realistic examples. Consider the adjacency graph in Figure 12a. Initial genome A, has a string of three genes (2, 3 and 4) to be deleted, leaving outer genes (1 and 5) in the target, genome B. This may be compared with Figure 12b where the extra genes occur in the target instead of the initial genome. This last example may be compared with the example in Figure 6, an insertion of a single gene.

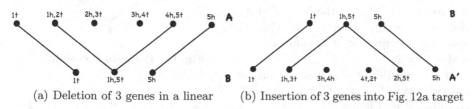

(a) Deletion of 3 genes in a linear   (b) Insertion of 3 genes into Fig. 12a target

Fig. 12. Insertion and Deletion in Linear Chromosomes

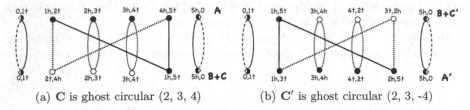

(a) **C** is ghost circular (2, 3, 4)   (b) **C′** is ghost circular (2, 3, -4)

Fig. 13. Examples of Fig. 12 with caps and ghosts added

Initial genome A has two vertices for which neither gene end occurs in the target, genome B, of Figure 12a. These can each be doubly coupled with a target ghost to produce two 1- cycles as shown in Figure 13a. After capping, the telomeric vertices become part of two bystander 1- cycles, leaving two remaining vertices, (1h,2t) and (4h,5t) containing two gene ends (2t and 4h) not present in the final genome which will require an additional single ghost between them. (See Figure 13a; compare with Figure 8 which has fewer ghosts.) With six adjacencies and five cycles: $N'' - C'' = 6 - 5 = 1$, so only one DCJ is required to sort this problem with ghosts in place, similar to the passage of Figure 8 to Figure 9.

Now, however, consider Figure 12b, in which the initial genome, B, is the same as the target genome of Figure 12a, but the target genome, A′, has three genes (2, 3, -4) inserted. Introducing ghosts and closing cycles in Figure 12b, we again find (Figure 13b) 5 cycles and a net $N'' - C''$ of one unit.

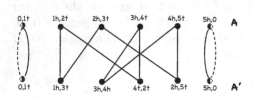

Fig. 14. Direct comparison of A and A′

A paradox emerges when we consider the distance between genomes A and A′ directly. These have equal gene content but are rearranged from one another. If we form the adjacency graph between them in the normal way and close the paths terminating in caps we arrive at Figure 14 which has six adjacencies and three cycles. Sorting it requires $6 - 3 = 3$ DCJs to perform the rearrangement.

But now the triangle inequality, expected of "reasonable" distance metrics and illustrated in Figure 15, is violated, as Figures 13a (AB) and 13b (BA′) each show a distance of only 1. That is, the $N''-C''$ distance $AA' = 3$ is *greater* than (rather than being less than or equal to) the sum of $AB = 1$ and $BA' = 1$.

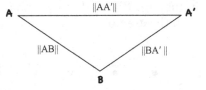

Fig. 15. Triangle Inequality
$AA' \leq AB + BA'$

It is possible to arrive at an alternative way of connecting A to A′ by concatenating Figures 13a and 13b. Here, $N'' - C'' = 9 - 7 = 2$. The first DCJ converts the graph into Figure 13b, plus some 1-cycles; the second

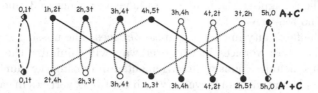

Fig. 16. Alternative comparison of A and A′ using ghosts. C and C′ are ghost circulars as in Fig. 13.

completes the sorting. It appears that use of ghosts has shortened the distance from 3 arrived at by rearrangement scenario (Figure 14) to 2 (Figure 16).

4.2 The Free Lunch Problem

Thus the use of ghosts, introduced in order to bring the unequal content problem into our scheme, threatens to disrupt the distance structure already established for the equal content problem. This disruption is drastic in the case of an equal content rearrangement involving $N >> 1$genes. The DCJ distance without ghosts can easily be comparable to N; but with ghosts, following the example of Figure 16, it is reduced to $O[1]$. Moreover, this reduced distance is independent of the complexity of the rearrangement, since it is based on the possibility of simply removing all the genes and replacing them in the new order at nearly zero cost. We call this the *free lunch* problem.

One's first thought may be that Figure 16 should not be allowed as it might be considered an improper use of ghosts to have the same gene ends (2h, 3t, 3h, 4t) appear as ghosts in both genomes. Perhaps this should be forbidden. But that will not change the fact that Figures 13a and 13b taken together with Figure 14 violate the triangle inequality. Moreover, it permits us to change a genome (1, 2, 3, 4, 5) into (1, 6, 7, 8, 5) by simply removing the three middle genes and putting in new ones. Thus it leaves us asserting that two genomes with utterly dissimilar content can be closer together than two with the same content rearranged. Such a solution is certainly unsatisfactory.

4.3 Surcharge Rule

Having pursued several alternatives, we arrive at the following proposal to compute genomic distance uniquely in the case of unequal gene content via DCJ so as to circumvent the free lunch problem.

The *distance* for a particular adjacency graph shall be the minimum possible number of DCJs required to sort the graph after necessary nulls and ghosts have been added, plus a surcharge of one for every 1-cycle (cycle having 1 vertex per genome) containing a ghost vertex.

Applying these ideas to Figure 13 we see that there are two "ghostly 1-cycles" (G1C) in Figure 13a so that the distance becomes $6 - 5 + 2 = 3$. Likewise in

Figure 13b. Thus Figures 13a, b and 14 all have a distance of 3, and the triangle inequality is no longer violated.

We note the flexibility in choosing ghost vertices. The missing gene ends may be paired into ghost vertices in whatever way will minimize the distance. In particular, this includes choosing arrangements minimizing the number of ghosts in 1-cycles by placing them in 2-cycles (which contain 2 vertices in each genome).

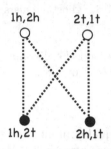

For example, with the surcharge rule, the Figure 10 transformation has a distance 2, due to the two G1C in Figure 10c. The transformation shown in Figure 11, however, can be performed with less cost by setting up the ghosts as shown in Figure 17, in which the two G1C have been replaced by a single ghostly 2-cycle or *bow tie*. This costs 1 cycle but avoids the surcharge, so the distance becomes $2 - 1 = 1$, whereas in Figure 11c it would be 2.

Since each bow tie costs $N'' - C'' = 2 - 1 = 1$, it can add two ghost vertices for the price of one G1C. It is conducive to pair up ghost vertices whenever they exist in G1C and combine them into bow ties. In particular, we observe, that Figure 13a is no longer the optimal way to treat Figure 12a. It is better to avoid the surcharge by combining the two G1C into a 2-cycle as in Figure 18. The distance now

Fig. 17. *Bow Tie* results after rearranging Fig. 11c ghosts by DCJ

becomes $6 - 4 = 2$. Likewise Figure 13b can be replaced by a figure with distance $6 - 4 = 2$. Since $2 + 2 > 3$, the triangle inequality is satisfied.

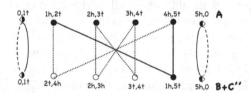

Fig. 18. The two G1C in Fig. 13c are rearranged into a *"bow tie"* by performing DCJ which result in *ghost circular* $C'' = (2, -3, 4)$

Similarly, when the surcharge rule is applied to Figure 16, the distance becomes $9 - 7 + 4 = 6$, but the surcharge can be avoided as shown in Figure 19 by using 2-cycles, so that the distance becomes $9 - 5 = 4$. The "free lunch" is not offered, though, since Figure 14, with no ghosts, still gives a shorter distance.

The proof of the pudding is in the deletion (or insertion) of a very long string of, say M, genes. Without the surcharge this could be done at nearly zero cost, since $M - 1$ G1C can appear in the adjacency graph so that the large number M is cancelled in $N'' - C''$.

One might argue that insertion of a long string is still only a single rearrangement operation, and should therefore only incur a cost of one. Without the surcharge rule, a

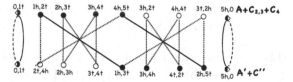

Fig. 19. Two pairs of G1C in Fig. 16 are rearranged into bow ties by performing 2 DCJs. Ghost circular $C_{2,3}$ contains genes 2 and 3, C_4, gene 4, and C'' is the same as in Fig. 18.

large string could in effect be inserted (or deleted) at a cost of only 1. However, this leads to the "free lunch" paradoxes associated with the triangle inequality. The most complex transformations could be performed at a cost of only 2, by first deleting everything and then inserting what is desired.

With the surcharge rule, it initially appears (not using bow ties) that the cost of inserting or deleting a string of length M will be $\sim M$. This solves the free lunch problem with overkill, as the triangle inequality $M + M \geq M$ is satisfied with a large margin. But it makes the insertion of a large string cost no more than M individual insertions in different places.

With the distance rule we propose, and for an insertion size M, the surcharge will add back the contribution $M - 1$ so that the distance is $O[M]$. This, however, is not the optimal distance; the surcharge can be avoided by using 2-cycles so that one obtains roughly $M/2$ cycles with no surcharge, and the distance is $O[M - M/2]$ or $O[M/2]$. Now if the M genes are reinserted in a different order, the distance by similar reasoning will again be $O[M/2]$. But if the reordering of genes were done directly without ghosts and without first deleting and then reinserting them, the distance at most would be $O[M]$. Since $M/2 + M/2 \geq M$, the triangle inequality is not violated.

We may go through this explicitly with a particular M, say $M = 23$. Imagine Figure 13a with 20 additional "middle" genes, so that the initial genome has 25 genes and $N' = 26$. Instead of two G1C there are now twenty-two, so that the distance is $26 - 25 + 22 = 23$. But a shorter distance is obtained by using 2-cycles as in Figure 18; instead of one 2-cycle with ghosts, there are now eleven, plus one without ghosts and two 1-cycles without ghosts. Thus $C'' = 14$, and the distance is $26 - 24 = 12$.

Thus the 23 middle genes have been deleted at a cost of 12. By reinserting them in arbitrary order we would incur another cost of 12, making a total of 24. This is to be compared with the "direct" (ghostless) reordering analogous to Figure 14, in which there would be at least three cycles (two telomeric 1-cycles and one large cycle) so that the cost would be at most $26 - 3 = 23$. Again, no free lunch.

In conclusion, one sees that with the surcharge rule in place, the cost of inserting or deleting a string of length M is only $\sim M/2$, achieved by using bow ties (by combining 1-cycles into 2-cycles after performing DCJ). The triangle inequality is satisfied efficiently, free lunch is avoided, but nevertheless the insertion of a long string is less costly than inserting an equal number of genes in separate operations.

4.4   Hybrid Paths

Returning to the Bergeron *et al* [7] language of "paths", we note that before *nulls* (two caps joined together) and ghosts have been added, there are paths ending in *telomeres*, gene ends which terminate linear chromosomes, and paths ending in gene ends unmatched in the other genome. In our discussion so far we have used nulls to close the first kind of path and ghosts to close the second. But there can also exist a *hybrid path*, which terminates at one end in a telomere and at the other an unmatched gene end.

Consider Figure 20. The initial genome has one linear chromosome with three genes. It is to be converted into a circular chromosome with only two genes, one of which is reversed. One may expect that the sorting will require three steps, since it is necessary to delete gene 2, reverse gene 3, and circularize the linear chromosome.

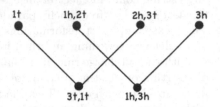

Fig. 20. Linear turning into a circular with a deletion

The "raw" adjacency graph displays two paths, both *even* in that they start and end in the same genome, and both *hybrid* in that each one has a telomere at one end and an unmatched gene end at the other. Proceeding as in the previous examples, we introduce (Figure 21) a null vertex in the target genome to accommodate the two caps in the initial genome, and a ghost vertex to accommodate the unmatched gene ends. The result is a single cycle, and a distance of $4 - 1 = 3$ as expected.

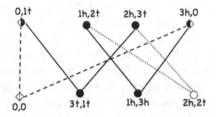

Fig. 21. Fig. 20 after adding null, ghost and caps

The distance can be shortened, however, if one is allowed to introduce a new entity, a *nugh*, which is half null (ie a cap) and half ghost. It can receive one green line from a cap in the other genome, and one from an unmatched gene end. As shown in Figure 22, two nughs can be connected into the graph in such a way that the number of cycles is 2 rather than 1. Thus the distance becomes $4 - 2 = 2$.

We have not reached a definitive judgment on whether nughs ought to be allowed. For the rest of this paper we assume they are forbidden.

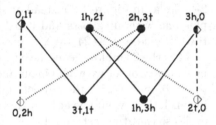

Fig. 22. Fig. 20 after adding **nughs** ie a cap (white triangle) combined with a half ghost (white semicircle)

5   Multiple Copies

5.1   Equal Genomic Content

Suppose that some genes occur multiply in each genome, with each gene occurring the same number of times in each. Once we have decided which copy of a gene in the initial genome corresponds to which copy in the target, it is perfectly straightforward to draw the adjacency graph accordingly, and the distance can be found by the formula of Bergeron *et al* [7] or by closing paths just as if all genes were distinct. No ghosts are needed, but nulls may be needed if there are capped even paths.

The decision, how to match copies in the initial genome to those of the same gene in the target, should depend on which matching gives the shortest distance. For example, Figure 23 shows an adjacency graph for the initial genome, linear chromosome [1, -1, 2], and final genome, linear chromosome [-1, 2, -1]. There are two odd paths and no cycles. With three genes, the Bergeron *et al* [7] distance: $D_B = N - C - (\#odd\ paths)/2$ gives the distance as $= 3 - 0 - (2/2) = 2$. Alternatively we may close the odd paths, each to itself, obtaining two new cycles; the same distance is given as $4 - 2 = 2$.

Figure 24, however, shows the adjacency graph of an alternative way of matching copies. There are now two odd paths and one cycle. The distance by the Bergeron *et al* formula is $3-1-(2/2) = 1$. By closing paths it is $4 - 3 = 1$. Since this distance is less than the one from Figure 23, the second way of matching copies, by which the left end gene in the initial genome is matched to the right end gene in the target, is preferred.

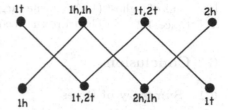

Fig. 23. A way of matching multiple copies of gene 1 in this graph yields 2 **odd paths** (beginning and ending in different genomes), and no cycles

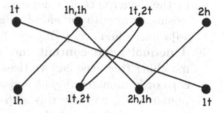

Fig. 24. This alternative matching strategy is preferred since it results in a smaller distance

5.2 Unequal Genomic Content

Now suppose that gene 1 is duplicated in the initial but not in the target genome. Figure 25 shows an adjacency graph in which the initial genome is [1, -1, 2] and the target is [1, 2]. The left-hand gene 1 has been matched to the target 1 and the right-hand gene -1 has been left unmatched. Caps have been matched in the obvious (best) way, yielding two cycles and one even path. When we close this path by means of a ghost (1h,1t) in the target genome, we shall have three cycles, giving a distance $4 - 3 = 1$.

Figure 26 shows an alternative adjacency graph for the same problem, in which the gene -1 in the initial genome has been matched to the target 1, and the gene 1 in the initial genome left hanging. Again the caps are connected optimally. Here there is only one cycle, and one even path. When the path is closed

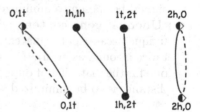

Fig. 25. Left-hand gene 1 is matched to target, right hand gene -1 can be matched to a ghost

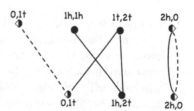

Fig. 26. A way of matching multiple copies of gene 1 in this graph yields 2 odd paths and no cycles

by means of a ghost (1h,1t) in the target genome, there will be two cycles, giving a distance 4-2 = 2. Therefore the second matching (Figure 26) is preferred.

6   Conclusion

6.1   Summary of Rules

We may summarize the proposed rules for drawing graphs and finding distances by considering four cases:

i) **Equal gene content, no duplication.** This case has been treated in previous literature [4]. The adjacency graph is unique. Distance may be found by the Bergeron *et al* [7] formula by counting genes, paths, and cycles, or by closing paths with the aid of nulls where needed, and counting the resulting adjacencies and cycles [10].

ii) **Unequal gene content, no duplication.** Again the adjacency graph matching true gene end vertices is unique. Paths may terminate either in caps or in unmatched gene ends. Ghosts are introduced so as to match the unmatched gene ends; this step is not unique in general. Distance is then found as in (i), except that one unit of distance is added for each 1-cycle containing a ghost (surcharge rule). The manner of introducing ghosts is to be chosen so as to minimize the resulting distance.

iii) **Equal gene content with duplication.** The adjacency graph is not unique, because the duplicated genes can be matched in more than one way. Once the matching is chosen, distance is found as in (i). The matching should be chosen to minimize the resulting distance.

iv) **Unequal gene content with duplication.** The adjacency graph is not unique because of the matching ambiguity. Once the matching is chosen, distance is found as in (ii). There are in general two ambiguities to be resolved, one for the matching of duplicates and one for the assignment of ghosts. The distance is to be minimized with respect to both ambiguities.

6.2   Questions to Be Answered

In this paper we have addressed only the proposed system of graphs and DCJ distance rules. We leave to later work the open questions associated with this system:

1) To find a formula analogous to that of Bergeron *et al* [7] which takes into account paths terminating in unmatched gene ends as well as those terminating in caps. We note here only that while odd paths terminating in caps contribute negatively to the distance, odd paths terminating in unmatched gene ends will contribute positively. In the language of the procedure of closing all paths, the reason for the difference is that an odd path terminating in caps can be closed efficiently, by connecting the caps at the two ends of the path, whereas an odd path terminating in unmatched gene ends cannot be closed in this way because the two ends are dissimilar.

2) To find an efficient algorithm for optimizing the choice of ghosts in cases (ii) and (iv).
3) To find an efficient algorithm for resolving the matching ambiguity in cases (iii) and (iv).
4) To give a rigorous proof that no case exists in which the surcharge rule does not eliminate the free lunch problem.
5) To arrive at a good reason why *nughs* should be either allowed or disallowed.

6.3 Concluding Remarks

Efficient algorithms exist for transforming one genome to another in the case of equal gene content. However when insertions, deletions and duplications are introduced, the problem becomes more challenging. The extensions of the DCJ paradigm in this paper allow us to consider genome rearrangements within this more general framework.

Acknowledgements

SY thanks Nicholas Chiorazzi for support and Robert Warren, Martin Bader and Michal Ozery-Flato for stimulating conversations during RECOMB CG 2007 that helped propel this inquiry.

References

1. Kent, W.J., Baertsch, R., Hinrichs, A., Miller, W., Haussler, D.: Evolutions cauldron: duplication, deletion, and rearrangement in the mouse and human genomes. Proc. Natl Acad. Sci. USA 100, 11484–11489 (2003)
2. El-Mabrouk, N.: Sorting signed permutations by reversals and insertions/deletions of contiguous segments. Journal of Discrete Algorithms 1(1), 105–122 (2001)
3. Marron, M., Swenson, K., Moret, B.: Genomic distances under deletions and insertions. Theoretical Computer Science 325(3), 347–360 (2004)
4. Yancopoulos, S., Attie, O., Friedberg, R.: Efficient sorting of genomic permutations by translocation, inversion and block interchange. Bioinformatics 21, 3340–3346 (2005)
5. Feuk, L., Carson, A.R., Scherer, S.W.: Structural variation in the human genome. Nat. Rev. Genet. Feb. 7(2), 85–97 (2006)
6. The Chimpanzee Sequencing and Analysis Consortium, Initial sequence of the chimpanzee genome and comparison with the human genome. Nature 437, 69–87 (2005)
7. Bergeron, A., Mixtacki, J., Stoye, J.: A unifying view of genome rearrangements. In: Bücher, P., Moret, B.M.E. (eds.) WABI 2006. LNCS (LNBI), vol. 4175, pp. 163–173. Springer, Heidelberg (2006)
8. Bergeron communication
9. Bafna, V., Pevzner, P.A.: Genome rearrangements and sorting by reversals. In: Proc. 34th Ann. IEEE Symp Found. Comp. Sci., pp. 148–157. IEEE Press, Los Alamitos (1993)
10. Friedberg, R., Darling, A.E., Yancopoulos, S.: Genome Rearrangement by the Double Cut and Join Operation. In: Keith, J.M. (ed.) Bioinformatics, Data, Sequence Analysis and Evolution, ch. 18. vol. I. Humana Press (2008)

A Fast and Exact Algorithm for the Median of Three Problem—A Graph Decomposition Approach

Andrew Wei Xu

Department of Mathematics and Statistics, University of Ottawa, Canada K1N 6N5

Abstract. In a previous paper, we have shown that adequate subgraphs can be used to decompose multiple breakpoint graphs, achieving a dramatic speedup in solving the median problem. In this paper, focusing on the median of three problem, we prove more important properties about adequate subgraphs with rank 3 and discuss the algorithms inventorying simple adequate subgraphs. After finding simple adequate subgraphs of small sizes, we incorporate them into ASMedian, an algorithm to solve the median of three problem. Results on simulated data show dramatic speedup so that many instances can be solved very quickly, even ones containing hundreds or thousands of genes.

1 Introduction

The median problem[3,7,8,6,2,1] for genomic rearrangement distances is NP-hard [4,9]. Algorithms have been developed to find exact solutions for small instances [4,6] and there are rapid heuristics of varying degrees of efficiency and accuracy [2,1,5]. In a previous paper [10], with the aim of finding a decomposition method to reduce the size of the problem, we introduced the notion of adequate subgraph and showed how they lead to such a decomposition. By applying this method recursively, the size of the problem is effectively reduced. In this paper, we focus on the median of three problem, which is to find a genome q with smallest total distance $\sum_{1 \le i \le 3} d(q, g_i)$ for any given three genomes g_1, g_2, g_3.

Because of its simple structure, we choose to work with DCJ distance[11] $d = n - c$ as most likely to yield non-trivial mathematical results, where n is the number of genes in each genome (assuming that they have the same gene content) and c is the number of cycles in the breakpoint graph. We require genomes to consist of one or more circular chromosomes, but our results could be extended to genomes with multiple linear chromosomes.

In Section 2 several related concepts are defined, such as *breakpoint graph* and *adequate subgraph*. In Section 3 some important properties about adequate subgraphs of rank 3 are proved. We discuss the problem of inventorying simple adequate subgraphs in Section 4. Then in Section 5, we give an algorithm AS-Median to solve the median problem. Results on simulated data are given and discussed in Section 6.

C.E. Nelson and S. Vialette (Eds.): RECOMB-CG 2008, LNBI 5267, pp. 184–197, 2008.
© Springer-Verlag Berlin Heidelberg 2008

2 Graphs, Subgraphs and More

2.1 Breakpoint Graph

We construct the breakpoint graph of two genomes by representing each gene by an ordered pair of vertices, adding coloured edges to represent the adjacencies between two genes, red edges for one genome and black for the other.

In a genome, every gene has two adjacencies, one incident to each of its two endpoints, since it appears exactly once in that genome. Then in the breakpoint graph, every vertex is incident to one red edge and one black one. Thus the breakpoint graph is a 2-regular graph which automatically decomposes into a set of alternating-colour cycles.

The edges of one colour form a perfect matching of the breakpoint graph, which we will simply refer to as a *matching*, unless otherwise specified. By the red matching, we mean the matching consisting of all the red edges.

The **size** of the breakpoint graph is defined as half the number of vertices it contains, which equals to the size of its matchings and the number of gens in each genome.

2.2 Multiple Breakpoint Graph and Median Graph

The breakpoint graph extends naturally to a multiple breakpoint graph (MBG), representing a set $\mathcal{G}$ of three or more genomes. The number of genomes[1] $N_{\mathcal{G}} \geq 3$ in $\mathcal{G}$ is called the **rank** of the MBG, which is also its edge chromatic number. The colours assigned to the genomes are labeled by the integers from 1 to $N_{\mathcal{G}}$. The **size** of an MBG or its subgraph is also defined as half the number of vertices it contains.

For a candidate median genome, we use a different colour for its matching E, namely colour 0. Adding E to the MBG results in the **median graph**. The set of all possible candidate matchings is denoted by $\mathcal{E}$.

The **0-i cycles** in a median graph with matching E, numbering $c(0, i)$ in all, are the cycles where 0-edges and i edges alternate. Let $c_E = \sum_{1 \leq i \leq 3} c(0, i)$. Then $c_{\max} = \max\{c_E : E \in \mathcal{E}\}$ is the maximum number of cycles that can be formed from the MBG. *Minimizing the total DCJ distance in the median problem is equivalent to finding an optimal 0-matching E, i.e., with $c_E = c_{\max}$.*

2.3 Subgraphs

Let $\mathbf{V}(G)$ and $\mathbf{E}(G)$ be the sets of vertices and edges of a regular graph G. A **proper subgraph** H of G is one where $\mathbf{V}(H) = \mathbf{V}(G)$ and $\mathbf{E}(H) = \mathbf{E}(G)$ do not both hold at the same time. An **induced subgraph** H of G is the subgraph which satisfies the property that if $x, y \in \mathbf{V}(H)$ and $(x, y) \in \mathbf{E}(G)$, then $(x, y) \in \mathbf{E}(H)$.

In this paper, we will focus on the induced proper subgraphs of MBGs, with even numbers of vertices. Through this paper, the size of a subgraph is denoted

---

[1] For the median of three problem, this number is just 3.

by m. For a proper induced subgraph H, $\mathcal{E}(H)$ is the set of all its perfect 0-matchings $E(H)$. The number of cycles determined by H and $E(H)$ is $c_{E(H)}(H)$, and $c_{\max}(H)$ is the maximum number of cycles that can be formed from H. A 0-matching $E^{\star}(H)$ with $c_{E^{\star}(H)}(H) = c_{\max}(H)$ is called an optimal partial 0-matching, and $\mathcal{E}^{\star}(H)$ is the set of such 0-matchings.

2.4   Non-crossing 0-Matchings and Decomposers

For a subgraph H of an MBG G, a potential 0-edge would be H-**crossing** if it connected a vertex in $\mathbf{V}(H)$ to a vertex in $\mathbf{V}(G) - \mathbf{V}(H)$. A candidate matching containing one or more H-crossing 0-edges is an H-crossing.

An MBG subgraph H is called a **decomposer** if for any MBG containing it, there is an optimal matching that is not H-crossing. It is a **strong decomposer** if for any MBG containing it, all the optimal matchings are not H-crossing.

2.5   Adequate and Strongly Adequate Subgraphs

A connected MBG subgraph H of size m is an **adequate subgraph** if $c_{\max}(H) \geq \frac{1}{2}mN_G$; it is **strongly adequate** if $c_{\max}(H) > \frac{1}{2}mN_G$. For the median of three problem, an adequate subgraph of rank 3 is a subgraph with $c_{\max}(H) \geq \frac{3m}{2}$ and a strongly adequate subgraph of rank 3 is one with $c_{\max}(H) > \frac{3m}{2}$.

A (strongly) adequate subgraph H is **simple** if it does not contain another (strongly) adequate subgraph as an induced subgraph; deleting any vertex from H will destroy its adequacy.

Adequate subgraphs enable us to decompose the MBG into a set of smaller ones, as in the next theorem.

Theorem 1. *[10] Any adequate subgraph is a decomposer. Any strongly adequate subgraph is a strong decomposer.*

3   The Properties of Simple Adequate Subgraphs of Rank 3

In this section, we prove other important properties about simple adequate subgraphs of rank 3. Multiple edges in MBGs are the simple adequate subgraphs of size one, which are the only exceptions to many of the properties stated below.

3.1   More Properties about Adequate Subgraphs of Rank 3

Lemma 1. *The vertices of simple adequate subgraphs of rank 3 have degrees either 2 or 3.*

Proof. Since the MBG for the median of three problem is 3-regular, the vertex degrees of its induced subgraphs can only be 1, 2 or 3.

The lemma is true for parallel edges (the smallest simple adequate subgraphs), where the vertex degrees are 2 or 3. For simple adequate subgraphs of size two or more, we prove by contradiction that they can not contain vertices of degree 1.

Assume there is a simple adequate subgraph H of size m containing a vertex x of degree 1. In one of the optimal 0-matchings of H, x is connected to vertex y by a 0-edge e, and e appears only in one of the colour-alternating cycles. By deleting edge e and vertices x, y, only that cycle is destroyed. Because of its adequacy, the maximum number of cycles formed with H is at least $\frac{3m}{2}$. So for the resultant subgraph F of size $m - 1$, the maximum number of cycles can be formed is at least $\frac{3m}{2} - 1 = \frac{3(m-1)}{2} + \frac{1}{2} > \frac{3(m-1)}{2}$. Therefore F, as a subgraph of H, is also an adequate subgraph, which contradicts the assumption that H is simple.

So the vertex degrees in a simple adequate subgraph can only be 2 or 3. $\square$

Lemma 2. *Except for multiple edges, the size of a simple adequate subgraph of rank 3 is even.*

Proof. Suppose there is an odd-sized simple adequate subgraph H of size $2k+1$. Because of its adequacy, the maximum number of cycles formed with H is at least $\left\lceil \frac{3(2k+1)}{2} \right\rceil = 3k + 2$. Since H is an proper subgraph, there exists a vertex x with degree 2. Suppose 0-edge e is incident to x in one of H's optimal 0-matchings. By deleting e and the corresponding vertices, two colour-alternating cycles are destroyed. Then for the resultant subgraph F of size $2k$, the maximum number of cycles formed with F is at least $3k = \frac{3}{2} \times 2k$. Hence F, as a subgraph of H, is also an adequate subgraph, which contradicts the simplicity of H. $\square$

Lemma 3. *Except for multiple edges, the maximum number of cycles of a simple adequate subgraph of rank 3 is exactly $\frac{3m}{2}$, where m is its size.*

Proof. Because of Lemma 2, we only need to consider even-sized simple adequate subgraphs. Suppose H is a simple adequate subgraph of size $2k$, with which the maximum number of cycles formed is at least $3k + 1$. Then by deleting a 0-edge connecting to a degree 2 vertex, the size of the subgraph decreases by 1 and the number of cycles decreases by 2. So H contains another adequate subgraph of size $2k - 1$ whose maximum number of cycles is at least $3k - 1 = \left\lceil \frac{3}{2}(2k - 1) \right\rceil$, which contradicts the simplicity assumption for H. $\square$

3.2 There Are Infinite Many Simple Adequate Subgraphs

In this subsection we show that there are infinitely many adequate subgraphs, by proving the number of simple adequate subgraphs is infinite, which follows from the infinite size of a special family of simple adequate subgraphs—the mirrored-tree graphs.

Definition 1 *An **mirrored-tree graph:** two identical 3-edge-coloured binary trees with corresponding pairs of leaf vertices connected by simple edges. Being an MBG subgraph, the size of an mirrored-tree graph is defined as half the number of its vertices, which also is the number of vertices contained in each tree.*

Proposition 1   *1. Any binary tree containing more than one vertex must have even size;*

2. *For a binary tree with m vertices, there are $\frac{m}{2}+1$ leaf vertices (with degree 1) and $\frac{m}{2}-1$ inner vertices (with degree 3).*

3. *The total number of edges is $m-1$.*

Proposition 2   *For a mirrored-tree graph of size m, there are $\frac{5m}{2}-1$ edges in total; $\frac{m}{2}+1$ of them connect the two binary trees and $2m-2$ of them lie in the trees.*

Definition 2   *Double-Y end: A mirrored-tree graph of size 4, with one connecting edge missing, as illustrated by Figure 1(a). Being a part of an MBG subgraph, it is connected to the remaining graph through the two vertices of degree one.*

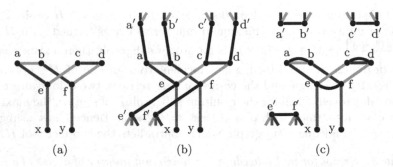

(a)   (b)   (c)

Fig. 1. (a) Illustration of a double-Y end and it is connected to the remaining subgraph only through vertices x and y; (b) shows a 0-matching not containing 0-edges $(a,b),(c,d),(e,f)$; (c) shows another 0-matching obtained by applying 3 DCJ operations to the 0-matching in (b), that does contain those three 0-edges and forms more colour-alternating cycles

Lemma 4. *If a double-Y end appears in an MBG subgraph H, then the 0-edges $(a,b),(c,d)$ and (e,f) connecting corresponding vertices of the two identical trees, must exist in any optimal 0-matching of H, as illustrated by Figure 1(c).*

Proof. In an optimal 0-matching of H, if any of the three 0-edges (a,b), (c,d) and (e,f) appears, the other two 0-edges must also exist. Then only one case is left to disprove—that none of these 0-edges appears in some optimal 0-matching, as illustrated by Figure 1(b).

Figure 1(c) is obtained by three DCJ operations on the 0-edges of Figure 1(b), creating 0-edges (a,b), (c,d) and (e,f). By comparison we can see that: a', b' are connected by a green-black alternating path and c', d' are connected by a blue-black alternating path in both figures. So they are involved in the same number of cycles in both figures.

Apart from these, Figure 1(b) contains another 6 paths, which can form at most 6 colour-alternating cycles; Figure 1(c) contains 4 cycles as well as another 6 paths of 3 different colours which will form at least 3 cycles, summing up to a total of 7 cycles or more. So Figure 1(c) forms more cycles than Figure 1(b).

For the cases where vertices a', b', c', d', e', f' are incident to different set of edges, the same result still holds.

Since 0-matchings of H containing 0-edges (a, b), (c, d) and (e, f) have more cycles than 0-matchings not containing them, these three 0-edges must exist in any optimal 0-matchings.   □

Theorem 2. *With a mirrored-tree graph of size m, we can form a maximum of $\frac{3m}{2}$ colour alternating cycles, hence it is an adequate subgraph; Furthermore, it does not contain any smaller adequate subgraphs, so it is a simple adequate subgraph.*

Proof. We first prove that there is a 0-matching of the mirrored-tree graph, forming $\frac{3m}{2}$ colour alternating cycles. This is just the set of 0-edges connecting the corresponding vertices of the two trees. With this 0-matching, each non-0 edge connecting two trees makes a cycle by itself; and the edges on the tree form cycles of size 2 with the corresponding edges on the other tree. From Proposition 2, there are $\frac{m}{2} + 1$ edges connecting trees and $2m - 2$ edges on the trees, the total number of cycles is $\frac{3m}{2}$.

Next we show that is the only optimal 0-matching. For any binary tree, since the number of leaf vertices is larger than the number of inner vertices by 2, there is always an inner vertex being connected to two leaf vertices. In the corresponding mirrored-tree graph, this gives a double-Y end.

For a mirrored-tree graph H, we add two 0-edges parallel to the connecting edges of its double-Y end as 0-edges (a, b) and (c, d) in Figure 1(c). Then by shrinking them, H becomes a quasi mirrored-tree graph[2] of smaller size, containing double-Y ends or quasi double-Y ends[3]. By applying this procedure of adding and shrinking 0-edges to the (quasi) double-Y ends recursively, H finally becomes a three-parallel edge. Since in each step the new added 0-edges must appear in all optimal 0-matchings, the resultant perfect 0-matching is the only optimal 0-matching of H.

The symmetrical structure of mirrored-tree graphs leads to colour-alternating cycles of smallest sizes—1 and 2 only. In detecting whether a mirrored-tree graph H contains any smaller adequate subgraphs, it is sufficient to only consider its subgraphs with symmetrical structures. Using reasoning similar to the above paragraphs, it can be shown that the optimal 0-matchings for these symmetrical subgraphs of H are the subsets of the 0-edges in the optimal 0-matching of H. However none of these symmetrical subgraphs of H can form cycles of $\frac{3}{2}$ times their sizes. So the mirrored-tree graphs are simple adequate subgraphs.   □

Theorem 3. *There are infinitely many simple adequate subgraphs.*

Proof. Since there are binary trees of arbitrary large size, which give mirrored-tree graphs with arbitrary large (even) size. Also because mirrored-tree graphs are simple adequate subgraphs, there exist simple adequate subgraphs with arbitrary large (even) size.   □

---

[2] In which, there may be multiple edges connecting the two identical trees.
[3] The ones whose connecting edges might be multiple edges. Obviously the conclusion in Lemma 4 also applies to quasi double-Y ends.

4 Inventorying Simple Adequate Subgraphs

4.1 It Is Practical to Use Simple Adequate Subgraphs of Small Sizes

Before using the adequate subgraphs to reduce the search space for finding an optimal 0-matching, we need to inventory the adequate subgraphs. Theorem 3 states that there are infinitely many simple adequate subgraphs, hence infinitely many adequate subgraphs, so it is impossible to inventory all of them and use them to decompose the median problems. However, it is practical to work on simple adequate subgraphs of small sizes, as justified by the following:

1. There are much fewer simple adequate subgraphs. And many non-simple adequate subgraphs can be decomposed into several simple adequate subgraphs embedded in each other. Hence many non-simple ones can be detected through the constituent simple ones.
2. The algorithms to inventory simple adequate subgraphs for a given size require more than exponential time in their size.
3. The total number of simple adequate subgraphs increases dramatically as the size increases. The complexity of the algorithm to detect the existence of a given simple adequate subgraph also increases accordingly. Combining these two factors, we conclude that it is prohibitively expensive to detect the existence of simple adequate subgraphs of large sizes.
4. Simple adequate subgraphs of small sizes exist with much higher probability than subgraphs of greater size on random MBGs. The details will be given in the full version of this paper.

4.2 Algorithms to Inventory Simple Adequate Subgraphs

To enumerate the simple adequate subgraphs, we need to search among all the MBG subgraphs, which consist of (perfect or non-perfect) matchings of three colours. In order to count the number of cycles, the perfect 0-matchings must also be enumerated. So the algorithms need to work on graphs consisting of 4 matchings, hence the problem is computationally costly.

Our simple adequate subgraph inventorying algorithm uses a depth-first search method. The graph grows by adding an edge at each step. It is backtracked whenever the current graph contains a smaller simple adequate subgraph and then restrained on another path to grow the graph until all subgraphs have been searched.

To speed up the algorithm, we adopt several useful methods and techniques:

1. Only inventory simple adequate subgraphs of even sizes, as a result of Lemma 2.
2. Fix the 0-matching. Any median subgraph is isomorphic to $\frac{(2m)!}{2^m\, m!} - 1$ other median subgraphs by permuting the 0-edges.
3. Only allow the graphs whose number of 1-edges is no less than the number of 2-edges and the number of 2-edges is no less than the number of 3-edges, because of the isomorphism associated with the permutation of colours.
4. Every vertex must be incident to 2 or 3 non-0-edges, because of Lemma 1.

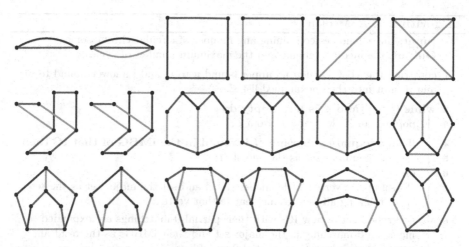

Fig. 2. Simple adequate subgraphs of size 1, 2 and 4 for MBGs on three genomes

4.3 Simple Adequate Subgraph Enumerated

In Figure 2, the simple adequate subgraphs of size 1, 2 and 4 are listed. Each subgraph represent a class of subgraphs isomorphic under the permutations of vertices and colours.[4]

5 Solving the Median of Three Problem by Recursively Detecting Simple Adequate Subgraphs

Our algorithm using adequate subgraphs to decompose the median problems is called **ASMedian**. It adopts a branch-and-bound method to find an optimal 0-matching for any given MBG. At any intermediate step during the branch-and-bound search, an intermediate configuration (IC for short) is constructed, containing a partial 0-matching and an intermediate MBG (iMBG for short) resulted by a process of edge-shrinking [10] of that partial 0-matching from the original MBG. The algorithm keeps a list of unexamined ICs $\mathcal{L}$, initially just consisting of the original MBG.

At each step, from $\mathcal{L}$ an unexamined IC with the largest upper bound is selected to examine. According to whether an inventoried simple adequate subgraph exists in that iMBG and what simple adequate subgraph it is, a number of new ICs are generated, containing smaller and non-empty iMBGs and expanded partial 0-matchings. Then we update U the largest upper bound of all unexamined ICs and c^* the largest cycle number encountered so far. We prune the ICs whose upper bounds are no larger than c^*. The algorithm stops when $c^* \geq U$ or no unexamined ICs remain. Then c^* is the maximum cycle number for the original MBG, and the corresponding 0-matching is an optimal 0-matching.

---

[4] Strictly speaking, Figures (a) and (f) are not proper subgraphs of a connected MBG.

Algorithm 1. ASMedian

    Input: three genomes containing any number of circular chromosomes
    Output: the median genome and the maximum number of cycles c

1 construct the MBG, assign its upper bound u to U and its lower bound to c^*
    and push it into the unexamined list $\mathcal{L}$;

2 **while** $U > c^*$ **and** $\mathcal{L}$ **is not empty do**
3   pop out an IC with $u = U$ from $\mathcal{L}$;

4   **if an adequate subgraph H is found in the iMBG of that IC then**
5   set the major set as the one of H;
6   **else**
7   select the vertex with smallest label and set the major set as the set
            containing all 0-edges incident to that vertex;

8   generate a set of new ICs with their partial 0-matchings are expanded to
      include a 0-matching in the major set and their iMBGs as the resultant
      graphs of shrinking these partial 0-matchings;

9   update U and c^*;
10   **if c^* gets updated then** Remove all the ICs with $u \le c^*$ in $\mathcal{L}$;
11   push the new generated ICs with $u > c^*$ into $\mathcal{L}$;

    `// the maximum cycle number has been found;`
12 set c as c^* and construct the median genome from the optimal 0-matching
    obtained;
13 **return** c and the median genome;

5.1 Examining the Intermediate MBGs

Definition 3 *The inquiry set is the set of simple adequate subgraphs (of small sizes) for which the ASMedian algorithm looks on the intermediate MBGs. For a specific algorithm, the inquiry set is given as a parameter.*

The iMBG of the selected IC is examined to see the existence of any simple adequate subgraph in the inquiry set. If one of such subgraphs H exists, then we know there is an optimal 0-matchings of iMBG which is non-H-crossing. This 0-matching can be divided into two parts: a 0-matching of H and a partial 0-matching of the remaining intermediate MBG.

    A **major set** of H is the minimal set of 0-matchings of H, which guarantees that at least one of them must appear in an optimal 0-matchings of the MBG, without the knowledge of the remaining part of the MBG (as will be shown else where). The size of the major set is denoted by μ. Since the inquiry set is given in advance, the major sets for the simple adequate subgraphs are also known in advance. Then according to this major set, μ new ICs will be generated with smaller iMBGs, each resulting from the shrinking of a 0-matching of H in the major set from the iMBG of the currently selected IC.

    When no simple adequate subgraph in the inquiry set exists, the vertex v with the smallest remaining label is selected. The nominal major set of size $2\tilde{n} - 1$ is constructed, where $\tilde{n}$ is the size of the iMBG — just the set of 0-edges incident to v (here each partial 0-matching is just a 0-edge). Then $2\tilde{n} - 1$ new ICs are created accordingly.

If the inquiry set is chosen as all simple adequate subgraphs of sizes 1, 2 and 4, then the sizes of their major sets are just one, except for one case where it is 2. It can be seen that whenever a simple adequate subgraph is detected, the search space is roughly reduced by a factor of $\tilde{n}$, $\tilde{n}^2$ or $\tilde{n}^4$.

5.2   The Lower Bound and the Upper Bound

For each intermediate configuration, the ASMedian algorithm calculates its upper bound and prunes it if the value is no larger than c^*—the maximum number of cycles encountered so far.

Because of the search schema we use (see next subsection), it takes a while for the algorithm to reach any perfect 0-matching. Due to the fact that the number of cycles formed by partial 0-matchings are small, to calculate c^* from them will make the pruning procedure very inefficient. Instead, for each intermediate configuration, a tight lower bound is calculated and c^* takes the maximum of these lower bounds and the encountered cycle numbers.

Since the DCJ distance is a metric measure, for any median of three problem, there is an associated lower bound for the total distance. Assume the three known genomes are labeled as 1, 2, 3, and $d_{1,2}, d_{1,3}, d_{2,3}$ denote the pairwise distances and $c_{1,2}, c_{1,3}, c_{2,3}$ denote the cycle numbers between any two pairs. The lower bound for the total distance is $d \geq \frac{d_{1,2}+d_{1,3}+d_{2,3}}{2}$. Because $d_{i,j} = n - c_{i,j}$, then we get an upper bound for the total cycle number,

$$c \leq \frac{3n}{2} + \frac{c_{1,2} + c_{1,3} + c_{2,3}}{2}. \tag{1}$$

To find a lower bound for the total cycle number, we can set the 0-matching to any of the matchings representing the three known genomes and take largest total cycle number of the three as the lower bound, so that

$$c \geq c_{1,2} + c_{1,3} + c_{2,3} - \min\{c_{1,2}, c_{1,3}, c_{2,3}\}. \tag{2}$$

For any IC, by adding $\tilde{c}$, the number of cycles formed by its partial 0-matching, to the lower bound and upper bound of its intermediate MBG, we get the upper bound and lower bound of this IC, denoted by u and l correspondingly.

$$u = \tilde{c} + \frac{3\tilde{n}}{2} + \frac{\tilde{c}_{1,2} + \tilde{c}_{1,3} + \tilde{c}_{2,3}}{2} \tag{3}$$

$$l = \tilde{c} + \tilde{c}_{1,2} + \tilde{c}_{1,3} + \tilde{c}_{2,3} - \min\{\tilde{c}_{1,2}, \tilde{c}_{1,3}, \tilde{c}_{2,3}\}. \tag{4}$$

The IC and all ICs derived from it, are referred as the **parent** IC and the **child** ICs. A non-increasing property holds between the upper bounds of the parent IC and the child ICs.

Lemma 5. *The upper bounds of the child ICs are never larger than the upper bound of their parent IC.*

Proof. Suppose a child IC is obtained from the parent IC by adding a 0-edge e. We first inspect the possible effects on $\tilde{c}$ and $\tilde{c}_{1,2}$ of adding e to the iMBG of the parent IC.

  a If e connects two 1-2 cycles, then the two cycles will be merged into one. Then $\tilde{c}$ remains the same and $\tilde{c}_{1,2}$ decreases by 1;
  b if e parallels a 1-edge (or a 2-edge), then one 0-1 cycle of size 1 is formed and the 1-2 cycle containing that edge becomes a shorter one. So that $\tilde{c}$ increases by 1 and $\tilde{c}_{1,2}$ remains the same;
  c if e connects two vertices of the same 1-2 cycle, not paralleling any edges, then this 1-2 cycle may be split into two or remain with a smaller size. Therefore $\tilde{c}$ remains the same and $\tilde{c}_{1,2}$ increases by 0 or 1.

Since the size of the iMBG in the child IC decreases by 1, $\frac{3\tilde{n}}{2}$ decreases by $\frac{3}{2}$. As long as $\tilde{c} + \frac{\tilde{c}_{1,2}+\tilde{c}_{1,3}+\tilde{c}_{2,3}}{2}$ does not increase more than $\frac{3}{2}$, u never increases.

  1 If e does not parallel any edge, then $\tilde{c}$ remains the same, and each of $\tilde{c}_{1,2}, \tilde{c}_{1,3}, \tilde{c}_{2,3}$ increases at most by 1. So u does not increase;
  2 if e parallels one edge, $\tilde{c}$ increases by 1 and only one of $\tilde{c}_{1,2}, \tilde{c}_{1,3}, \tilde{c}_{2,3}$ increases at most by 1. So u does not increase;
  3 if e parallels two edges, $\tilde{c}$ increases by 2 and the cycle formed by the parallel edges is destroyed and the other two terms of $\tilde{c}_{1,2}, \tilde{c}_{1,3}, \tilde{c}_{2,3}$ remain the same. So u does not change;
  4 e parallels three edges, $\tilde{c}$ increases by 3 and the three cycles formed by the parallel edges are destroyed. So u remains the same.

So the upper bound of the child ICs are never larger than the upper bound of their parent IC. □

The algorithm maintains an overall upper bound U which is the maximum upper bound of all unexamined ICs. Another global variable, as mentioned before, is the largest total cycle number or lower bound c^* found so far. Obviously the maximum total cycle number c of the original MBG lies between c^* and U.

5.3   The Optimistic Search Schema

Our algorithm is neither a strict depth-first nor a strict breadth-first search schema, but follows an "optimistic" search strategy . From the list of all unexamined ICs, we select the one with the largest upper bound. The intuition behind this is, the ICs with larger upper bounds are more likely to lead to perfect 0-matchings with larger cycle numbers. Beside the intuitive aspect, we can prove that this optimistic search schema has a smallest search space in terms of the number of ICs it examines.

Theorem 4. *The set of ICs the optimistic search schema examines includes all ICs with $u > c$, plus a subset of ICs with $u = c$. Further more, since the search space of every branch-and-bound method includes all ICs with $u > c$, the optimistic search schema has the smallest search space possible.*

Proof. Obviously every IC with $u > c$ should be examined by the algorithm, otherwise, the possibility of having a maximum total cycle number with $c + 1$ or more can not be eliminated.

Because of Lemma 5, for any IC with $u \geq c$, all the ICs lying on the path from the original of the search to this IC have their upper bounds larger than or equal to c. So the algorithm with optimistic search schema never needs to examine any IC with $u < c$ to find the ones with $u \geq c$, i.e., this algorithm finds all ICs with $u \geq c$ without examining any ones with smaller upper bounds. By the time that the ones with $u \geq c$ have been examined, an optimal 0-matching with c cycles has been found and the algorithm stops. And the search space for the optimistic schema includes all the ICs with $u > c$ and a subset of the ICs with $u = c$.

Hence the optimistic search schema has the smallest search space possible.   □

The exact algorithm in [4] consists of cascaded runs of depth-first branch-and-bound search, with the first run seeking a solution whose cycle number is equal to the upper bound of the original MBG and the subsequent runs seeking solutions with one cycle less than the previous ones, until a solution is found. The cascaded branch-band-bound algorithm and our optimistic branch-and-bound algorithm are similar in terms of the search spaces. The intermediate configurations may be examined more than once in the former algorithm. In our optimistic algorithm, some intermediate configurations with smaller upper bounds need to be stored temporarily. Although storing huge amount of these intermediate configurations can be a challenge to physical memories or even hard disks, the problem is dramatically improved with the adequate subgraph decomposition method and it can be further improved by finding better pruning methods, such as finding a better lower bound or running a heuristic before the main exact algorithm starts.

6   Results on Simulated Data

ASMedian algorithm is implemented in Java and runs on a MacBook, using only one 2.16GHz CPU. Sets of data are simulated with varying parameters n and $\pi = \rho/n$, where n is the number of gene in each genome and ρ is the number of random reversals applied to the ancestor $I = 1, \ldots, n$ independently to derive each of the three different genomes. n ranges among 10, 20, 30, 40, 50, 60, 80, 100, 200, 300, 500, 1000, 2000, 5000 and π starts from 0.1 and increases by intervals of 0.1. For each data set, 10 instances are generated.

6.1   The Running Time for Simulated Data Sets with Varying n and $\pi = \rho/n$

Table 1 shows the average running time in seconds for all data sets whose 10 instances can all be solved within one hour or the number of solved instances in parenthesis for the remaining data sets. It can be seen that relatively large instances can be solved if ρ/n remains at 0.3 or less. It also shows that for small n, the median is easy to find even if ρ/n is large enough to effectively scramble the genomes.

Table 1. For each data set, if its ten instances all finish in 1 hour, then their average running time is shown in seconds; otherwise the number of finished instances is shown with parenthesis

n	ρ/n	0.1	0.2	0.3	0.4	0.5	0.6	0.7	0.8	0.9	1.0
10		$4\,10^{-4}$	$1\,10^{-4}$	$2\,10^{-4}$	$8\,10^{-4}$	$4\,10^{-4}$	$2\,10^{-3}$	$2\,10^{-3}$	$8\,10^{-4}$	$4\,10^{-4}$	$5\,10^{-4}$
20		$2\,10^{-4}$	$2\,10^{-4}$	$3\,10^{-4}$	$6\,10^{-4}$	$9\,10^{-4}$	$6\,10^{-4}$	$2\,10^{-2}$	$7\,10^{-3}$	$2\,10^{-2}$	$5\,10^{-3}$
30		$2\,10^{-3}$	$2\,10^{-3}$	$3\,10^{-4}$	$7\,10^{-4}$	$5\,10^{-3}$	$3\,10^{-2}$	$5\,10^{-2}$	$1\,10^{-1}$	$4\,10^{-1}$	1
40		$1\,10^{-4}$	$2\,10^{-4}$	$3\,10^{-4}$	$6\,10^{-4}$	$4\,10^{-2}$	1	6	$6\,10^{1}$	$6\,10^{1}$	$5\,10^{1}$
50		0	$4\,10^{-4}$	$5\,10^{-4}$	$2\,10^{-3}$	$7\,10^{-2}$	$7\,10^{1}$	(9)	(7)	(7)	
60		$2\,10^{-3}$	$1\,10^{-3}$	$5\,10^{-3}$	$3\,10^{-2}$	$5\,10^{1}$	(7)				
80		$3\,10^{-4}$	$4\,10^{-4}$	$7\,10^{-4}$	$8\,10^{-2}$	(9)					
100		$3\,10^{-4}$	$7\,10^{-4}$	$1\,10^{-3}$	$7\,10^{1}$	(1)					
200		$7\,10^{-3}$	$1\,10^{-2}$	$3\,10^{-2}$	(0)						
300		$5\,10^{-3}$	$5\,10^{-3}$	$2\,10^{-2}$	(0)						
500		$2\,10^{-2}$	$2\,10^{-2}$	$1\,10^{-1}$	(0)						
1000		$9\,10^{-2}$	$7\,10^{-2}$	$8\,10^{1}$	(0)						
2000		$9\,10^{-2}$	$3\,10^{-1}$	(3)							
5000		2	2	(0)							

Table 2. Speedup due to adequate subgraph (AS) discovery. Three genomes are generated from the identity genome with $n = 100$ by 40 random reversals. Time is measured in seconds. Runs were halted after 10 hours. AS1, AS2, AS4, AS0 are the numbers of edges in the solution median constructed consequent to the detection of adequate subgraphs of sizes 1, 2, 4 and at steps where no adequate subgraphs were found, respectively.

run	speedup factor	run time with AS	no AS	AS1	AS2	AS4	AS0
1	41,407	4.5×10^{-2}	1.9×10^{3}	53	39	8	0
2	85,702	3.0×10^{-2}	2.9×10^{3}	53	34	12	1
3	2,542	5.4×10^{0}	1.4×10^{4}	56	26	16	2
4	16,588	3.9×10^{-2}	6.5×10^{2}	58	42	0	0
5	$> 10^{6}$	5.9×10^{2}	stopped	52	41	4	3
6	199,076	6.0×10^{-3}	1.2×10^{3}	56	44	0	0
7	6,991	2.9×10^{-1}	2.1×10^{3}	54	33	12	1
8	$> 10^{6}$	4.2×10^{1}	stopped	57	38	0	5
9	1,734	8.7×10^{0}	1.5×10^{4}	65	22	8	5
10	855	2.1×10^{0}	1.8×10^{3}	52	38	8	2

6.2 The Effect of Adequate Subgraph Discovery on Speed-Up

Table 2 shows how the occurrence of adequate subgraphs can dramatically speed up the solution to the median problem, generally from more than a half an hour to a fraction of a second.

7   Conclusion

In this paper, several important properties about the adequate subgraphs of rank 3 are proved. We show that there are infinitely many adequate subgraphs, hence it is not possible to list all these subgraphs. By showing that the simple adequate subgraphs of small sizes have the largest occurrence probability on random MBGs and the algorithms of detecting them are simple and fast, it is practical and efficient to solve the median of three problem by only using simple adequate subgraphs of small sizes. This is confirmed by the dramatic speedup shown in the results on simulated data. Whether it is worth exploring simple adequate subgraphs of size 6 is not clear. It depends on many factors, such as the size of the problem (number of genes genomes contained) and the algorithms for detecting subgraphs and their implementations.

References

1. Adam, Z., Sankoff, D.: The ABCs of MGR with DCJ. Evol. Bioinform. 4, 69–74 (2008)
2. Bourque, G., Pevzner, P.: Genome-scale evolution: Reconstructing gene orders in the ancestral species. Genome Res. 12, 26–36 (2002)
3. Bryant, D.: The complexity of the breakpoint median problem. TR CRM-2579. Centre de recherches mathématiques, Université de Montréal (1998)
4. Caprara, A.: The reversal median problem. Informs J. Comput. 15, 93–113 (2003)
5. Lenne, R., Solnon, C., Stützle, T., Tannier, E., Birattari, M.: Reactive stochastic local search algorithms for the genomic median problem. In: van Hemert, J., Cotta, C. (eds.) EvoCOP 2008. LNCS, vol. 4972, pp. 266–276. Springer, Heidelberg (2008)
6. Moret, B.M.E., Siepel, A.C., Tang, J., Liu, T.: Inversion medians outperform break-point medians in phylogeny reconstruction from gene-order data. In: Guigó, R., Gusfield, D. (eds.) WABI 2002. LNCS, vol. 2452. Springer, Heidelberg (2002)
7. Pe'er, I., Shamir, R.: The median problems for breakpoints are NP-complete. the Electronic Col loquium of Computational Complexity Report, number TR98-071 (1998)
8. Sankoff, D., Blanchette, M.: Multiple genome rearrangement and breakpoint phy-logeny. J. Comput. Biol. 5, 555–570 (1998)
9. Tannier, E., Zheng, C., Sankoff, D.: Multichromosomal median and halving prob-lems. In: Crandall, K.A., Lagergren, J. (eds.) WABI 2008. LNCS (LNBI), vol. 5251. Springer, Heidelberg (2008)
10. Xu, A.W., Sankoff, D.: Decompositions of multiple breakpoint graphs and rapid exact solutions to the median problem. In: Crandall, K.A., Lagergren, J. (eds.) WABI 2008. LNCS (LNBI), vol. 5251. Springer, Heidelberg (2008)
11. Yancopoulos, S., Attie, O., Friedberg, R.: Efficient sorting of genomic permutations by translocation, inversion and block interchange. Bioinform. 21, 3340–3346 (2005)

A Phylogenetic Approach to Genetic Map Refinement

Denis Bertrand[1], Mathieu Blanchette[2], and Nadia El-Mabrouk[3]

[1] DIRO, Université de Montréal, H3C 3J7, Canada
bertrden@iro.umontreal.ca
[2] McGill Centre for Bioinformatics, McGill University, H3A 2B4, Canada
blanchem@mcb.mcgill.ca
[3] DIRO
mabrouk@iro.umontreal.ca

Abstract. Following various genetic mapping techniques conducted on different segregating populations, one or more genetic maps are obtained for a given species. However, recombination analyses and other methods for gene mapping often fail to resolve the ordering of some pairs of neighboring markers, thereby leading to sets of markers ambiguously mapped to the same position. Each individual map is thus a partial order defined on the set of markers, and can be represented as a Directed Acyclic Graph (DAG). In this paper, given a phylogenetic tree with a set of DAGs labeling each leaf (species), the goal is to infer, at each leaf, a single combined DAG that is as resolved as possible, considering the complementary information provided by individual maps, and the phylogenetic information provided by the species tree. After combining the individual maps of a leaf into a single DAG, we order incomparable markers by using two successive heuristics for minimizing two distances on the species tree: the breakpoint distance, and the Kemeny distance. We apply our algorithms to the plant species represented in the Gramene database, and we evaluate the simplified maps we obtained.

1 Introduction

Similarly to a road map indicating landmarks along a highway, a genetic map indicates the position and approximate genetic distances between markers along chromosomes. Genetic mapping using DNA markers is a key step towards the discovery of regions within genomes containing genes associated with particular quantitative traits (QTLs). This is particularly important in crops and other grasses, where the localization of markers linked to genes playing major roles in traits such as yield, quality, and disease resistance, can be harnessed for agricultural purposes [3].

In order to fulfill their purpose of locating QTLs as precisely as possible, ideal genetic maps should involve as many markers as possible, evenly distributed over the chromosomes, and provide precise orders and distances between markers. In reality, recombination analysis, physical imaging and the other methods used for

C.E. Nelson and S. Vialette (Eds.): RECOMB-CG 2008, LNBI 5267, pp. 198–210, 2008.

genetic mapping only give an approximate evaluation of genetic distances between markers, and often fail to order some pairs of neighboring markers, leading to partial orders, with sets of incomparable markers, that is, set of markers affected to the same locus. Moreover, to identify a specific marker locus, one requires polymorphisms at that locus in the considered population. As different populations do not contain polymorphisms for all the desired loci, the different genetic maps obtained for the same species on the basis of different segregating populations generally contain different markers. However, as long as some common markers are used, individual maps can be combined into a single one.

Various approaches have been considered to integrate different maps of a single species. As genetic distances are poorly comparable between maps, a standard approach has been to reduce each map to the underlying partial order between markers. This simplification allows representing a map as a Directed Acyclic Graph (DAG), where nodes correspond to markers, and paths between nodes to the ordering information [18]. Combining DAGs from different maps may lead to cycles, corresponding to conflicts (two markers A and B that are ordered $A \to B$ in one map, and $B \to A$ in another). Different approaches have been considered to cope with such conflicts. In [18], a DAG is recovered by simply "condensing" the subgraph corresponding to a maximum subset of "conflicting" vertices into a single vertex. In [6,7] the authors find a median by removing a minimum number of conflicts.

In contrast to the work that has been done for combining information of different maps of a single species, no similar effort has been expended to improve the markers' partial order information on one species on the basis of the genetic information of related species. In this context, the only comparative genomic study for genetic mapping is the one that we have conducted [1] for linearizing a DAG representing the map of a given species, with respect to a related species for which a total order of markers is known. In the context of computing the rearrangement distance between two maps, a more general study has been conducted by Zheng et. al [19] for inferring the minimal sequence of reversals transforming one DAG into another. Another study by the same authors [20] has considered the problem of reconstructing synthenic blocks between two gene maps by eliminating as few noisy markers as possible.

In this paper, starting from a species tree and a set of DAGs (individual maps) labeling each leaf (species), the goal is to infer, at each leaf, a single combined DAG that is as resolved as possible, considering the complementary information provided by individual maps, and the phylogenetic information provided by the species tree. Ideally (assuming sufficient complementary information between maps and sufficient phylogenetic information), a complete linearization of the DAGs is desirable. However, as ideal situations are rarely encountered, we will consider the more restricted, but more biologically relevant problem, of integrating maps and reducing pairs of incomparable and conflicting markers of each DAG, given the phylogenetic signal.

We proceed as follows. After combining the individual maps of a leaf into a single DAG by using a method similar to that of Yap et al. [18], we resolve incomparable pairs of markers by using two successive heuristics for minimizing two distances on the species tree: the breakpoint distance, and the Kemeny distance [10] defined as the total number of pairwise ordering conflicts over the branches of the tree.

The second heuristic is based on the previous work of Ma *et.al* [11] for reconstructing ancestral gene orders. The developed algorithm is guaranteed to identify a most parsimonious scenario for the history of each incomparable pair of markers, although it provides no guarantee as to the optimality of the global solution. The paper is organized as follows. We introduce all concepts and notations in Section 2. We then describe our methodology in Section 3, and present our two heuristics in Section 4. In Section 5, we apply our method to the Gramene database [8] and evaluate the simplified maps we obtain.

2   Gene Order Data and Representation as Graphs

Experimental methods used for genetic mapping give rise to individual maps, generally represented by lines upon which are placed individual loci (Figure 1, Map1 and Map2). Each locus represents the position of a specific marker that might appear at several positions in the genome. However, in this paper, we

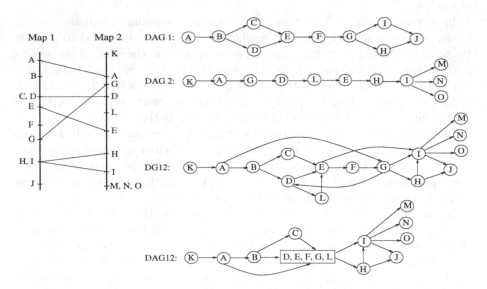

Fig. 1. Modeling maps as Directed Acyclic Graphs (DAGs). Left: maps as they are represented in a gene map database such as Gramene. Map1 and Map2 correspond to two mapping studies of the same chromosome. Letters correspond to markers placed at positions proportional to their distance from each other. Common markers between the two maps are linked. Right, from up to bottom: DAG representing Map1; DAG representing Map2; Directed Graph (DG) corresponding to the union of DAG1 and DAG2; DAG corresponding to map integration, after simplifying the strongly connected components.

assume that each marker exhibits a single polymorphism along the genome, allowing to treat the concept of marker and locus synonymously. In other words, marker duplications are not allowed.

Modeling a map as a DAG. Following the notations of [18], maps may be represented as Directed Acyclic Graphs (DAGs), where each marker is represented by a vertex, and each pair of adjacent markers are connected by an edge (Figure 1, DAG1 and DAG2). Often due to the lack of recombination between two loci, a number of different markers may appear at the same position on the map (for example, markers C and D in Map1). It follows that a single marker may be connected to a set of other markers.

Such DAGs represent partial orders between markers. Two markers A and B are *comparable* iff there is a directed path from A to B (in which case we write $A < B$) or from B to A (we write $A > B$), and *incomparable* otherwise. A *conflict* between two maps is a pair of markers A and B that are ordered $A < B$ on one map, and $A > B$ on the other. The *Kemeny distance* [10] between two DAGs (or partial orders) is the number of conflicts between them. For example, in Map1 (Figure 1), (A, C) is a pair of comparable markers ($A < C$, or similarly $C > A$) and (C, D) is a pair of incomparable markers. Moreover, the Kemeny distance between Map1 and Map2 is 20 since markers D, E, F, G and L are in conflict with each other.

Map integration. Different studies on the same species conducted on different populations give rise to different maps involving different markers. As long as some identical markers are shared between studies, maps can be merged to a Directed Graph (DG) with a single connected component, by performing the union of the individual maps [18]. More precisely, let $\mathcal{D}_1, \mathcal{D}_2, \ldots, \mathcal{D}_n$ be n DAGs corresponding to n maps, and $\mathcal{M}$ be the set of markers represented in at least one $\mathcal{D}_i$, for $1 \le i \le n$. Then the *union DG* is the directed graph $\mathcal{G}$ defined as follows: a vertex is in $\mathcal{G}$ iff it is in at least one $\mathcal{D}_i$, and an edge is in $\mathcal{G}$ iff it is in at least one $\mathcal{D}_i$, for $1 \le i \le n$ (Figure 1, DG12).

Due to conflicts between maps, such union DG may contain cycles (for example, (D, L, E, F, G) is a cycle in DG12). Markers involved in such cycles cannot be ordered relative to each other without yielding a contradiction. Two main approaches have been used in the literature to cope with cycles.

1. A *Strongly Connected Component* (SCC) of a DG $\mathcal{G}$ refers to a maximum subset $\mathcal{V}$ of vertices of $\mathcal{G}$ such that, for each $(v_1, v_2) \in \mathcal{V}^2$, there is a directed path in $\mathcal{G}$ from v_1 to v_2 and from v_2 to v_1. For example, $\{D, E, F, G, L\}$ is an SCC in DG12. A number of very efficient algorithms are able to find the SCCs in a graph [12,16]. This yields to the possibility of simplifying a DG to a DAG by "condensing" the subgraph that comprises an SCC into a single vertex (DAG12 in Figure 1). Markers belonging to such a vertex are considered pairwise incomparable.

2. Based on the hypothesis that conflicts are due to mapping errors, Jackson *et.al* [7] considered the problem of inferring a consensus map leading to a minimum number of such errors. Their method is based on finding a median

order for the Kemeny distance, which is an NP-hard problem [17]. They
proved that inferring a median order according to this distance is equivalent
to finding an acyclic subgraph of minimum weight in a weighted directed
graph (i.e. a minimum feedback arc set), and designed an exact algorithm
and a heuristic to solve it.

3   Methodology

Given a phylogenetic tree T for a set of n species and a set of DAGs (set of
individual maps) at each leaf of T, our goal is to produce a single DAG at
each leaf of T that is as resolved as possible considering the shared information
between maps and the phylogenetic information provided by T.

In the rest of this paper, *resolving a pair of incomparable markers* will refer
to fixing an order between the two markers, and *simplifying a DAG* will refer to
resolving a number of pairs of incomparable markers in the DAG.

3.1   Integrating Maps

We integrate the set of DAGs labeling each leaf of T into a single DAG as
follows. We first construct the DAG's union DG as described in Section 2. Then,
in contrast to [7], we do not try to solve *conflicts of the union DG*, that is the
pairs of markers involved in a cycle of the DG, at this stage. Rather, conflicts
are reduced to SCCs, as in [18], and resolving such SCCs is delayed to the next
phase considering the phylogenetic information of the species tree.

3.2   Marker Content of Internal Nodes

We would like to account for the phylogenetic information represented by the
species tree T. Considering a most parsimonious model of evolution, the goal
is to infer marker orders that minimize a given distance on T. Preliminary to
computing any distance on T is the assignment of marker content at each internal
node. We will proceed as follows, assuming a model with no convergent evolution.
Let M be a marker, $\mathcal{L}$ be the set of leaves that contain the marker M, and v be
the node of T representing the least common ancestor of $\mathcal{L}$. Then, we assign M
to each node belonging to a path from v to an element of $\mathcal{L}$.

3.3   Minimizing an Evolutionary Distance

In contrast to gene order data, maps do not provide information on adjacencies,
but rather on relative orders between markers: an edge $A \to B$ in a DAG does not
mean that A is adjacent to B, but rather that A precedes B on the chromosome.
Indeed another DAG for the same species may contain an edge $A \to C$, leading
to two possible total orders for the three markers: $A\,B\,C$ or $A\,C\,B$. Therefore,
a classical gene order distance such as the inversion distance, or its reduction
to the breakpoint distance, is not directly applicable to such data. In this case,
a more natural distance is the number of conflicts between two maps, that is
the Kemeny distance. In the case of a species tree, the Kemeny distance can be
generalized as follows:

Definition 1. *Given a species tree T with a total order assigned to each node, the Kemeny distance on T is the sum of Kemeny distances of each pair of adjacent nodes (nodes connected by an edge of T).*

For the purpose of introducing our optimization problems, we recall the classical notion of a linear extension.

Definition 2. *Let $\mathcal{D}$ be a DAG on a set $\mathcal{M}$ of markers. A linear extension of $\mathcal{D}$ is a total order $\mathcal{O}$ of $\mathcal{M}$ such that if $A < B$ in $\mathcal{D}$ then $A < B$ in $\mathcal{O}$.*

Now, consider the following optimization problems, where "Given" should be replaced by either Kemeny, Breakpoint or Inversion:

MINIMUM-"GIVEN" LINEARIZATION PROBLEM

Given: A species tree T with a DAG at each leaf and a set of markers at each internal node;

Find: A total order at each internal node of T, and at each leaf of T, a linear extension of its DAG, minimizing the "Given" distance on T.

Notice that this problem is proved to be NP-hard for the breakpoint distance [13], for the inversion distance [2], and for the Kemeny distance [17].

The Minimum-Kemeny Linearization Problem is the one most directly applicable to partial orders. This problem is equivalent to the Minimum-Breakpoint Linearization Problem and the Minimum-Inversion Linearization Problem in the case of a marker set restricted to the same two markers $\mathcal{M} = \{A, B\}$ at each node and leaf of T. Moreover, it is equivalent to the Minimum-Inversion Linearization Problem with inversions restricted to segments of size 2 [14]. However, in the general case, a solution to the Minimum-Kemeny Linearization Problem is not guaranteed to minimize the inversion or breakpoint distance. Using this distance only allows combining the information obtained on closely related species, in case of no large genome rearrangements.

Simplifying DAGs: Following the above observations, we will present, in the next section, two algorithms aiming to simplify each leaf's DAG as follows:

1. Simplify the DAG based on the breakpoint distance. Although the resulting DAG $\mathcal{D}$ is not a total order, the developed algorithm can be seen as a heuristic for the Minimum-Breakpoint Linearization Problem, as any linear extension of $\mathcal{D}$ can be seen as a (possibly suboptimal) solution to this problem;
2. Simplify the resulting DAG based on the Kemeny distance. Similarly to the above step, the developed algorithm can be seen as a heuristic for the Minimum-Kemeny Linearization Problem.

4   Algorithms

Our two heuristics are inspired from the general methodology used by Ma *et. al* [11] for inferring ancestral gene orders, which in turn is inspired by the Fitch algorithm for substitution parsimony [4].

4.1   A Heuristic for the Minimum-Kemeny Linearization Problem

Considering the assumption of no convergent mutation, the Fitch algorithm infers the DNA sequences at the internal nodes of a phylogenetic tree based on the DNA sequences at the leaves [4]. The sequences are treated site-by-site. Although nucleotide assignment is not unique, any assignment gives an evolutionary history with the minimum number of substitutions.

A similar idea has been considered in [11] for inferring ancestral gene orders on the basis of minimizing the number of breakpoints (or maximizing the number of adjacencies). The Ma *et. al* algorithm [11] proceeds in two steps. First, using a bottom-up traversal, it determines the potential adjacencies of each individual gene. This step results in a graph at each internal node, potentially with cycles. Then, in a top-down traversal, the information obtained on a node's parent is used to simplify the node's graph. The whole algorithm is guaranteed to identify a most-parsimonious scenario for the history of each individual adjacency. However, in contrast to the case of DNA sequences for which individual nucleotides are independent, adjacencies are not, and thus the whole-genome prediction is not guaranteed to minimize the number of breakpoints.

As DAGs provide information on relative orders between markers, rather than immediate adjacencies, we aim at inferring the ancestral order ($A < B$ or $B < A$) for each pair of markers (A, B). The *Kemeny-Simplification* algorithm described in Figure 2 is guaranteed to identify a most-parsimonious scenario for the history of each individual pair of markers. However, as pairs of markers are not independent, this does not guarantee the optimality of whole-map predictions.

Algorithm Kemeny-Simplification (T)
1.   In a bottom-up traversal of T,
For each internal node v of T *do*
   For each pair (A, B) of markers of $\mathcal{M}_v^2$ *do*
     If $A < B$ (resp. $A > B$) in both children of v *then*
        Set $A < B$ (resp. $A > B$) in v;
     Else If $A > B$ in one child and $A < B$ in the other, *then*
        Set (A, B) incomparable in v;
     Else If $A < B$ (resp. $A > B$) in one child and incomparable in the other, *then*
        Set $A < B$ (resp. $A > B$) in v;
     Else If (A, B) are incomparable in both children of v *then*
        Set (A, B) incomparable in v;

2.   In a top-down traversal of T,
For each node v of T that is not the root *do*
   For each pair (A, B) of markers of $\mathcal{M}_v^2$ *do*
     If (A, B) are incomparable in v but ordered $A < B$ (resp. $A > B$) in v's parent *then*
        Set $A < B$ (resp. $A > B$) in v;

Fig. 2. For each node v, $\mathcal{M}_v$ is the marker content at v

During the second step of the algorithm (top-down traversal), conflicts may be created. For example, let A, B, C be three markers such that $A > B$ and (A, C) and (B, C) are incomparable. Resolving this two pairs by $B > C$ and $C > A$ results in transforming the comparable pair (A, B) into a conflicting pair. This may lead to a loss of order information at the leaves of T. To avoid this problem, we weight each order between two markers based on the number of times it appears in all species. Then for each leaf v, orders between pairs of markers in the parent of v are sorted according to their weight, and added successively in the DAG of v if they do not create a conflict.

4.2 A Heuristic for the Minimum-Breakpoint Linearization Problem

As the Kemeny distance is not guaranteed to provide a good evaluation of the evolutionary distance in the case of large inversions, before applying the *Kemeny-simplification* algorithm on T, we first simplify DAGs by using a heuristic for the Minimum-Breakpoint Linearization Problem. This heuristic is based on the third step of the Ma *et al.* [11] algorithm aiming to recover a "partially linearized" gene order at a particular node of the tree. This step proceeds by first weighting each edge by an estimate of the likelihood of its presence in the ancestor, and then choosing adjacency paths of maximum weight.

Based on this idea, we develop the *Breakpoint-Simplification* algorithm that proceeds as follows:

For each leaf v of T,

1. Convert v's DAG into an *extended DG* $\mathcal{G}$ (possibly containing cycles) as follows: (1) expand each vertex corresponding to an SCC into the set of vertices of this SCC; (2) add an edge between each vertex connected to an SCC and each vertex of this SCC; (3) add an edge between each pair of markers that are potentially adjacent in a linear extension of the DAG (i.e. between all incomparable markers). For example, in DAG12 of Figure 1, the SCC $\mathcal{S} = \{D, E, F, G, L\}$ is replaced by five vertices labeled D, E, F, G, L; each pair of vertices (X, Y) belonging to $\{B, C\} \times \mathcal{S}$, $\mathcal{S} \times \{I, H\}$ and $\mathcal{S}^2$ is connected by an edge.

2. Weight each edge (A, B) of $\mathcal{G}$ by an estimate $w(A, B)$ of the probability of having B following A in the species f. This estimate is computed as follows:

$$w(A, B) = \frac{\sum_{i=i_1}^{i_k} \frac{1}{ADJ(A,i)}}{n}$$

where $i_1, \ldots, i_k$ represent the k leaves of T (including v) containing (A, B) as an edge in their corresponding extended DG, and $ADJ(A, i)$, for $i_1 \leq i \leq i_k$, is the number of edges adjacent to A in the extended DG of i. Recall that n is to the number of leaves (species) of T.

3. Construct a set of paths of maximum weight that cover all nodes of $\mathcal{G}$. This problem is known to be NP-hard [5] and we propose a simple greedy heuristic

to resolve it. Our heuristic proceeds by sorting all the edges of $\mathcal{G}$ by weight, and then adding them in order to a new graph, initially restricted to the set of vertices of $\mathcal{G}$ and no edges, until each vertex has a unique predecessor and successor.

4. Incorporate the obtained set of adjacency paths into the original v's DAG. This is done by applying the heuristic that we have developed in [1] for simplifying a DAG with respect to a given total order. In our case, the heuristic is applied successively to the total order represented by each adjacency path.

4.3   The General Method

In summary, our methodology can be subdivided into three main steps:

- **Step 1.** Perform map integration at each leaf of T;
- **Step 2.** Apply the Breakpoint-Simplification algorithm on T;
- **Step 3.** Apply the Kemeny-Simplification algorithm on T.

In the following section, we will analyse the efficiency of each step of the general method.

5   Experiments on the Gramene Database

Gramene [8] is an important comparative genomics mapping database for crop grasses. It uses the completely sequenced rice genome to organize information on maize, sorghum, wheat, barley, and other gramineae (see Figure 3 for a phylogenetic tree of the species present in Gramene, excluding rice). It provides curated information on genetic and genomic datasets related to maps, markers, genes, genomes and quantitative trait loci, as well as invaluable tools for map comparison.

Correlating information from one map to another and from one species to another requires to have common markers, i.e. markers that are highly polymorphic among several populations. Such markers, also called "anchor markers" are typically SSRs (Simple Sequence Repeats, or microsatellites) or RFLPs (Restriction Fragment Length Polymorphism). In our study, we selected exclusively RFLP markers, as they appeared to be the most shared among all crop species present in Gramene, and thus those most likely to gain additional order information following a phylogenetic analysis. Moreover, they represent the largest family of DNA markers present in Gramene (17,715 different markers).

In order to consider only non-duplicated markers, we select, in each species, those appearing at a single locus. Moreover, as only markers shared between species may gain additional order information from a phylogenetic study, we further restrict ourselves, for each species s, to the set of "valid markers" defined as follows: a *valid marker* in s is a non-duplicated marker in s that appears as a non-duplicated marker in at least one other species.

Figure 3 gives the distribution of total and valid RFLPs among species, and also the number of incomparable and conflicting (markers involved in a cycle)

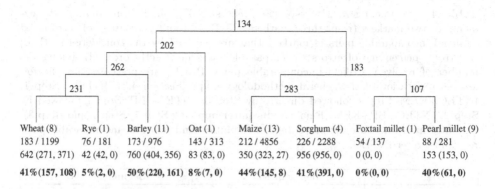

Fig. 3. Species included in the Gramene database (excluding the rice genome), with the phylogeny provided by [9]. Each internal node is labeled by the cardinality of its marker set. Labels of each leaf are defined from line 1 to line 4 as follows: (1) the species name followed, in brackets, by the number of map sets used in our study (each map set involves one map for each chromosome); (2) the number of valid RFLPs followed by the total number of RFLPs; (3) the total number of incomparable and conflicting pairs of markers in the union DG resulting from map integration (in brackets, the number of incomparable pairs, followed by the number of conflicting pairs); (4) the percentage of resolved incomparable and conflicting pairs of markers (in brackets, the number of resolved incomparable pairs , followed by the number of resolved conflicting pairs).

pairs of markers in the union DG obtained after the first step of of map integration. The total set of incomparable and conflicting pairs are those we hope to resolve following a phylogenetic analysis.

Results of applying our methodology (Section 4.3) to the Gramene database are given in the last line of leaf labels in Figure 3. The percentage of resolved incomparable and conflicting pairs of markers is given, followed in brackets by the actual number of resolved pairs. Overall, for species with a number of map set greater than one, the resolution rate ranges from 40% to 50%.

Results evaluation. To test the efficiency of our methodology, we perform the following experiments. We randomly choose 50 segments of two or three adjacent genes, each from a randomly chosen genetic map; the markers of each segment are made incomparable. We then apply our methodology, and check the percentage of incomparable pairs correctly resolved after each step (Section 4.3). This process is repeated 500 times.

Results are presented in Table 1. Performing the union of individual maps allows the integration, in a single map, of the complementary information interspersed in these maps. As conflicts between individual maps are usually due to mapping errors rather than to real rearrangement events that would have affected one particular population, they are expected to be rare. This observation is confirmed by our results. Indeed, the step of integrating maps (INTEG in Table 1) allows to resolve a large proportion of incomparable pairs, with high resolution power ($\sim$ 2% errors).

Table 1. "Segment size 2" (resp. "Segment size 3"): simulations done with segments of two markers (resp. three markers); % Resolution: percentage of introduced artificial incomparable pairs of markers that are resolved by the considered method; % Errors: percentage of errors (incomparable pairs incorrectly ordered) among the number of resolved artificial incomparable pairs; Results are presented for the following application of the general methodology steps (Section 4.3). INTEG: Step 1; INTEG+KEM: Step 1 followed directly by Step 3; INTEG+BP: Step 1 followed by Step 2; INTEG+BP+KEM: Final results (after applying Step 1, Step 2 and Step 3). Numbers in brackets are the percentage of resolution and error, for incomparable pairs remaining after INTEG.

	Segment size 2				Segment size 3			
	% Resolution		% Errors		% Resolution		% Errors	
INTEG	36.7		2.3		37.4		2.6	
INTEG+KEM	51.8	(15.1)	12.1	(36.0)	52.9	(15.6)	11.7	(34.0)
INTEG+BP	48.5	(11.8)	9.0	(30.0)	50.1	(12.7)	8.7	(26.7)
INTEG+BP+KEM	54.4	(17.7)	11.5	(30.6)	54.9	(17.5)	10.6	(27.5)

Following this step, the Kemeny-Simplification algorithm (KEM) has a higher resolution rate than the Breakpoint-Simplification algorithm (BP), but with a lower level of efficiency ($\sim$ 12% errors for KEM, versus 9% for BP). Applying the complete methodology (BP followed by KEM) leads to a good compromise. However, it should be noted that less confidence should be given to incomparable pairs resolved from the phylogenetic information in comparison to those resolved from combining individual maps of a given species. This is indicated by the percentage of error ($\sim$ 30%) for incomparable pairs remaining after step INTEG (number in brackets in Table 1).

6   Conclusion

This paper is a first effort towards accounting for the phylogenetic information of a species tree to increase the resolution of genetic maps. The main assumption is that individual maps of one species may gain additional order information by considering the complementary information obtained from closely related species. In the case of species that are close enough to preserve a high degree of gene order conservation, minimizing the Kemeny distance on the species tree is an appropriate way of increasing the resolution of individual maps. However, the Kemeny distance is not appropriate anymore for species that have diverged from each other by large rearrangement events. In this case, using a genomic rearrangement measure (e.g. inversions or breakpoints) is more appropriate. Based on this idea, we have designed a two-step methodology: resolve a number of incomparable markers by considering a rearrangement distance (namely the breakpoint distance), and then increase the resolution rate by considering the Kemeny distance.

Another more accurate heuristic for the Minimum-Breakpoint Linearization Problem may be designed by using a Median Branch-and-Bound approach, similar to the one developed for inferring ancestral gene orders of a species tree [15]. The general idea would be to begin with an arbitrary order at each internal node of the species tree, and then, in a bottom-up traversal, consider each triplet, and improve the order of the median by minimizing the breakpoint or inversion distance. However, as leaves are labeled by partial orders, instead of a linear-time algorithm for computing the breakpoint distance between two orders, an exponential-time algorithm, as the one that we have developed in [1], would be needed for computing a distance between a partial and a total order. The resulting complete heuristic is therefore likely to be intractable for reasonably large datasets. Moreover, as the number of possible solutions is likely to be huge, evaluating the obtained resolutions may be much more difficult.

Results obtained on the Gramene database are encouraging, as a high level of resolution is reached. However, our preliminary simulations performed to evaluate the method reveal a lack of specificity. These simulations may be improved, for example by removing an individual map and checking whether the order information it contains can be recovered by our methodology. Additional work should also be done to improve the various steps of the methodology, and better adapt it to the gramineae species.

References

1. Blin, G., Blais, E., Hermelin, D., Guillon, P., Blanchette, M., El-Mabrouk, N.: Gene maps linearization using genomic rearrangement distances. Journal of Computational Biology 14(4), 394–407 (2007)
2. Caprara, A.: The reversal median problem. Journal on Computing 15(1), 93–113 (2003)
3. Collard, B.C.Y., Jahufer, M.Z.Z., Brouwer, J.B., Pang, E.C.K.: An introduction to markers, quantitative trait loci (QTL) mapping and marker-assisted selection for crop improvement: The basic concepts. Euphytica 142, 169–196 (2005)
4. Fitch, W.M.: Toward defining the course of evolution: Minimum change for a specific tree topology. Systematic Zoology 20, 406–416 (1971)
5. Garey, M.R., Johnson, D.S.: Computers and Intractability: A Guide to the Theory of NP-Completeness. Freeman, San Francisco (1979)
6. Jackson, B.N., Aluru, S., Schnable, P.S.: Consensus genetic maps: a graph theoretic approach. In: IEEE Computational Systems Bioinformatics Conference (CSB 2005), pp. 35–43 (2005)
7. Jackson, B.N., Schnable, P.S., Aluru, S.: Consensus genetic maps as median orders from inconsistent sources. IEEE/ACM Transactions on Computational Biology and Bioinformatics 5(2), 161–171 (2008)
8. Jaiswal, P., et al.: Gramene: a bird's eye view of cereal genomes. Nucleic Acids Research 34, D717–D723 (2006)
9. Kellogg, E.A.: Relationships of cereal crops and other grasses. Proceedings of the National Academy of Sciences of USA 95(5), 2005–2010 (1998)
10. Kemeny, J.P.: Mathematics without numbers. Daedelus 88, 577–591 (1959)
11. Ma, J., Zhang, L., Suh, B.B., Raney, B.J., Burhans, R.C., Kent, W.J., Blanchette, M., Haussler, D., Miller, W.: Reconstructing contiguous regions of an ancestral genome. Genome Research 16(12), 1557–1565 (2006)

12. Nuutila, E., Soisalon-Soininen, E.: On finding the strongly connected components in a directed graph. Information Processing Letters 49, 9–14 (1993)
13. Pe'er, I., Shamir, R.: The median problems for breakpoints are NP-complete. In: Electronic Colloquium on Computational Complexity (ECCC), Report 71 (1998)
14. Saari, D., Merlin, V.: A geometric examination of Kemeny's rule. Social Choice and Welfare 7, 81–90 (2000)
15. Sankoff, D., Blanchette, M.: Multiple genome rearrangement and breakpoint phylogeny. Journal of Computational Biology 5, 555–570 (1998)
16. Tarjan, R.E.: Depth-first search and linear graph algorithms. SIAM Journal of Computing 1(2), 146–160 (1972)
17. Wakabayashi, Y.: The complexity of computing medians of relations. Resenhas 3, 323–349 (1998)
18. Yap, I.V., Schneider, D., Kleinberg, J., Matthews, D., Cartinhour, S., McCouch, S.R.: A graph-theoretic approach to comparing and integrating genetic, physical and sequence-based maps. Genetics 165, 2235–2247 (2003)
19. Zheng, C., Lenert, A., Sankoff, D.: Reversal distance for partially ordered genomes. Bioinformatics 21(supp. 1), 502–508 (2005)
20. Zheng, C., Zhu, Q., Sankoff, D.: Removing noise and ambiguities from comparative maps in rearrangement analysis. IEEE/ACM Transactions on Computational Biology and Bioinformatics 4(4), 515–522 (2007)

Sorting Cancer Karyotypes by Elementary Operations

Michal Ozery-Flato and Ron Shamir

School of Computer Science, Tel-Aviv University, Tel Aviv 69978, Israel
{ozery,rshamir}@post.tau.ac.il

Abstract. Since the discovery of the "Philadelphia chromosome" in chronic myelogenous leukemia in 1960, there is an ongoing intensive research of chromosomal aberrations in cancer. These aberrations, which result in abnormally structured genomes, became a hallmark of cancer. Many studies give evidence to the connection between chromosomal alterations and aberrant genes involved in the carcinogenesis process. An important problem in the analysis of cancer genomes, is inferring the history of events leading to the observed aberrations. Cancer genomes are usually described in form of *karyotypes*, which present the global changes in the genomes' structure. In this study, we propose a mathematical framework for analyzing chromosomal aberrations in cancer karyotypes. We introduce the problem of sorting karyotypes by elementary operations, which seeks for a shortest sequence of elementary chromosomal events transforming a normal karyotype into a given (abnormal) cancerous karyotype. Under certain assumptions, we prove a lower bound for the elementary distance, and present a polynomial-time 3-approximation algorithm. We applied our algorithm to karyotypes from the Mitelman database, which records cancer karyotypes reported in the scientific literature. Approximately 94% of the karyotypes in the database, totalling 57,252 karyotypes, supported our assumptions, and each of them was subjected to our algorithm. Remarkably, even though the algorithm is only guaranteed to generate a 3-approximation, it produced a sequence whose length matches the lower bound (and hence optimal) in 99.9% of the tested karyotypes.

Introduction

Cancer is a genetic disease, caused by genomic mutations leading to the aberrant function of genes. Those mutations ultimately give cancer cells their proliferative nature. Hence, inferring the evolution of these mutations is an important problem in the research of cancer. Chromosomal mutations that shuffle/delete/duplicate large genomic fragments are common in cancer. Many methods for detection of chromosomal mutations use chromosome painting techniques, such as G-banding, to achieve a visualization of cancer cell genomes. The description of the observed genome organization is called a *karyotype* (see Fig. 1). In a karyotype, each chromosome is partitioned into continuous genomic regions called *bands*, and the total number of bands is the *banding resolution*. Over the last

C.E. Nelson and S. Vialette (Eds.): RECOMB-CG 2008, LNBI 5267, pp. 211–225, 2008.

decades, a large amount of data has been accumulated on cancer karyotypes. One of the largest depositories of cancer karyotypes is the Mitelman database of chromosomal aberrations in cancer [9], which records cancer karyotypes reported in the scientific literature. These karyotypes are described using the ISCN nomenclature [8], and thus can be parsed automatically.

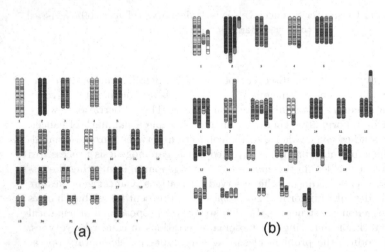

(a)   (b)

Fig. 1. A schematic view of two real karyotypes: a normal female karyotype (a), and the karyotype of MCF-7 breast cancer cell-line (b) [1]. In the normal karyotype, all chromosomes, except X and Y, appear in two identical copies, and each chromosome has a distinct single color. In the cancer karyotype presented here, only chromosomes 11,14, and 21 show no chromosomal aberrations.

Cancer karyotypes exhibit a wide range of chromosomal aberrations. The common classification of these aberrations categorizes them into a variety of specific types, such as translocations, iso-chromosomes, etc. Inferring the evolution of cancer karyotypes using this wide vocabulary of complex alteration patterns is a difficult task. Nevertheless, the entire spectrum of chromosomal alterations can essentially be spanned by four elementary operations: breakage, fusion, duplication, and deletion (Fig. 2). A *breakage*, formally known as a "double strand break", cuts a chromosomal fragment into two. A *fusion* ligates two chromosomal fragments into one. Genomic breakages, which occur quite frequently in our body cells, are normally repaired by the corresponding inverse fusion. Mis-repair of genomic breakages is believed to be a major cause of chromosomal aberrations in cancer [4]. Other prevalent chromosomal alterations in cancer genomes are *duplications* and *deletions* of chromosomal fragments. These four elementary events play a significant role in carcinogenesis: fusions and duplications can activate oncogenes, while breakages and deletions can eliminate tumor suppressor genes.

Based on the four elementary operations presented above, we introduce a new model for analyzing chromosomal aberrations in cancer. We study the problem of finding a shortest sequence of operations that transforms a normal karyotype

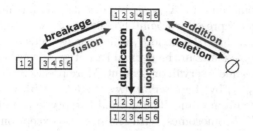

Fig. 2. Illustrations of elementary operations: breakage, fusion, duplication, and deletion. The inverse elementary operations are: fusion, breakage, c-deletion, and addition.

into a given cancer karyotype. This is the problem of *karyotype sorting by elementary operations* (KS), and the length of a shortest sequence is the *elementary distance* between the normal and cancer karyotypes. The elementary distance indicates how far, in terms of number of operations, a cancer karyotype is from the normal one, and is *not* a metric in the mathematical sense. The elementary distance corresponds to the complexity of the cancer karyotype, which may give an indication of the tumor phase [6]. The reconstructed elementary operations can be used to detect common events for a set of cancer karyotypes, and thus point out to genomic regions suspect of containing genes associated with carcinogenesis.

Under certain assumptions, which are supported by most cancer karyotypes, the KS problem can be reduced in linear time to a simpler problem, called RKS. For the latter problem we prove a lower bound for the elementary distance, and present a polynomial-time 3-approximation algorithm. We show that approximately 94% of the karyotypes in the Mitelman database (57,252) support our assumptions, and each of these was subjected to our algorithm. Remarkably, even though the algorithm is only guaranteed to generate a 3-approximation, it produced a sequence whose length matches the lower bound (and hence optimal) in 99.9% of the tested karyotypes. Manual inspection of the remaining cases reveals that the computed sequence for each of these cases is also optimal.

The paper is organized as follows. In Section 1 we give the combinatorial formulation of the KS problem and its reduced variant RKS. In the rest of the paper we focus on the RKS problem. In Section 2 we prove a lower bound for the elementary distance for RKS. Section 3 describes our 3-approximation algorithm for RKS. Finally, in Section 4 we present the results of the application of our algorithm to the karyotypes in the Mitelman database. Due to space limits, most proofs are omitted.

1 Problem Formulation

1.1 The KS Problem

The KS problem receives two karyotypes as an input: K_{normal}, and the cancer karyotype, K_{cancer}. We represent each of the two karyotypes by a multi-set of chromosomes. Every chromosome in K_{normal} is presented as an interval of B integers, where each integer represents a *band*. For simplicity we assume that all the chromosomes in K_{normal} share the same B, which corresponds to the banding resolution. Every two chromosomes in the normal karyotype are either identical, i.e. are represented by the same interval, or disjoint. More precisely, we represent every chromosome in K_{normal} by the interval $[(k-1)B+1, kB]$, where k is an integer that identifies the chromosome. The normal karyotype usually contains exactly two copies of each chromosomes, with the possible exception of the sex chromosomes. Every chromosome in K_{cancer} is either a fragment or a concatenation of several fragments, where a *fragment* is a maximal sub-interval, with two bands or more, of a chromosome in the normal karyotype. More formally, a fragment is a maximal interval of the karyotype of the form $[i, j] \equiv [i, i+1, \ldots, j]$, or $[j, i] \equiv [j, j-1, \ldots, i]$, where $i < j$, $i, j \in \{(k-1)B+1, \ldots, kB\}$, and $[(k-1)B+1, kB] \in K_{\text{normal}}$. Note that in particular, a chromosome in K_{cancer} can be identical to a chromosome in K_{normal}. We use the symbol "::" to denote a concatenation between two fragments, e.g., $[i, j] :: [i', j']$. Every chromosome, in both K_{normal} and K_{cancer}, is orientation-less, i.e. reversing the order of the fragments, along with their orientation, results in an equivalent chromosome. For example, $X = [i, j] :: [i', j'] \equiv [j', i'] :: [j, i] = \overline{X}$.

We refer to the concatenation point of two intervals as an *adjacency* if the union of their intervals is equivalent to a larger interval in K_{normal}. In other words, two concatenated intervals that form an adjacency can be replaced by one equivalent interval. For example, the concatenation point in $[5, 3] :: [3, 1] \equiv [5, 1]$ is an adjacency. Typically, a breakage occurs within a band, and each of the resulting fragments contains a visible piece of this broken band. For example, if $[5, 1]$ is broken within band 3, then the resulting fragments are generally denoted the by $[5, 3]$ and $[3, 1]$. For this reason, we do *not* consider the concatenation $[5, 3] :: [2, 1]$ as an adjacency. A concatenation point that is *not* an adjacency, is called a *breakpoint*[1]. Further examples of concatenation points that are breakpoints are: $[1, 3] :: [5, 6]$ and $[2, 4] :: [4, 3]$.

We assume that the cancer karyotype, K_{cancer}, has evolved from the normal karyotype, K_{normal}, by the following four *elementary operations*:

I.   **Fusion.** A concatenation of two chromosomes, X_1 and X_2, into one chromosome $X_1 :: X_2$.

II. **Breakage.** A split of a chromosome into two chromosomes. A split can occur within a fragment, or between two previously concatenated fragments, i.e. in a breakpoint. In the former case, where the break is in a fragment $[i, j]$, the fragment is split into two complementing fragments: $[i, k]$ and $[k, j]$, where $k \in \{i+1, i+2, \ldots, j-1\}$.

III. **Duplication.** A whole chromosome is duplicated, resulting in two identical copies of the original chromosome.

IV. **Deletion.** A complete chromosome is deleted from the karyotype.

---

[1] Formally, since the broken ends of a chromosome are not considered breakpoints here, the term "fusion-point" may seem more appropriate. However, we kept the name "breakpoint" due to its prior use and brevity.

Given K_{normal} and K_{cancer}, we define the KS problem as finding a short-est sequence of elementary operations that transforms K_{normal} into K_{cancer}. An equivalent formulation of the KS problem is obtained by considering the inverse direction: find a shortest sequence of *inverse* elementary operations that trans-forms K_{cancer} into K_{normal}. Clearly, fusion and breakage operations are inverse to each other. The inverse to a duplication is a *constrained deletion* (abbreviated *c-deletion*), where the deleted chromosome is one of two or more identical copies. In other words, a c-deletion can delete a chromosome only if there exists another identical copy of it. The inverse of a deletion is an *addition* of a chromosome. Note that in general, the added chromosome need not be a duplicate of an ex-isting chromosome and can contain any number of fragments. For the rest of the paper, we analyze KS by sorting in reverse order, i.e. starting from K_{cancer} and going back to K_{normal}. The sorting sequences will also start from K_{cancer}.

1.2   Reducing KS to RKS

In this section we present a basic analysis of KS, which together with two ad-ditional assumptions, allows the reduction of KS to a simpler variant in which no breakpoint exists (RKS). As we shall see, our assumptions are supported by most analyzed cancer karyotypes.

We start with several definitions. A sequence of inverse elementary operations is *sorting*, if its application to K_{cancer} results in K_{normal}. We shall refer to a shortest sorting sequence as *optimal*. Since every fragment contains two or more bands, we can present any band i within it by an ordered pair of its two ends, i^0, which is the end closer to the minimal band in the fragment, and i^1, the end closer to the maximal band in the fragment. More formally, we map the fragment $[i, j]$, $i \neq j$, to $[i^1, j^0] \equiv [i^1, (i + 1)^0, (i + 1)^1, \ldots, j^0]$ if $i < j$, and oth-erwise to $[i^0, j^1] \equiv [i^0, (i - 1)^1, (i - 1)^0, \ldots, j^1]$. We say that two fragment-ends, a and a', are *complementing* if $\{a, a'\} = \{i^0, i^1\}$. The notion of viewing bands as ordered pairs is conceptually similar to considering genes / synteny blocks as oriented, as is standard in the computational studies of genome rearrangements in evolution [3]. In this study we consider bands as ordered pairs to well identify breakpoints: as mentioned previously, a breakage usually occurs within a band, say i, and the two ends of i, i^0 and i^1, are separated between the two new result-ing fragments. Thus, a fusion of two fragment-ends forms an adjacency iff these ends are complementing. We identify a breakpoint, and a concatenation point in general, by the two corresponding fragment-ends that are fused together. More formally, the concatenation point in $[a, b]::[a', b']$ is identified by the (unordered) pair $\{b, a'\}$. For example, the breakpoint in $[1, 2] :: [4, 3] \equiv [1^1, 2^0] :: [4^0, 3^1]$ is identified by $\{2^0, 4^0\}$. Note that the two other fragment-ends, 1 and 3, do not matter for that breakpoint's identity. Having defined breakpoint identities, we refer to a breakpoint as *unique* if no other breakpoint shares its identity, and otherwise we call it *recurrent*. In particular, a breakpoint in a non-unique chro-mosome (i.e., a chromosome with another identical copy) is recurrent. Last, we say that a chromosome X is *complex* if it contains at least one breakpoint, and *simple* otherwise. In other words, chromosome X is simple if it consists of one fragment. Analogously, an addition is *complex* if the chromosome added is complex, and *simple* otherwise.

Observation 1. *Let S be an optimal sorting sequence. Suppose K_{cancer} contains a breakpoint, p, that is not involved in a c-deletion in S. Then there exists an optimal sorting sequence S', in which the first operation is a breakage of p.*

Proof. Since K_{normal} does not contain any breakpoint, p must be eventually eliminated by S. A breakpoint can be eliminated either by a breakage or by a c-deletion. Since p is not involved in a c-deletion, p is necessarily eliminated by a breakage. Moreover, this breakage can be moved to the beginning of S since no other operation preceding it involves p.   □

Corollary 1. *Let S be an optimal sorting sequence. Suppose S contains an addition of chromosome $X = f_1::f_2:: \ldots ::f_k$, where $f_1, f_2, \ldots, f_k$ are fragments, and none of the $k-1$ breakpoints in X is involved in any subsequent c-deletion in S. Then the sequence S', obtained from S by replacing the addition of X with the additions of $f_1, f_2, \ldots, f_k$ (overall k additions), is an optimal sorting sequence.*

Proof. By Observation 1, the breakpoints in X can be immediately broken after its addition. Thus replacing the addition of X, and the $k-1$ breakages following it, by k additions of $f_1, f_2, \ldots, f_k$, yields an optimal sorting sequence.   □

It appears that complex additions, as opposed to simple additions, makes KS very difficult to analyze. Moreover, based on Corollary 1, complex additions can be truly beneficial only in complex scenarios in which c-deletions involve recurrent breakpoints that were formerly created by complex additions. An analysis of a large collection of cancer karyotypes reveals that only 6% of the karyotypes contain recurrent breakpoints (see Section 4). Therefore, for the rest of this paper, we make the following assumption:

Assumption 1. *Every addition is simple, i.e., every added chromosome consists of one fragment.*

Using the assumption above, the following observation holds:

Observation 2. *Let p be a unique breakpoint in K_{cancer}. Then there exists an optimal sorting sequence in which the first operation is a breakage of p.*

Proof. If p is not involved in a c-deletion, then by Observation 1, p can be broken immediately. Suppose there are k c-deletions involving p or other breakpoints identical to it. Note that following Assumption 1, from the four inverse elementary operations, only fusion can create a new breakpoint. Now, any c-deletion involving p requires another fusion that creates a breakpoint p', identical to p. Thus we can obtain an optimal sorting sequence, S', from S, by: *(i)* first breaking p, *(ii)* not creating any breakpoint point p' identical to p, *(iii)* replacing any c-deletion involving p, or one of its copies, with two c-deletions of the corresponding 4 unfused chromosomes, and *(iii)* not having to break the last instance of p (since it was already broken). In summary, we moved the breakage of p to the beginning of the sorting sequence and replaced k fusions and k c-deletions (i.e. 2k operations) with 2k c-deletions.   □

Observation 3. *In an optimal sequence, every fusion creates either an adjacency, or a recurrent breakpoint.*

Proof. Let S be an optimal sorting sequence. Suppose S contains a fusion that creates a new unique breakpoint p. Then, following Observation 2, p can be immediately broken after it was formed, a contradiction to the optimality of S. □

In this work, we choose to focus on karyotypes that do not contain recurrent breakpoints. According to our analysis of the Mitelman database, 94% of the karyotypes satisfy this condition. Thus we make the following additional assumption:

Assumption 2. *The cancer karyotype, K_{cancer}, does not contain any recurrent breakpoint.*

Assumption 2 implies that *(i)* we can immediately break all the breakpoints in K_{cancer} (due to Observation 2), and *(ii)* consider fusions only if they create an adjacency (due to Observation 3). Hence, for each unique normal chromosome, its fragments can be used separated from all the other fragments and used to solve a simpler variant of KS: In this variant, *(i)* $K_{normal} = \{[1, B] \times N\}$, *(ii)* there are no breakpoints in K_{cancer}, and *(iii)* neither fusions, nor additions, form breakpoints. Usually, $N = 2$. Exceptions are $N = 1$ for the sex chromosomes, and $N > 2$ for cases of global changes in the ploidy. We refer to this reduced problem as RKS (restricted KS). For the rest of the paper, we shall limit our analysis to RKS only.

2 A Lower Bound for the Elementary Distance

In this section we analyze RKS and define several combinatorial parameters that affect the elementary distance between K_{normal} and K_{cancer}, denoted by $d \equiv d(K_{normal}, K_{cancer})$. Based on these parameters, we prove a lower bound on the elementary distance. Though theoretically our lower bound is not tight, we shall demonstrate in Section 4 that in practice, for the vast majority (99.9%) of the real cancer karyotypes analyzed, the elementary distance to the appropriate normal karyotype achieves this bound.

2.1 Extending the Karyotypes

For the simplicity of later analysis, we extend both K_{normal} and K_{cancer} by adding each $2N$ "tail" intervals:

$$\widehat{K}_{normal} = K_{normal} \cup \{[0, 1] \times N, [B, B + 1] \times N\}$$

$$\widehat{K}_{cancer} = K_{cancer} \cup \{[0, 1] \times N, [B, B + 1] \times N\}$$

These new "tail" intervals do not take part in elementary operations: breakage and fusion are still limited to $\{2, 3, \ldots, B - 1\}$, and intervals added/c-deleted are contained in $[1, B]$. Hence $d(K_{normal}, K_{cancer}) \equiv d(\widehat{K}_{cancer}, \widehat{K}_{cancer})$. Their only role is to simplify the definitions of parameters given below.

2.2   The Histogram

We define the *histogram* of $\widehat{K}_{cancer}$, $H \equiv H(\widehat{K}_{cancer}) : \{[i-1,i] \mid i = 1,2,\ldots,B+1\} \rightarrow \mathbb{N} \cup \{0\}$, as follows. Let $H([i-1,i])$ be the number of fragments in $\widehat{K}_{cancer}$ that contain the interval $[i-1,i]$. See Fig. 3(b) for an example. From the definition of $\widehat{K}_{cancer}$, it follows that $H([0,1]) = H([B,B+1]) = N$. For simplicity we refer to $H([i-1,i])$ as $H(i)$. The histogram H has a *wall* at $i \in \{1,\ldots,B\}$ if $H(i) \neq H(i+1)$. If $H(i+1) > H(i)$ (respectively, $< H(i)$) then the wall at i is called a *positive* wall (respectively, a *negative* wall). Intuitively, a wall is a vertical boundary of H. We define w to be the total size of walls in H. More formally,

$$w = \sum_{i=1}^{B} |H(i+1) - H(i)|$$

Since $H(1) = H(B+1) = N$, the total size of positive walls is equal to the total size of negative walls, and hence w is even. Note that if $\widehat{K}_{cancer} = \widehat{K}_{normal}$ then $w = 0$. The pair $(i,h) \equiv (i,[h-1,h])$, $h \in \mathbb{N}$, is a *brick* in the wall at i if $H(i)+1 \leq h \leq H(i+1)$ or $H(i+1)+1 \leq h \leq H(i)$. A brick (i,h) is *positive* (respectively, *negative*) if the wall at i is positive (respectively, negative). Note that the number of bricks in a wall is equal to its total size. Hence w corresponds to the total number of bricks in H.

Observation 4. *For a breakage/fusion,* $\Delta w = 0$; *For a c-deletion/addition,* $\Delta w = \{-2,0,2\}$.

2.3   Counting Complementing End Pairs

Consider the case where $w = 0$. Then there are no gains and no losses of bands, and the number of fragments in $\widehat{K}_{cancer}$ is greater or equal to the number of fragments in $\widehat{K}_{normal}$. Note that each of the four elementary operations can decrease the total number of fragments by at most one. Hence when $w = 0$, an optimal sorting sequence would be to fuse pairs of complementing fragment-ends, not including the tails. Let us define $f \equiv f(\widehat{K}_{cancer})$ as the maximum number of disjoint pairs of complementing fragment-ends. Note there could be many alternative choices of complementing pairs. Nevertheless, any maximal disjoint pairing is also maximum. It follows that if $w = 0$, then $d(\widehat{K}_{normal}, \widehat{K}_{cancer}) = f - 2N$. Also, when $w \neq 0$, a c-deletion may need to be preceded by some fusions of complementing ends, to form two identical fragments. In general, the following holds:

Observation 5. *For breakage* $\Delta f = 1$; *For fusion,* $\Delta f = -1$; *For c-deletion,* $\Delta f \in \{0,-1,-2\}$; *For addition,* $\Delta f \in \{0,1,2\}$.

Lemma 1. *For breakage/addition,* $\Delta(w/2 + f) = 1$; *For fusion/c-deletion,* $\Delta(w/2 + f) = -1$.

2.4 Simple Bricks

A brick (i, h) is *simple* if: *(i)* $(i, h - 1)$ is not a brick, and *(ii)* $\widehat{K}_{cancer}$ does not contain a pair of complementing fragment-ends in i. Thus, in particular, a simple brick cannot be eliminated by a c-deletion. On the other hand, for a non-simple brick, (i, h), there are two fragments ending in the corresponding location (i.e. i). Nevertheless, it may still be impossible to eliminate (i, h) by a c-deletion if these two fragments are not identical. We define $s \equiv s(\widehat{K}_{cancer})$ as the number of simple bricks.

Observation 6. *For breakage, $\Delta s \in \{0, -1\}$; For fusion, $\Delta s \in \{0, 1\}$; For c-deletion, $\Delta s = 0$; For addition, $|\Delta s| \leq 2$.*

2.5 The Weighted Bipartite Graph of Bricks

We now define the last parameter in the lower bound formula for the elementary distance. It is based upon matching pairs of bricks, where one is positive and the other is negative. Note that in the process of sorting $\widehat{K}_{cancer}$, the histogram is flattened, i.e., all bricks are eliminated, which can be done only by using c-deletion/addition operations. Observe that if a c-deletion/addition eliminates a pair of bricks, then one of these bricks is positive and the other is negative. Thus, roughly speaking, every sorting sequence defines a matching between pairs of positive and negative bricks that are eliminated together.

Given two bricks, $v = (i, h)$ and $v' = (i', h')$, we write $v < v'$ (resp. $v = v'$) if $i < i'$ (resp. $i = i'$). Let V^+ and V^- be the sets of positive and negative bricks respectively. We say that v and v' have the same *sign*, if either $v, v' \in V^+$, or $v, v' \in V^-$. Two bricks have the same *status* if they are either both simple, or both non-simple. Let $BG = (V^+, V^-, \delta)$ be the weighted complete bipartite graph, where $\delta : V^+ \times V^- \to \{0, 1, 2\}$ is an edge-weight function defined as follows. Let $v^+ \in V^+$ and $v^- \in V^-$ then:

$$\delta(v^+, v^-) = \begin{cases} 0 & v^+ \text{ and } v^- \text{ are both simple and } v^- < v^+ \\ 0 & v^+ \text{ and } v^- \text{ are both non-simple and } v^+ < v^- \\ 1 & v^+ \text{ and } v^- \text{ have opposite status} \\ 2 & \text{otherwise} \end{cases}$$

For an illustration of BG see Fig 3(c). A *matching* is a set of vertex-disjoint edges from $V^+ \times V^-$. A matching is *perfect* if it covers all the vertices in BG (recall that $|V^+| = |V^-|$). Thus a perfect matching is in particular a maximum matching. Given a matching M, we define $\delta(M)$ as the total weight of its edges. Let $m \equiv m(\widehat{K}_{cancer})$ denote the minimum weight of a perfect matching in BG. The problem of finding a minimum-weight perfect matching can be solved in polynomial-time (see, e.g., [7, Theorem 11.1]). We note there exists a simple efficient algorithm for computing m, which relies heavily on the specific weighting scheme, δ.

Let K' be obtained from K by an elementary operation (a move). For a function F defined on karyotypes, define $\Delta(F) = F(K') - F(K)$.

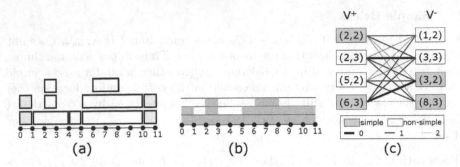

Fig. 3. An example of an (extended) cancer karyotype $\widehat{K}_{\text{cancer}}$ and its combi-natorial parameters. (a) The (extended) cancer karyotype is $\widehat{K}_{\text{cancer}} = \{[0, 1] \times 2, [1, 4], [4, 5], [5, 10] \times 2, [10, 11] \times 2, [2, 3] \times 2, [6, 8]\}$. Here $N = 2$, $B = 10$. The number of disjoint pairs of complementing fragment-ends, f, is 5. (b) The histogram $H \equiv H(\widehat{K}_{\text{cancer}})$, which has walls at 1, 2, 3, 5, 6, and 8. There are four positive bricks: $(2, 2)$, $(2, 3)$, $(5, 2)$, and $(6, 3)$, and four negative bricks: $(1, 2)$, $(3, 3)$, $(3, 2)$, and $(8, 3)$. Hence $w = 8$. Four of the eight bricks are simple: $(2, 2)$, $(3, 2)$, $(6, 3)$, and $(8, 3)$, thus $s = 4$. (c) The weighted-bipartite graph of BG. It is not hard to verify that $M = \{ ((2, 3), (3, 3)), ((6, 3), (3, 2)), ((2, 2), (1, 2)), ((5, 2), (8, 3)) \}$ is a minimum-weight perfect matching and hence $m = 2$.

Lemma 2. $d \geq w/2 + f - 2N + s + m \geq 0$.

3   The 3-Approximation Algorithm

Algorithm 1 below is a polynomial procedure for the RKS problem. We shall prove that it is a 3-approximation, and then describe a heuristic that aims to improve it.

Lemma 3. *Algorithm 1 transforms $\widehat{K}_{normal}$ into $\widehat{K}_{normal}$ using at most $3w/2 + f - 2N + s + m$ inverse elementary operations.*

Theorem 1. *Algorithm 1 is a polynomial-time 3-approximation algorithm for RKS.*

Note that the same result applies to multi-chromosomal karyotypes, by summing the bounds for the RKS problem on each chromosome. Note also that the results above imply also that $d \in [w/2 + f - 2N + s + m, 3w/2 + f - 2N + s + m]$.

We now present Procedure 2, a heuristic that improves the performance of Algorithm 1, by replacing Steps 12-21. The procedure assumes that *(i)* all bricks are non-simple, and *(ii)* $v^+ < v^-$, for every $(v^+, v^-) \in M$, $v^- \in V^-$. In this case, $m = 0$, and the lower bound is reached only if no additions are made. Thus, Procedure 2 attempts to minimize the number of extra addition operations performed. For an interval I, let $L(I)$ and $R(I)$ be the left and right endpoints of I respectively.

Algorithm 1. Elementary Sorting (RKS)

1: $M \leftarrow$ a minimum-weight perfect matching in BG
2: **for all** $(v^-, v^+) \in M$ where $v^- < v^+$ **do**
3:     Add the interval $[v^-, v^+]$.
4: **end for** /* *Now $v^+ < v^-$ for every $(v^+, v^-) \in M$, where $v^+ \in V^+$, $v^- \in V^-$* */
5: **for all** $v \in V^+ \cup V^-$, v is simple, **and** $v \neq 1, B$ **do**
6:     **if** $v \in V^+$ **then**
7:         Add the interval $[1, v]$
8:     **else**
9:         Add the interval $[v, B]$
10:     **end if**
11: **end for** /* *Now $v^+ < v^-$ for every $(v^+, v^-) \in M$, where $v^+ \in V^+$, $v^- \in V^-$ **and** all the bricks are non-simple* */
12: **for all** $v^- \in V^-$, $v^- < B$ **do**
13:     Add the interval $[v^-, B]$
14: **end for** /* *Now all the bricks are non-simple, **and** $v^- = B, \forall v^- \in V^-$* */
15: **while** $V^+ \neq \emptyset$ **do**
16:     $v^+ \leftarrow \max V^+$
17:     **for all** $p > v^+$, $p < B$ **do**
18:         Fuse any pair of intervals complementing at p.
19:     **end for**
20:     C-delete an interval $[v^+, B]$
21: **end while**

---

4 Experimental Results

In this section we present the results of sorting real cancer karyotypes, using Algorithm 1, combined with the improvement heuristic in Procedure 2.

4.1 Data Preprocessing

For our analysis, we used the Mitelman database (version of February 27, 2008), which contained 56,493 cancer karyotypes, collected from 9,088 published studies. The karyotypes in the Mitelman database (henceforth, MD) are represented in the ISCN format and can be automatically parsed and analyzed by the software package CyDAS [5]. We refer to a karyotype as *valid* if it is parsed by CyDAS without any error. According to our processing, 49,622 (88%) of the records were valid karyotypes. Since some of the records contain multiple distinct karyotypes found in the same tissue, the total number of simple (valid) karyotypes that we deduced from MD is 61,137.

A karyotype may contain uncertainties, or missing data, both represented by a '?' symbol. We ignored uncertainties and deleted any chromosomal fragments that were not well defined.

4.2 Sorting the Karyotypes

Out of the 61,137 karyotypes analyzed, only 3,885 karyotypes (6%) contained recurrent breakpoints. Our analysis focused on the remaining 57,252 karyotypes.

Procedure 2. Heuristic for eliminating non-simple bricks

1: **while** $V^+ \neq \emptyset$ **do**
2:     $v^+ \leftarrow \max V^+$
3:     **for all** $p > v^+, p < B, p \notin V^-$ **do**
4:         Fuse any pair of intervals complementing at p.
5:     **end for**
6:     **if** $\exists I_1, I_2$, where $I_1 = I_2$ and $L(I_1) = v^+$, and $R(I_1) \in V^-$ **then**
7:         Let I_1, I_2 be a pair of intervals with minimal length satisfying the above.
8:         C-delete I_1
9:     **else if** $\exists I_1, I_2$, where $L(I_1) = L(I_2) = v^+$ and $R(I_1) < R(I_2) \in V^-$ **then**
10:         Let I_1, I_2 be a pair of intervals with minimal length satisfying the above.
11:         Add the interval $[R(I_1), R(I_2)]$
12:     **else**
13:         Let $u^- = \min\{v^- \in V^- | v^- > v^+\}$
14:         Add the interval $[u^-, B]$
15:     **end if**
16: **end while**

We note that 37% (21,315) of these karyotypes do not contain any breakpoint at all. (In these karyotypes, no bands that are not adjacent in normal chromosomes are fused, but some chromosome tails as well as full chromosomes may be missing or duplicated). Following our assumptions (see Section 1.2), we broke all the breakpoints in each karyotype. We then applied Algorithm 1, combined with Procedure 2, to the fragments of each of the chromosomes in these karyotypes. We used the ploidy of each karyotype, as the normal copy-number (N) of each chromosome. In 99.9% (57,223) of the analyzed karyotypes our algorithm achieved the lower-bound, and thus the produced sequences are optimal. Each of the remaining 29 karyotypes contained a chromosome for which the computed sequence was larger in 2 than the lower-bound. Manual inspection revealed that for each of these cases the elementary distance was indeed 2 above the lower bound. Hence the computed sequences were found to be optimal in 100% of the analyzed cases.

4.3 Operations Statistics

We now present statistics on the (direct) elementary operations performed by our algorithm. The 57,252 analyzed karyotypes, contained 84,601 (unique) breakpoints in total. Hence the average number of fusions (eq. breakpoints) per karyotype is approximately 1.5. The distribution of the number of breakpoints per a karyotypes, including the non-sorted karyotypes (i.e karyotypes with recurrent breakpoints), is presented in Fig. 4. The most frequent number of breakpoints after zero is 2, which may point to the prevalence of reciprocal translocations in the analyzed cancer karyotypes. Table 1 summarizes the average number of operations per sorted karyotype.

Table 1. Average number of elementary operations per (sorted) cancer karyotype

breakage	fusion	deletion	addition	all
2.4	1.5	2.6	1.1	7.6

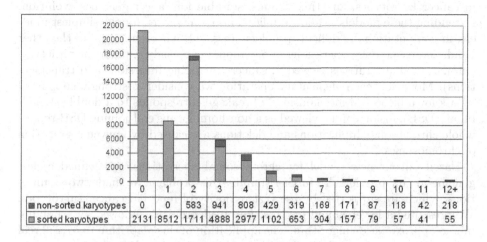

Fig. 4. The distribution of number of breakpoints (i.e. fusions of non-adjacent bands) per karyotype. "Sorted karyotypes" correspond to karyotypes with *no* recurrent breakpoints. "Non-sorted karyotypes" correspond to karyotypes with recurrent breakpoints. About 35% of all the karyotypes do not contain any breakpoint.

5 Discussion

In this paper we propose a new mathematical model for the evolution of cancer karyotypes, using four simple operations. Our model is in some sense the result of an earlier work [10], where we showed that chromosome gain and loss are dominant events in cancer. The analysis in [10] relied on a purely heuristic algorithm that reconstructed events using a wide catalog of complex rearrangement events, such as inversions, tandem-duplication etc. Here we make our first attempt to reconstruct rearrangement events in cancer karyotypes in a more rigorous, yet simplified, manner.

The fact that we model and analyze bands and karyotypes may seem out of fashion. While modern techniques today allow *in principle* detection of chromosomal aberrations in cancer at an extremely high resolution, the clinical reality is that karyotyping is still commonly used for studying cancer genomes, and to date it is the only abundant data resource for cancer genomes structure. Moreover, our framework is not limited to banding resolution as the "bands" in our model may represent any DNA blocks.

Readers familiar with the wealth of computational works on evolutionary genome rearrangements (see [3] for a review), may wonder why we have not used traditional operations, such as inversions and translocations, as has been previously done, e.g., by Raphael et al. [12]. The reason is that while inversions and translocations are believed to dominate the evolution of species, they form less than 25% of the rearrangement events in cancer karyotypes [10], and 15% in malignant solid tumors in particular. The extant models for genome rearrangements do not cope with duplications and losses, which are frequently observed in cancer karyotypes, and thus are not suitable for cancer genomes evolution. Extending these models to allow duplications results, even for the simplest models, in computationally difficult problems (e.g. [11, Theorem 10]). On the other hand, the elementary operations in our model can easily explain the variety of chromosomal aberrations viewed in cancer (including inversions and translocations). Moreover, each elementary operation we consider is strongly supported by a known biological mechanism [2]: breakage corresponds to a double-strand-break (DSB); fusion can be viewed as a non-homologous end-joining DSB-repair; whole chromosome duplications and deletions are caused by uneven segregation of chromosomes.

Based on our new model for chromosomal aberrations, we defined a new genome sorting problem. To further simplify this problem, we made two assumptions, which are supported by the vast majority of reported cancer karyotypes. We presented a lower bound for this simplified problem, followed by a polynomial 3-approximation algorithm. The application of this algorithm to 57,252 real cancer karyotypes yielded solutions that achieve the lower bound (and hence an optimal solution) in almost all cases (99.9%). This is probably due to the relative simplicity of reported karyotypes, especially after removing ones with repeated breakpoints (cf. Fig. 4).

In the future, we would like to extend this preliminary work by weakening our assumptions in a way that will allow the analysis of the remaining non-analyzed karyotypes (6% of the data), which due to their complexity, are likely to correspond to more advanced stages of cancer. Our hope is that this study will lead to further algorithmic research on chromosomal aberrations, and thus help in gaining more insight on the ways in which cancer evolves.

Acknowledgements

We thank the referees for their careful and critical comments. This study was supported in part by the Israeli Science Foundation (grant 385/06).

References

1. NCI and NCBI's SKY/M-FISH and CGH Database (2001),
   http://www.ncbi.nlm.nih.gov/sky/skyweb.cgi
2. Albertson, D.G., Collins, C., McCormick, F., Gray, J.W.: Chromosome aberrations in solid tumors. Nature Genetics 34, 369–376 (2003)
3. Bourque, G., Zhang, L.: Models and methods in comparative genomics. Advances in Computers 68, 60–105 (2006)

4. Ferguson, D.O., Frederick, W.A.: DNA double strand break repair and chromosomal translocation: Lessons from animal models. Oncogene 20(40), 5572–5579 (2001)
5. Hiller, B., Bradtke, J., Balz, H., Rieder, H.: CyDAS: a cytogenetic data analysis system. BioInformatics 21(7), 1282–1283 (2005), http://www.cydas.org
6. Höglund, M., Frigyesi, A., Säll, T., Gisselsson, D., Mitelman, F.: Statistical behavior of complex cancer karyotypes. Genes, Chromosomes and Cancer 42(4), 327–341 (2005)
7. Korte, B., Vygen, J.: Combinatorial optimization: theory and algorithms. Springer, Berlin (2002)
8. Mitelman, F. (ed.): ISCN: An International System for Human Cytogenetic Nomenclature. S. Karger, Basel (1995)
9. Mitelman, F., Johansson, B., Mertens, F. (eds.): Mitelman Database of Chromosome Aberrations in Cancer (2008), http://cgap.nci.nih.gov/Chromosomes/Mitelman
10. Ozery-Flato, M., Shamir, R.: On the frequency of genome rearrangement events in cancer karyotypes. In: The first annual RECOMB satellite workshop on computational cancer biology (2007)
11. Radcliffe, A.J., Scott, A.D., Wilmer, E.L.: Reversals and transpositions over finite alphabets. SIAM J. Discret. Math. 19(1), 224–244 (2005)
12. Raphael, B.J., Volik, S., Collins, C., Pevzner, P.: Reconstructing tumor genome architectures. Bioinformatics 27, 162–171 (2003)

On Computing the Breakpoint Reuse Rate in Rearrangement Scenarios

Anne Bergeron[1], Julia Mixtacki[2], and Jens Stoye[3]

[1] Dépt. d'informatique, Université du Québec à Montréal, Canada
bergeron.anne@uqam.ca
[2] International NRW Graduate School in Bioinformatics and Genome Research,
Universität Bielefeld, Germany
julia.mixtacki@uni-bielefeld.de
[3] Technische Fakultät, Universität Bielefeld, Germany
stoye@techfak.uni-bielefeld.de

Abstract. In the past years, many combinatorial arguments have been made to support the theory that mammalian genome rearrangement scenarios rely heavily on breakpoint reuse. Different models of genome rearrangements have been suggested, from the classical set of operations that include inversions, translocations, fusions and fissions, to more elaborate models that include transpositions. Here we show that the current definition of *breakpoint reuse rate* is based on assumptions that are seldom true for mammalian genomes, and propose a new approach to compute this parameter. We explore the formal properties of this new measure and apply these results to the human-mouse genome comparison. We show that the reuse rate is intimately linked to a particular rearrangement scenario, and that the reuse rate can vary from 0.89 to 1.51 for scenarios of the same length that transform the mouse genome into the human genome, where a rate of 1 indicates no reuse at all.

1 Introduction

There has been ample evidence, since the birth of modern genetics, that the genomes of species have been often reshuffled, with large chunks of genetic material being moved [13]. Rearrangements such as inversions within one chromosome, translocations, fusions, and fissions of chromosomes have been regularly observed through the comparison of the genomes of close species. In the past two decades it became possible to compare more distant species, and the problem of explaining how to transform one genome into another with a sequence of rearrangements became a central problem of computational biology [11].

Computing the minimal number of inversions, translocations, fusions and fissions necessary to transform one genome into another is a solved and well understood mathematical problem [4,5,6,8,14]. A sequence of rearrangements that achieves this minimum is called a *parsimonious sorting scenario*.

The *breakpoint reuse rate* [1,9,12] measures how often regions of chromosomes are *broken* in these rearrangement scenarios. This parameter has traditionally a

C.E. Nelson and S. Vialette (Eds.): RECOMB-CG 2008, LNBI 5267, pp. 226–240, 2008.

value between 1 and 2. Low or high values of the reuse rate have been used in the literature to support different models of distributions of breaks along chromosomes [9,12]. However, in this paper, we show that the reuse rate is extremely sensitive to both genome representation, and the rearrangement model.

In [9], the reuse rate is defined as a function of the length of rearrangement scenarios, but not the particular rearrangement operations used in it. Moreover, its computation is based on the assumptions that the compared genomes have the same number of chromosomes, and that they share the same set of telomere markers. In order to compute the reuse rate in the comparison of genomes that do not fit in this model, such as the mouse and human genomes, 'empty' chromosomes and additional telomere markers are added as necessary.

In this paper, we first argue that this approach necessarily yields abnormally high values of breakpoint reuse, and we then suggest a new approach to compute the reuse rate that uses all the information contained in a particular rearrangement scenario. We give lower and upper bounds for this measure, demonstrating that, depending on the particular scenario, a wide range of reuse rates can be inferred for the same data-set. Finally we apply our results to the human-mouse data-set studied in [9] and give a particular rearrangement scenario where the lower bound is actually achieved.

2   Preliminaries

Whole genomes are compared by identifying homologous segments along their DNA sequences, called *blocks*. These blocks can be relatively small, such as gene coding sequences, or very large fragments of chromosomes. The order and orientation of the blocks may vary in different genomes. In this paper, we assume that the genomes are compared with sufficiently large blocks such that each block occurs once and only once in each genome. For genomes that have linear chromosomes, a simple representation is to list the blocks in each chromosome, using minus signs to model relative orientation, such as in the following example:

Genome S: ○ -2 -4 -5 -1 7 3 ○   ○ 8 6 10 9 11 12 ○
Genome T: ○ 1 2 3 4 5 ○   ○ 6 7 8 9 ○   ○ 10 11 12 ○

Here, genome S has two chromosomes, and genome T has three. The unsigned symbol '○' is used to mark ends of chromosomes. An *adjacency* in a genome is a sequence of two consecutive blocks, or a block and a '○'. For example, in the above genomes, $(-2\ -4)$ is an adjacency of genome S, and $(5\ ○)$ is an adjacency, also called a *telomere*, of genome T. Since a whole chromosome can be flipped, we always have $(a\ b) = (-b\ -a)$, and $(○\ a) = (-a\ ○)$. Chromosomes can be represented by the set of their adjacencies.

Two genomes are said to be *co-tailed* if they have the same set of telomeres. Genomes that have only circular chromosomes are always co-tailed, since they do not have telomeres. Co-tailed genomes whose chromosomes are all linear have always the same number of chromosomes.

The *adjacency graph* of two genomes A and B is a graph whose vertices are the adjacencies of A and B, respectively called A-vertices and B-vertices, and such that for each block b there is an edge between adjacency $(b\ c)$ in genome A and $(b\ c')$ in genome B, and an edge between $(a\ b)$ in genome A, and $(a'\ b)$ in genome B. For example, the adjacency graph of genomes S and T from above is shown in Figure 1.

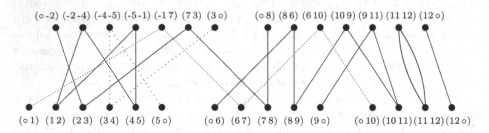

Fig. 1. The adjacency graph of genomes S and T. The S-vertices are on the top of the figure, and the T-vertices are on the bottom.

Since each vertex has at most two incident edges, the adjacency graph can be decomposed into connected components that are either cycles or paths. We classify paths as AA-paths, BB-paths and AB-paths depending on the type of vertices, A or B, of its extremities. For example, the adjacency graph of genomes S and T has the following connected components, described by the list of their adjacencies in genome T:

>   Two cycles, [(1 2)(4 5)] and [(11 12)].
>   One SS-path, [(2 3)(7 8)].
>   Two TT-paths, [(○ 1)(6 7)(○ 10)] and [(○ 6)(8 9)(10 11)(9 ○)].
>   Two ST-paths, [(3 4)(5 ○)] and [(12 ○)].

When two genomes share an adjacency, such as (11 12) or (12 ○) for genomes S and T, then they belong to cycles of length 2 or paths of length 1 in the adjacency graph. We have:

Proposition 1 ([3]). *Two genomes are equal if and only if their adjacency graph has only cycles of length 2 and paths of length 1.*

We will refer to cycles of length more than 2, and paths of length more than 1, as *long* cycles and paths.

A *double-cut-and-join* (DCJ) rearrangement operation [3,16] on genome A acts on two adjacencies $(a\ b)$ and $(c\ d)$ to produce either $(a\ d)$ and $(c\ b)$, or $(a\ -c)$ and $(-b\ d)$. When a, b, c and d are all blocks, the DCJ operation makes two cuts in genome A; these correspond, for linear chromosomes, to *inversions*, *translocations*, and *circularizations* when the two cuts act on the same chromosome and produce a circular chromosome. When one of a, b, c or d is a telomere, the operation makes one cut; these are special cases of inversions, translocations and circularizations that we will refer to as *semi*-operations. When there are two telomeres, the operation makes no cut, and it corresponds to a *fusion* of chromosomes, its inverse operation is a *fission* that makes one cut in a chromosome.

For a fixed set of rearrangement operations, the *distance* between genomes A and B is the minimum number of operations needed to rearrange – or *sort* – genome A into genome B. When all DCJ operations are allowed, we will denote this distance by $d_{DCJ}(A, B)$. The DCJ distance is easily computed from the adjacency graph [3]. It is given by:

$$d_{DCJ}(A, B) = N - (C + I/2) \tag{1}$$

where N is the number of blocks, C is the number of cycles, and I is the number of AB-paths. For example, the DCJ distance between our genomes S and T is given by $12 - (2 + 2/2) = 9$.

Rearrangement operations that do not create circular chromosomes are called *HP operations*, from the names of the authors of the first algorithm to compute the distance between genomes using these operations [5]. The corresponding distance is denoted by $d_{HP}(A, B)$. It is always greater or equal to the DCJ distance, and the difference is generally very small in real data. For example, the difference is 0 when the human and mouse genomes are compared with 281 blocks; and the difference is 1 when the human and chicken genomes are compared with 586 blocks.

A rearrangement operation is *sorting* if it lowers the distance by 1, and a sequence of sorting operations of length $d(A, B)$ is called a *parsimonious sorting scenario*. For the DCJ distance, it is easy to detect sorting operations since, by the distance formula (1), any operation that increases the number of cycles by 1, or the number of AB-paths by 2, is sorting. Note that no operation can modify simultaneously those two parameters [3].

3   Breakpoint Reuse Rate

Rearrangement scenarios have been used in the past years to assess whether some regions of chromosomes were particularly susceptible to break. The measure used to infer the existence of these regions is called the *breakpoint reuse rate*. Before formally defining this measure, it is important to note that most previous definitions relied on the assumption that *the compared genomes are co-tailed*. This assumption implies that genomes A and B have the same number of linear chromosomes, and that there are no long paths. Since all adjacencies are in cycles or short paths, it also implies that the adjacency graph has the same number of A-vertices and B-vertices.

Let b be the number of A- or B-vertices that are in long cycles, then the *breakpoint reuse rate r* for co-tailed genomes A and B is traditionally defined as:

$$r = 2d(A, B)/b.$$

This definition reflects the fact that, with co-tailed genomes, the only sorting operations are inversions and translocations that must make two *cuts* in the chromosomes since all adjacencies are in cycles.

A well known result from sorting theory [2,5,7] asserts that at least $\ell - 1$ inversions or translocations are needed in order to sort a cycle of length 2ℓ. We thus have the following lower bound on the breakpoint reuse rate for co-tailed genomes:

Proposition 2. *If the adjacency graph of genomes A and B has c long cycles of total length $L_c = 2\ell_1 + \ldots + 2\ell_c$ and no long paths, then the breakpoint reuse rate is at least*
$$2 - 4c/L_c.$$

Proof. The number of A- or B-vertices that are in long cycles is $b = \ell_1 + \ldots + \ell_c = L_c/2$, and the distance $d(A, B) \geq (\ell_1 - 1) + \ldots + (\ell_c - 1) = L_c/2 - c$, implying that

$$
\begin{aligned}
r = 2d(A,B)/b &\geq 2(L_c/2 - c)/(L_c/2) \\
&= (L_c - 2c)/(L_c/2) \\
&= 2 - 4c/L_c. \qquad \square
\end{aligned}
$$

Proposition 2 establishes a link between the number of cycles and the breakpoint reuse rate: a few long cycles imply a high reuse rate, and many shorter cycles imply a low reuse rate. This fact is crucial in evaluating the breakpoint reuse of genomes that have been artificially made co-tailed. The procedure eliminates all paths in the adjacency graph by closing them into cycles [5,14]. Unfortunately, this process transforms all rearrangement operations that make less than two cuts into operations that makes two cuts. Clearly, this can lead to an overestimation of reuse rates.

In order to obtain more realistic measures, we first extend the definition of breakpoint reuse rate to arbitrary linear genomes.

Definition 1. *Consider a rearrangement scenario that transforms genome A into genome B. Let C be the total number of cuts made by the operations of the scenario, and b the number of B-vertices that are in long cycles or paths, then the breakpoint reuse rate r is defined by:*

$$r = C/b.$$

Note that this definition corresponds to the traditional breakpoint reuse rate when genomes are co-tailed. The reuse rate also depends on a particular scenario, and it is not necessarily symmetric. Indeed, for example, if genome B differs from genome A by a fusion of two chromosomes, then no cuts are necessary to transform A into B, yielding a breakpoint reuse rate of 0, but one cut is necessary to transform B into A, yielding a breakpoint reuse rate of 0.5. This example also shows that the breakpoint reuse rate can be less than 1 for general genomes.

4   Bounding the Breakpoint Reuse Rate

Since the breakpoint reuse rate depends on a particular rearrangement scenario, it is interesting to produce lower and upper bounds that are independent of a particular parsimonious scenario. The existence of long paths in an adjacency graph yields opportunities to choose operations that use less than two cuts. On the other hand, long cycles often impose mandatory reuse, as we saw in Proposition 2. For lower bounds, the strategy will thus be to avoid operations that create long cycles in the adjacency graph, while the opposite strategy will yield upper bounds. In Section 5, we will show that these bounds are effectively reached with real data.

4.1   Lower Bounds

We already saw that the number of cuts necessary to sort a cycle of length 2ℓ is at least $2(\ell - 1)$, and there is no possibility to lower this number. In fact, in the DCJ model this bound is tight, while in the HP model certain sets of paths or cycles may require extra operations to be sorted [4,5].

The situation is very different for paths. Consider first an AB-path of length $2\ell+1$ that begins with a telomere g_1 of genome B and ends with a telomere $g_{2\ell+1}$ of genome A, where we list all adjacencies of the path, with those of genome A underlined:

$$[(\circ \ g_1)\underline{(g_2 \ g_1)}(g_2 \ g_3)\cdots\underline{(g_{2\ell} \ g_{2\ell-1})}(g_{2\ell} \ g_{2\ell+1})\underline{(\circ \ g_{2\ell+1})}]$$

Consider the DCJ operation that acts on the adjacencies $(g_{2\ell} \ g_{2\ell-1})$ and $(\circ \ g_{2\ell+1})$ to produce the adjacency $(g_{2\ell} \ g_{2\ell+1})$ of genome B. This operation is sorting since it creates a cycle of length 2, and shortens the AB-path by 2. It also requires only one cut. We thus have:

Proposition 3. *The minimum number of cuts necessary to sort an AB-path of length $2\ell + 1$ is ℓ.*

Proof. Applied iteratively, the above strategy clearly sorts the path with ℓ cuts. It thus remains to prove that it is impossible to reduce the number of cuts below ℓ. However, the path contains ℓ adjacencies $(g_{2i} \ g_{2i-1})$ of A, $1 \le i \le \ell$, that are not adjacencies of B. Removing these requires at least ℓ cuts. □

Note that, in order to have a sorting sequence that minimizes the number of cuts, it is necessary to create the adjacency of genome B that is next to the genome A telomere. Any other sorting DCJ operation on the AB-path will increase the number of cuts.

In the case of a BB-path of length 2ℓ, we have more choices. Indeed, any path of the form

$$[(\circ \ g_1)\underline{(g_2 \ g_1)}(g_2 \ g_3)\cdots(g_{2\ell-2} \ g_{2\ell-1})\underline{(g_{2\ell} \ g_{2\ell-1})}(g_{2\ell} \ \circ)]$$

can be cut anywhere between adjacencies of the form $(g_{2k} \ g_{2k-1})$. This is always a sorting operation since it creates two new AB-paths of lengths $2k + 1$ and $2\ell - (2k + 1)$. Using Proposition 3, we easily have:

Proposition 4. *The minimum number of cuts necessary to sort a BB-path of length 2ℓ is ℓ.*

Finally, AA-paths can be sorted with a minimum number of cuts using the following procedure. Consider an AA-path of the form

$$[(g_1 \circ)(g_1 \ g_2)(g_3 \ g_2)\cdots(g_{2\ell-1} \ g_{2\ell-2})(g_{2\ell-1} \ g_{2\ell})(\circ \ g_{2\ell})].$$

Creating the adjacency $(g_1 \ g_2)$ of genome B is possible by acting on the adjacencies $(g_1 \circ)$ and $(g_3 \ g_2)$ of genome A. It is sorting since it creates a cycle, and it requires only one cut. The same holds for the last adjacency of genome B, $(g_{2\ell-1} \ g_{2\ell})$. Note that if the length of the path is 2, then no cut is necessary and we just do a fusion. Thus we have:

Proposition 5. *The minimum number of cuts necessary to sort an AA-path of length 2ℓ is $\ell - 1$.*

When the full spectrum of DCJ operations is allowed, it is always possible to construct a sorting sequence that achieves the minimum number of cuts, but this is not guaranteed when we restrict the operations to HP operations, as the following example shows. Let genomes E and F be the following:

Genome E: $\circ$ -1 -2 3 $\circ$
Genome F: $\circ$ 1 2 3 $\circ$

The adjacency graph has two EF-paths, one of length 1, and one of length 5. Thus the minimal number of cuts is 2. However, creating the adjacency 1 2 in genome F requires the creation of a circular chromosome, which is forbidden in the HP model. Thus the only allowable operation is the inversion of block number 2, which requires two cuts, followed by the inversion of block number 1, which requires one cut.

4.2   Upper Bounds

In order to compute the maximal number of cuts, we need to reverse the strategy of the preceding section and create cycles as large as possible.

Since sorting operations that act on an AB-path must split the path into a cycle and an AB-path, we choose to construct the largest possible cycle on a path of length $2\ell + 1$, splitting it in a path of length 1, and a cycle of length 2ℓ. We thus have:

Proposition 6. *The maximum number of cuts necessary to sort an AB-path of length $2\ell + 1$ is $2\ell - 1$.*

A BB-path of length 2ℓ can always be sorted with $2\ell - 1$ cuts, by creating a cycle of length $2\ell - 2$, requiring 2 cuts, and a path of length 2 that can be sorted with a fission. When there are only AA-paths in the adjacency graph, then the only sorting operations that can be applied to those must create a cycle, since splitting the path always yields two AA-paths. Creating a cycle can be easily done by a fusion of the two telomeres of genome A that are at the ends of the path. This creates a cycle of length 2ℓ, thus requiring $2(\ell - 1)$ cuts. On the other hand, it is possible to do better, in terms of maximizing the number of cuts, by pairing AA-paths with BB-paths in the following sense.

Consider an AA-path of length $2\ell_a$, and a BB-path of length $2\ell_b$:

$$[(g_1 \circ)(g_1\ g_2)(g_3\ g_2)\cdots(g_{2\ell_a-1}\ g_{2\ell_a-2})(g_{2\ell_a-1}\ g_{2\ell_a})(\circ\ g_{2\ell_a})]$$

$$[(\circ\ h_1)(h_2\ h_1)(h_2\ h_3)\cdots(h_{2\ell_b-2}\ h_{2\ell_b-1})(h_{2\ell_b}\ h_{2\ell_b-1})(h_{2\ell_b}\ \circ)]$$

The strategy is to act on the adjacencies $(h_2\ h_1)$ and $(g_1\ \circ)$ of genome A to produce $(\circ\ h_1)$ and $(h_2\ -g_1)$, resulting in two AB-paths, one of length 1 that corresponds to the new common telomere $(\circ\ h_1)$, and the other of length $2\ell_a + 2\ell_b - 1$. Using Proposition 6, the total number of cuts of this strategy would be $(2\ell_a - 1) + (2\ell_b - 1)$. We thus have:

Proposition 7. *The maximum number of cuts necessary to sort a BB-path of length 2ℓ is $2\ell - 1$. The maximum number of cuts necessary to sort an AA-path of length 2ℓ is $2\ell - 1$, when it can be paired with a BB-path, otherwise it is $2\ell - 2$.*

Again, the maximal values given here are for the DCJ model. In the HP model, the actual maxima could be a bit higher, but this generally represents a very small fraction in real data. For the mouse and human genome presented in the next section, this cost is null.

4.3   Bounds of the Reuse Rate

Using the various bounds on the number of cuts of the preceding sections, we can now derive closed formulas for bounding the breakpoint reuse rate. We have:

Theorem 1. *Suppose that the adjacency graph of two genomes A and B contains c long cycles of total length L_c, m long AB-paths of total length L_m, p AA-paths of total length L_p, and q BB-paths of total length L_q, then the number of B-vertices that are in long cycles or paths is given by:*

$$b = (L_c + L_m + L_p + L_q + m + 2q)/2$$

and the breakpoint resuse rate r for all parsimonious DCJ sorting scenarios is bounded by

$$1 + \frac{(L_c/2 - (2c + m + p + q))}{b} \leq r \leq 2 - \frac{2c + 3m + p + 3q + \delta(p - q)}{b},$$

where $\delta = 1$ if $p > q$ and 0 otherwise. Moreover there exist sorting scenarios that meet these bounds.

Proof. We show the detailed contributions for long AB paths. All the three other cases are treated similarly.

Suppose that there are m long AB-paths of lengths $2\ell_1+1, 2\ell_2+1, \ldots, 2\ell_m+1$, and let L_m be the sum of these lengths. Then the number of B-vertices that belong to these paths is:

$$v = \sum_{i=1}^{m} (\ell_i + 1) = \frac{(L_m + m)}{2}.$$

By Proposition 3, the minimum number of cuts required to sort these paths is:

$$\sum_{i=1}^{m} \ell_i = \frac{(L_m - m)}{2} = v - m$$

and, by Proposition 6, the maximum number of cuts required is:

$$\sum_{i=1}^{m} (2\ell_i - 1) = L_m - 2m = 2v - 3m.$$

Adding similar contributions in B-vertices and cuts for cycles, AA-paths and BB-paths, we can easily derive the bounds of the statement. The fact that there exist sorting scenarios that meet the bounds is due to the fact that all DCJ operations that were used to derive the minimal and maximal number of cuts were sorting operations.   □

5   Reuse Rates in the Mouse-Human Whole Genome Comparison

We now turn to test how these various computations behave with real data. We used the same mouse-human data that was analyzed in [9] to prove that there was extensive breakpoint reuse in mammalian evolution.

The data-set compares the order and orientation of 281 synteny blocks of the mouse and human genome, and is given in Appendix 1. The mouse genome M has 20 chromosomes, and the human genome H has 23. The adjacency graph has the following characteristics (see Appendix 2):

$c = 27$ long cycles: 24 of them of length 4 and one of length 6, 8 and 10 each, giving an overall length of $L_c = 120$,
$s = 4$ short MH-paths,
$m = 12$ long MH-paths: lengths ranging from 3 to 51,
$p = 12$ MM-paths: lengths ranging from 2 to 46,
$q = 15$ HH-paths: lengths ranging from 2 to 22.

By construction, the data-set has no short cycles since blocks adjacent in both genomes are merged together. The number of adjacencies in the human genome that are in long cycles or paths is $b = 300 = 281 + 23 - 4$, and the number in the mouse genome is $297 = 281 + 20 - 4$.

The HP distance, as computed by GRIMM [15] after closing all the paths and adding empty chromosomes, is $d(M, H) = 246$. There are 300 adjacencies in both the modified human and mouse genomes being in long cycles, yielding a breakpoint reuse rate for these co-tailed genomes of $2 \cdot 246/300 = 1.64$.

The DCJ distance is also $246 = 281 - (27 + 16/2)$, since there are 27 cycles and $4 + 12 = 16$ MH-paths. Using the bounds of Theorem 1 to estimate the breakpoint reuse rates of scenarios that transform the mouse into the human genome according to the definition given in this paper, we get a reuse rate r between $1 - 33/300$ and $2 - 147/300$, or:

$$0.89 \leq r \leq 1.51.$$

Compared to the value 1.64 obtained by forcing the genomes to be co-tailed, we obtain strictly lower values of reuse rate for all possible sorting scenarios.

We also verified whether there exist scenarios using HP operations that attain the minimum number of cuts. The answer is yes, and one such scenario can be found in Appendix 3. This scenario uses the following operations:

26 inversions to sort 20 long cycles, making 52 cuts,
7 translocations to sort 7 long cycles, making 14 cuts,
15 fissions and 48 semi-operations to sort 15 HH paths, making 63 cuts,
12 fusions and 57 semi-operations to sort 12 MM paths, making 57 cuts,
81 semi-operations to sort 12 long MH paths, making 81 cuts.

6 Conclusion

In this paper, we generalized the notion of breakpoint reuse rate to genomes that are not necessarily co-tailed. This new measure can be applied to circular or linear genomes, or even to genomes that have both types of chromosomes. We gave lower and upper bounds for the number of cuts in a rearrangement scenario, yielding lower and upper bounds to the reuse rate of all possible parsimonious scenarios that transform a genome into another. We also showed that transforming genome data-sets in order to make them co-tailed can yield to an overestimation of breakpoint reuse rate.

Acknowledgments

We kindly acknowledge Glenn Tesler for providing us the original data of [10] in a user friendly format. J. S. would like to thank his children Ferdinand, Leopold and Balduin for their help verifying the correctness of the mouse-human sorting scenario by coding the genomes by integers written on 281 post-it notes. The resulting video can be found at http://www.techfak.uni-bielefeld.de/~stoye rpublications/SvH.mov.

References

1. Alekseyev, M., Pevzner, P.A.: Are there rearrangement hotspots in the human genome? PLoS Comput. Biol. 3(11), e209 (2007)
2. Bafna, V., Pevzner, P.: Genome rearrangements and sorting by reversals. SIAM J. Computing 25(2), 272–289 (1996)
3. Bergeron, A., Mixtacki, J., Stoye, J.: A unifying view of genome rearrangements. In: Bücher, P., Moret, B.M.E. (eds.) WABI 2006. LNCS (LNBI), vol. 4175, pp. 163–173. Springer, Heidelberg (2006)

4. Bergeron, A., Mixtacki, J., Stoye, J.: HP distance via Double Cut and Join distance. In: Ferragina, P., Landau, G.M. (eds.) CPM 2008. LNCS, vol. 5029, pp. 56–68. Springer, Heidelberg (2008)
5. Hannenhalli, S., Pevzner, P.A.: Transforming men into mice (polynomial algorithm for genomic distance problem). In: Proceedings of FOCS 1995, pp. 581–592 (1995)
6. Jean, G., Nikolski, M.: Genome rearrangements: a correct algorithm for optimal capping. Inf. Process. Lett. 104(1), 14–20 (2007)
7. Kececioglu, J.D., Sankoff, D.: Exact and approximation algorithms for sorting by reversals with application to genome rearrangement. Algorithmica 13(1/2), 180–210 (1995)
8. Ozery-Flato, M., Shamir, R.: Two notes on genome rearrangements. J. Bioinf. Comput. Biol. 1(1), 71–94 (2003)
9. Pevzner, P., Tesler, G.: Human and mouse genomic sequences reveal extensive breakpoint reuse in mammalian evolution. Proc. Natl. Acad. Sci. USA 100(13), 7672–7677 (2003)
10. Pevzner, P., Tesler, G.: Transforming men into mice: The Nadeau-Taylor chromosomal breakage model revisited. In: Proceedings of RECOMB 2003, pp. 247–256 (2003)
11. Sankoff, D.: Edit distances for genome comparison based on non-local operations. In: Apostolico, A., Galil, Z., Manber, U., Crochemore, M. (eds.) CPM 1992. LNCS, vol. 644, pp. 121–135. Springer, Heidelberg (1992)
12. Sankoff, D., Trinh, P.: Chromosomal breakpoint reuse in genome sequence rearrangement. J. Comput. Biol. 12(6), 812–821 (2005)
13. Sturtevant, A.H.: A crossover reducer in *Drosophila melanogaster* due to inversion of a section of the third chromosome. Biologisches Zentralblatt 46(12), 697–702 (1926)
14. Tesler, G.: Efficient algorithms for multichromosomal genome rearrangements. J. Comput. Syst. Sci. 65(3), 587–609 (2002)
15. Tesler, G.: GRIMM: Genome rearrangements web server. Bioinformatics 18(3), 492–493 (2002)
16. Yancopoulos, S., Attie, O., Friedberg, R.: Efficient sorting of genomic permutations by translocation, inversion and block interchange. Bioinformatics 21(16), 3340–3346 (2005)

Appendix 1. The Dataset

The mouse genome:

```
 1: o -136 140 93 -95 -32 25 37 -38 39 -40 76 246 30 -29 33 -8 14 -11 10 -9 o
 2: o -161 162 -159 158 -157 156 -155 154 34 -35 36 -180 179 -178 -213 214 -24 28 259 -258 260 o
 3: o 141 139 -57 56 58 68 -201 55 -70 -7 -66 -5 o
 4: o 137 -142 -138 -97 146 153 148 145 4 -3 2 -1 o
 5: o 116 -115 120 124 18 62 -63 64 6 -267 195 -196 197 -113 -114 -119 105 118 200 o
 6: o 117 106 123 109 65 -67 -23 22 -21 -53 42 51 41 -167 -187 264 -188 189 o
 7: o 257 -255 254 -256 177 -210 212 211 -221 220 219 -218 -184 176 224 174 -175 -183 o
 8: o 250 205 126 -134 133 -132 -127 129 -71 130 -253 269 -69 -252 225 -226 227 12 -165 o
 9: o -185 251 110 -186 216 -215 -94 96 -217 -54 -48 -46 47 o
10: o 101 -100 -98 99 27 -170 -266 -263 248 194 -193 192 -191 o
11: o -268 112 -20 -85 -87 -80 84 231 -230 229 -228 -232 233 -234 237 -236 235 238 o
12: o -17 16 -15 -121 -107 -122 207 209 -125 -108 o
13: o -160 -13 -111 -89 88 -151 150 86 81 149 152 -72 -74 o
14: o 50 -45 171 -49 43 -168 -172 208 206 198 -199 203 -128 -131 -202 204 o
15: o -73 143 270 190 o
16: o 223 -135 -265 59 61 -60 -52 261 o
17: o -102 -103 104 -75 -222 91 262 -90 -92 44 -26 249 77 -240 19 239 o
18: o 164 163 -166 243 -31 78 82 79 -83 241 245 242 -244 -247 o
19: o 182 -181 -147 144 -169 173 o
 X: o -274 -275 273 281 -272 278 -279 280 -276 277 -271 o
```

The human genome:

```
 1: o 1 2 3 4 5 6 7 8 9 10 11 12 13 14 o
 2: o 15 16 17 18 19 20 21 22 23 24 25 26 27 28 29 30 31 32 33 34 35 36 37 38 39 40 o
 3: o 41 42 43 44 45 46 47 48 49 50 51 52 53 54 55 56 57 58 59 60 61 o
 4: o 62 63 64 65 66 67 68 69 70 71 o
 5: o 72 73 74 75 76 77 78 79 80 81 82 83 84 85 86 87 o
 6: o 88 89 90 91 92 93 94 95 96 97 98 99 100 101 102 103 104 o
 7: o 105 106 107 108 109 110 111 112 113 114 115 116 117 118 119 120 121 122 123 124 125 o
 8: o 126 127 128 129 130 131 132 133 134 135 136 137 138 139 140 141 142 143 o
 9: o 144 145 146 147 148 149 150 151 152 153 154 155 156 157 158 159 o
10: o 160 161 162 163 164 165 166 167 168 169 170 171 172 173 174 o
11: o 175 176 177 178 179 180 181 182 183 184 185 186 o
12: o 187 188 189 190 191 192 193 194 195 196 197 o
13: o 198 199 200 201 202 203 204 205 o
14: o 206 207 208 209 210 o
15: o 211 212 213 214 215 216 217 218 219 220 221 o
16: o 222 223 224 225 226 227 o
17: o 228 229 230 231 232 233 234 235 236 237 238 o
18: o 239 240 241 242 243 244 245 246 247 o
19: o 248 249 250 251 252 253 254 255 256 257 o
20: o 258 259 260 o
21: o 261 262 263 o
22: o 264 265 266 267 268 269 270 o
 X: o 271 272 273 274 275 276 277 278 279 280 281 o
```

Appendix 2. Cycles and Paths

We list the cycles and paths by their sequences of adjacencies in the human genome.

27 long cycles:

```
[(1 2) (2 3)]
[(9 10) (10 11)]
[(15 16) (16 17)]
[(21 22) (22 23)]
[(34 35) (35 36)]
```

[(37 38) (38 39)]
[(62 63) (63 64)]
[(132 133) (133 134)]
[(155 156) (156 157)]
[(157 158) (158 159)]
[(178 179) (179 180)]
[(191 192) (192 193)]
[(195 196) (196 197)]
[(225 226) (226 227)]
[(228 229) (229 230)]
[(278 279) (279 280)]
[(79 80) (83 84)]
[(80 81) (86 87)]
[(93 94) (95 96)]
[(105 106) (117 118)]
[(106 107) (122 123)]
[(127 128) (131 132)]
[(138 139) (141 142)]
[(277 278) (271 272)]
[(218 219) (219 220) (220 221)]
[(273 274) (275 276) (276 277) (280 281)]
[(233 234) (235 236) (236 237) (234 235) (237 238)]

16 MH-paths:

[(◦ 1)]
[(238 ◦)]
[(260 ◦)]
[(◦ 271)]
[(270 ◦) (189 190)]
[(281 ◦) (273 273) (274 275)]
[(◦ 72) (74 75) (222 223)]
[(◦ 62) (18 19) (239 240)]
[(159 ◦) (162 163) (164 165)]
[(◦ 206) (208 209) (207 208) (171 172) (49 50)]
[(186 ◦) (110 111) (89 90) (92 93) (140 141)]
[(263 ◦) (265 266) (134 135) (126 127) (128 129) (203 204) (201 202) (68 69) (252 253) (268 269)]
[(◦ 105) (118 119) (199 200) (198 199) (202 203) (130 131) (253 254) (254 255) (256 257)]
[(61 ◦) (60 61) (59 60) (52 53) (41 42) (167 168) (172 173) (168 169) (43 44) (91 92) (261 262)]
[(197 ◦) (113 114) (119 120) (114 115) (112 113) (20 21) (53 54) (48 49) (42 43) (50 51) (45 46)
(46 47) (47 48)]
[(◦ 144) (146 147) (152 153) (72 73) (142 143) (137 138) (97 98) (98 99) (99 100) (26 27) (44 45)
(170 171) (27 28) (23 24) (66 67) (6 7) (267 268) (111 112) (12 13) (165 166) (242 243) (244 245)
(241 242) (245 246) (76 77) (249 250)]

12 MM-paths:

[(73 74)]
[(190 191)]
[(136 137)]
[(182 183) (181 182)]
[(115 116) (116 117)]
[(161 162) (160 161)]
[(100 101) (101 102) (103 104) (102 103)]
[(200 201) (54 55) (216 217) (215 216) (185 186)]
[(17 18) (124 125) (108 109) (123 124) (120 121) (107 108)]
[(204 205) (250 251) (184 185) (217 218) (96 97) (145 146) (3 4) (4 5)]
[(163 164) (166 167) (187 188) (188 189) (264 265) (58 59) (67 68) (65 66) (5 6) (64 65) (109 110)
(251 252) (224 225) (173 174)]
[(8 9) (33 34) (154 155) (153 154) (147 148) (180 181) (36 37) (25 26) (248 249) (193 194)
(194 195) (266 267) (169 170) (144 145) (148 149) (81 82) (78 79) (82 83) (240 241) (77 78)
(30 31) (29 30) (246 247)]

15 HH-paths:

[(◦ 187) (◦ 264)]
[(◦ 126) (205 ◦)]

[(○ 41) (51 52) (○ 261)]
[(14 ○) (11 12) (227 ○)]
[(40 ○) (39 40) (75 76) (104 ○)]
[(○ 15) (121 122) (206 207) (○ 198)]
[(○ 222) (90 91) (262 263) (○ 248)]
[(174 ○) (175 176) (183 184) (○ 175)]
[(125 ○) (209 210) (211 212) (221 ○)]
[(210 ○) (177 178) (213 214) (212 213) (○ 211)]
[(143 ○) (269 270) (69 70) (7 8) (13 14) (○ 160)]
[(87 ○) (84 85) (230 231) (231 232) (232 233) (○ 228)]
[(○ 88) (88 89) (151 152) (149 150) (150 151) (85 86) (19 20) (○ 239)]
[(247 ○) (243 244) (31 32) (24 25) (214 215) (94 95) (32 33) (28 29) (258 259) (259 260) (○ 258)]
[(71 ○) (129 130) (70 71) (55 56) (56 57) (57 58) (139 140) (135 136) (223 224) (176 177) (255 256) (257 ○)]

Appendix 3. A Sorting Scenario with Minimum Number of Cuts

We list the adjacencies of the human genome in the order they are repaired.

16 inversions of one block:

(1 2) and (2 3)	(9 10) and (10 11)	(15 16) and (16 17)
(21 22) and (22 23)	(34 35) and (35 36)	(37 38) and (38 39)
(62 63) and (63 64)	(132 133) and (133 134)	(155 156) and (156 157)
(157 158) and (158 159)	(178 179) and (179 180)	(191 192) and (192 193)
(195 196) and (196 197)	(225 226) and (226 227)	(228 229) and (229 230)
(278 279) and (279 280)		

6 inversions to sort two cycles:

(218 219)	(219 220) and (220 221)	(233 234)	(235 236)
(236 237)	(234 235) and (237 238)		

4 inversions to sort the two interleaving cycles of chromosome X:

(273 274)	(275 276)	(277 278) and (271 272)	(276 277) and (280 281)

7 translocations that repair two adjacencies:

(79 80) and (83 84)	(80 81) and (86 87)	(93 94) and (95 96)
(105 106) and (117 118)	(106 107) and (122 123)	(127 128) and (131 132)
(138 139) and (141 142)		

15 fissions:

(○ 126) and (205 ○)	(○ 187) and (○ 264)	(14 ○)	(○ 15)	(40 ○)	(○ 41)	(71 ○)	(○ 88)
(143 ○)	(174 ○)	(○ 248)	(210 ○)	(221 ○)	(○ 228)	(○ 258)	

186 non-degenerate semi-operations:

(189 190) (164 165) (162 163) (239 240) (18 19)   (49 50)   (140 141) (256 257) (254 255)
(261 262) (91 92)   (47 48)   (46 47)   (45 46)   (50 51)   (249 250) (11 12)   (39 40)
(181 182) (115 116) (161 162) (100 101) (101 102) (103 104) (200 201) (17 18)   (204 205)
(250 251) (184 185) (163 164) (274 275) (272 273) (8 9)   (129 130) (70 71)   (55 56)
(56 57)   (57 58)   (139 140) (135 136) (223 224) (176 177) (255 256) (262 263) (88 89)
(151 152) (149 150) (150 151) (232 233) (231 232) (230 231) (84 85)   (85 86)   (19 20)
(177 178) (213 214) (212 213) (211 212) (209 210) (259 260) (258 259) (28 29)   (32 33)
(94 95)   (214 215) (24 25)   (31 32)   (243 244) (75 76)   (51 52)   (171 172) (207 208)
(208 209) (121 122) (206 207) (175 176) (183 184) (269 270) (69 70)   (7 8)   (13 14)
(92 93)   (89 90)   (110 111) (90 91)   (222 223) (74 75)   (253 254) (43 44)   (168 169)
(172 173) (167 168) (41 42)   (52 53)   (59 60)   (60 61)   (42 43)   (48 49)   (53 54)
(20 21)   (112 113) (114 115) (119 120) (113 114) (268 269) (252 253) (68 69)   (201 202)
(203 204) (128 129) (126 127) (134 135) (265 266) (130 131) (202 203) (198 199) (199 200)
(118 119) (76 77)   (245 246) (241 242) (244 245) (242 243) (165 166) (12 13)   (111 112)
(267 268) (6 7)   (66 67)   (23 24)   (27 28)   (170 171) (44 45)   (26 27)   (99 100)
(98 99)   (97 98)   (137 138) (142 143) (72 73)   (152 153) (146 147) (33 34)   (154 155)
(153 154) (147 148) (180 181) (36 37)   (25 26)   (248 249) (193 194) (194 195) (266 267)
(169 170) (144 145) (148 149) (54 55)   (216 217) (215 216) (217 218) (96 97)   (145 146)
(3 4)   (124 125) (108 109) (123 124) (120 121) (246 247) (29 30)   (30 31)   (77 78)
(240 241) (82 83)   (78 79)   (166 167) (187 188) (188 189) (264 265) (58 59)   (67 68)
(65 66)   (5 6)   (64 65)   (109 110) (251 252) (224 225)

12 fusions:

(182 183) (73 74)   (190 191) (136 137) (116 117) (160 161) (102 103) (185 186) (107 108)
(4 5)   (81 82)   (173 174)

Hurdles Hardly Have to Be Heeded

Krister M. Swenson, Yu Lin, Vaibhav Rajan, and Bernard M.E. Moret

Laboratory for Computational Biology and Bioinformatics
EPFL (Ecole Polytechnique Fédérale de Lausanne)
and Swiss Institute of Bioinformatics
Lausanne, Switzerland
{krister.swenson,yu.lin,vaibhav.rajan,bernard.moret}@epfl.ch

Abstract. As data about genomic architecture accumulates, genomic rearrangements have attracted increasing attention. One of the main rearrangement mechanisms, inversions (also called reversals), was characterized by Hannenhalli and Pevzner and this characterization in turn extended by various authors. The characterization relies on the concepts of breakpoints, cycles, and obstructions colorfully named hurdles and fortresses. In this paper, we study the probability of generating a hurdle in the process of sorting a permutation if one does not take special precautions to avoid them (as in a randomized algorithm, for instance). To do this we revisit and extend the work of Caprara and of Bergeron by providing simple and exact characterizations of the probability of encountering a hurdle in a random permutation. Using similar methods we, for the first time, find an asymptotically tight analysis of the probability that a fortress exists in a random permutation.

1 Introduction

The advent of high-throughput techniques in genomics has led to the rapid accumulation of data about the genomic architecture of large numbers of species. As biologists study these genomes, they are finding that genomic rearrangements, which move single genes or blocks of contiguous genes around the genome, are relatively common features: entire blocks of one chromosome can be found in another chromosome in another species. The earliest findings of this type go back to the pioneering work of Sturtevant on the fruit fly [10,11]; but it was the advent of large-scale sequencing that moved this aspect of evolution to the forefront of genomics.

The best documented type of rearrangement is the *inversion* (also called reversal), in which a block of consecutive genes is removed and put back in (the same) place in the opposite orientation (on the other strand, as it were). The most fundamental computational question then becomes: given two genomes, how efficiently can such an operation as inversion transform one genome into the other? Since an inversion does not affect gene content (the block is neither shortened nor lengthened by the operation), it makes sense to view these operations as being applied to a signed permutation of the set $\{1, 2, \ldots, n\}$.

Hannenhalli and Pevzner [6,7] showed how to represent a signed permutation of n elements as a *breakpoint graph* (also called, more poetically, a diagram of reality and

C.E. Nelson and S. Vialette (Eds.): RECOMB-CG 2008, LNBI 5267, pp. 241–251, 2008.
© Springer-Verlag Berlin Heidelberg 2008

desire), which is a graph on $2n + 2$ vertices (2 vertices per element of the permutation to distinguish signs, plus 2 vertices that denote the extremities of the permutation) with colored edges, where edges of one color represents the adjacencies in one permutation and edges of the other color those in the other permutation. In such a graph, every vertex has indegree 2 and outdegree 2 and so the graph has a unique decomposition into cycles of even length, where the edges of each cycle alternate in color. Hannenhalli and Pevzner introduced the notions of *hurdles* and *fortresses* and proved that the minimum number of inversions needed to convert one permutation into the other (also called "sorting" a permutation) is given by the number of elements of the permutation plus 1, minus the number of cycles, plus the number of hurdles, and plus 1 if a fortress is present. Caprara [5] showed that hurdles were a rare feature in a random signed permutation. Bergeron [2] provided an alternate characterization in terms of *framed common intervals* and went on to show that *unsafe inversions*, that is, inversions that could create new obstructions such as hurdles, were rare [3] when restricted to adjacency creating inversions. Kaplan and Verbin [8] capitalized on these two findings and proposed a randomized algorithm that sorts a signed permutation without paying heed to unsafe inversions, finding that, in practice, the algorithm hardly needed any restarts to provide a proper sorting sequence of inversions, although they could not prove that it is in fact a proper Las Vegas algorithm.

In this paper, we extend Bergeron's result about the possibility of creating a hurdle by doing an inversion. Her result is limited to inversions that create new adjacencies, but these are in the minority: in a permutation without hurdles, any inversion that increases the number of cycles in the breakpoint graph is a candidate. Using Sankoff's *randomness hypothesis* [9], we show that the probability that *any* cycle-splitting inversion is unsafe is $\Theta(n^{-2})$. We then revisit Caprara's complex proof and provide a simple proof, based on the framed intervals introduced by Bergeron, that the probability that a random signed permutation on n elements contains a hurdle is $\Theta(n^{-2})$. Finally, we show that this approach can be extended to prove that the probability such a permutation contains a fortress is $\Theta(n^{-15})$. Our results are elaborated for circular permutations, but simple (and by now standard) adaptations show that they also hold for linear permutations.

Framed common intervals considerably simplify our proofs; indeed, our proofs for hurdles and fortresses depend mostly on the relative scarcity of framed intervals. Our results add credence to the conjecture made by Kaplan and Verbin that their randomized algorithm is a Las Vegas algorithm, i.e., that it returns a sorting sequence with high probability after a constant number of restarts. Indeed, because our results suggest that the probability of failure of their algorithm is $O(1/n)$ when working on a permutation of n elements, whereas any fixed constant $0 < \varepsilon < 1$ would suffice, one could conceive taking advantage of that gap by designing an algorithm that runs faster by using a stochastic, rather than deterministic, data structure, yet remains a Las Vegas algorithm. Indeed, how fast a signed permutation can be sorted by inversions remains an open question: while we have an optimal linear-time algorithm to compute the number of inversions needed [1], computing one optimal sorting sequence takes subquadratic time—$O(n\sqrt{n \log n})$, either stochastically with the algorithm of Kaplan and Verbin or deterministically with the similar approach of Tannier and Sagot [12].

2   Preliminaries

Let Σ_n denote the set of signed permutations over n elements; a permutation π in this set will be written as $\pi = (\pi_1 \pi_2 \ldots \pi_n)$, where each element π_i is a signed integer and the absolute values of these elements are all distinct and form the set $\{1, 2, \ldots, n\}$. Given such a π, a pair of elements (π_i, π_{i+1}) or (π_n, π_1) is called an *adjacency* whenever we have $\pi_{i+1} - \pi_i = 1$ (for $1 \le i \le n-1$) or $\pi_1 - \pi_n = 1$; otherwise, this pair is called a *breakpoint*. We shall use Σ_n^0 to denote the set of permutations in which every permutation is entirely devoid of adjacencies. Bergeron *et al* [3] proved the following result about $|\Sigma_n^0|$.

Lemma 1. *[3] For all $n > 1$, $\frac{1}{2}|\Sigma_n| < |\Sigma_n^0| < |\Sigma_n|$.*

For any signed permutation π and the identity $I = (12 \ldots n)$, we can construct the breakpoint graph for the pair (π, I). Since there is one-to-one mapping between π and the corresponding breakpoint graph for (π, I), we identify the second with the first and so write that π contains cycles, hurdles, or fortresses if the breakpoint graph for (π, I) does; similarly, we will speak of other properties of a permutation π that are in fact defined only when π is compared to the identity permutation.

A *framed common interval* (FCI) of a permutation (made circular by considering the first and last elements as being adjacent) is a substring of the permutation, $a s_1 s_2 \ldots s_k b$ or $-b s_1 s_2 \ldots s_k$-a so that

– for each i, $1 \le i \le k$, $|a| < |s_i| < |b|$, and
– for each l, $|a| < l < |b|$, there exists a j with $|s_j| = l$.

So the substring $s_1 s_2 \ldots s_k$ is a signed permutation of the integers that are greater than a and less than b; a and b form the *frame*. The framed interval is said to be common, in that it also exists, in its canonical form, $+a+(a+1)+(a+2) \ldots +b$, in the identity permutation. Framed intervals can be nested. The *span* of an FCI is the number of elements between a and b, plus two, or $b - a + 1$. A *component* is comprised of all elements inside a framed interval that are not inside any nested subinterval, plus the frame elements. A *bad component* is a component whose elements all have the same sign.

In a circular permutation, a bad component A *separates* bad components B and C if and only if every substring containing an element of B and an element of C also has an element of A in it. We say that A *protects* B if A separates B from all other bad components. A *superhurdle* is a bad component that is protected by another bad component. A *fortress* is a permutation that has an odd number (larger than 1) of hurdles, all of which are superhurdles. The smallest superhurdles are equivalent to intervals $f = +(i)+(i+2)+(i+4)+(i+3)+(i+5)+(i+1)+(i+6)$ or the reverse $f' = -(i+6)-(i+1)-(i+5)-(i+3)-(i+4)-(i+2)-(i)$. A *hurdle* is a bad component that is not a superhurdle.

We will use the following useful facts about FCIs; all but fact 3 follow immediately from the definitions.

1. A bad component indicates the existence of a hurdle.
2. To every hurdle can be assigned a unique bad component.
3. FCIs never overlap by more than two elements [4].
4. An interval shorter than 4 elements cannot be bad.

3   The Rarity of Hurdles and Fortresses

In this section, we provide asymptotic characterizations in $\Theta(\ )$ terms of the probability that a hurdle or fortress is found in a signed permutation selected uniformly at random. Each proof has two parts, an upper bound and a lower bound; for readability, we phrase each part as a lemma and develop it independently. We begin with hurdles; the characterization for these structures was already known, but the original proof of Caprara [5] is very complex.

Theorem 1. *The probability that a random signed permutation on n elements contains a hurdle is* $\Theta(n^{-2})$.

Lemma 2 (Upper bound for shorter than $n - 1$**).** *The probability that a random signed permutation on n elements contains a hurdle spanning no more than* $n - 2$ *elements is* $O(n^{-2})$.

Proof. Fact 4 tells us that we need only consider intervals of at least four elements. Call $F_{\leq n-2}$ the event that a FCI spanning no more than $n - 2$ and no less than four elements exists. Call $F(i)_{\leq n-2}$ the event that such an FCI exists with a left endpoint at π_i. We thus have $F_{\leq n-2} = 1$ if and only if there exists an i, $1 \leq i \leq n$, with $F(i)_{\leq n-2} = 1$. Note that $F(i)_{\leq n-2} = 1$ implies either $\pi_i = a$ or $\pi_i = -b$ for some FCI. Thus we can write

$$Pr\big(F(i)_{\leq n-2} = 1\big) \leq \sum_{l=4}^{n-2} \frac{1}{2(n-1)} \binom{n-2}{l-2}^{-1} \tag{1}$$

since $\frac{1}{2(n-1)}$ is the probability the right endpoint matches the left endpoint (π_l is $-a$ or b if π_i is $-b$ or a respectively) of an interval of span l and $\binom{n-2}{l-2}^{-1}$ is the probability that the appropriate elements are inside the frame. We can bound the probability from (1) as

$$Pr\big(F(i)_{\leq n-2} = 1\big) \leq \frac{1}{2(n-1)} \sum_{l=2}^{n-4} \binom{n-2}{l}^{-1}$$

$$\leq \frac{1}{n-1} \sum_{l=2}^{\lceil n/2 \rceil - 1} \binom{n-2}{l}^{-1}$$

$$\leq \frac{1}{n-1} \left(\sum_{l=2}^{\sqrt{n}} \left(\frac{l}{n-2}\right)^l + \sum_{l=\sqrt{n}+1}^{\lceil n/2 \rceil - 1} \binom{n-2}{l}^{-1} \right) \tag{2}$$

where the second term is no greater than

$$\sum_{l=\sqrt{n}+1}^{\lceil n/2 \rceil - 1} \binom{n-2}{l}^{-1} \leq \sum_{l=\sqrt{n}+1}^{\lceil n/2 \rceil - 1} \left(\frac{1}{2}\right)^{\sqrt{n}+1} \in O(1/n^2) \tag{3}$$

and the first term can be simplified

$$\sum_{l=2}^{\sqrt{n}} \left(\frac{l}{n-2}\right)^l = \sum_{l=2}^{4} \left(\frac{l}{n-2}\right)^l + \sum_{l=5}^{\sqrt{n}} \left(\frac{l}{n-2}\right)^l$$

$$\leq \sum_{l=2}^{4} \left(\frac{l}{n-2}\right)^l + \sum_{l=5}^{\sqrt{n}} \left(\frac{n}{n-2}\frac{\sqrt{n}}{n}\right)^5$$

$$\in O\left(3 \times \frac{16}{(n-2)^2} + \sqrt{n}\, n^{-5/2}\right) = O(n^{-2}). \tag{4}$$

To compute $Pr(F_{\leq n-2})$ we use the union bound on $Pr(\bigcup_{i=1}^{n} F(i)_{\leq n-2})$. This removes the factor of $\frac{1}{n-1}$ from (2) yielding just the sum of (4) and (3) which is $O(n^{-2})$. The probability of observing a hurdle in some subsequence of a permutation can be no greater than the probability of observing a FCI (by fact 2). Thus we know the probability of observing a hurdle that spans no more than $n-2$ elements is $O(n^{-2})$.

We now proceed to bound the probability of a hurdle that spans $n-1$ or n elements. Call intervals with such spans n-*intervals*. For a bad component spanning n elements with $a = i$, there is only a single $b = (i-1)$ that must be a's left neighbor (in the circular order), and for a hurdle spanning $n-1$ elements with $a = i$, there are only two configurations ("+(i-2) +(i-1) +i" and its counterpart "+(i-2) -(i-1) +i") that will create a framed interval. Thus the probability that we see an n-interval with a particular $a = i$ is $O(1/n)$ and the expected number of n-intervals in a permutation is $O(1)$.

We now use the fact that a bad component is comprised of elements with all the same sign. Thus the probability that an n-interval uses all the elements in its span (i.e., there exist no nested subintervals) is $O(2^{-n})$. Call a bad component that does not use all of the elements in its span (i.e., there must exist nested subintervals) a *fragmented* interval.

Lemma 3 (Upper bound for n-intervals). *The probability that a fragmented n-interval is a hurdle is $O(n^{-2})$.*

Proof. We divide the analysis into three cases where the fragment-causing subinterval is of span

1. $n-1$,
2. 4 through $n-2$, and
3. less than 4.

The existence of a subinterval of span $n-1$ precludes the possibility of the frame elements from the larger n-interval being in the same component, so there cannot be a hurdle using this frame. We have already established that $Pr(F_{\leq n-2})$ is $O(n^{-2})$. Thus we turn to the third case. If an interval is bad, then the frame elements of any fragmenting subinterval must have the same sign as the frame elements of the larger one. If we view each such subinterval and each element not included in such an interval as single characters, we know that there must be at least $n/3$ signed characters. Since the signs of the characters are independent, the probability that all characters have the same sign is $1/2^{O(n)}$ and is thus negligible.

Thus the probability of a bad n-interval is $O(n^{-2})$. Now using fact 4 we conclude that the probability of existence of a hurdle in a random signed permutation on n elements is $O(n^{-2})$.

Lemma 4 (Lower bound). *The probability that a signed permutation on n elements has a hurdle with a span of four elements is* $\Omega(n^{-2})$.

Proof. Call h_k the hurdle with span four that starts with element $4k+1$. So the subsequence that corresponds to h_k must be $^+(4k+1)^+(4k+3)^+(4k+2)^+(4k+4)$ or $^-(4k+4)^-(4k+2)^-(4k+3)^-(4k+1)$. We can count the number of permutations with h_0, for instance. The four elements of h_0 are contiguous in $4!(n-3)!2^n$ permutations of length n. In $c = 2/(4!2^4)$ of those cases, the contiguous elements form a hurdle, so the total proportion of permutations with h_0 is

$$c\frac{4!(n-3)!2^n}{n!2^n} \in \Omega\left(\frac{1}{n^3}\right).$$

Similarly, the proportion of permutations that have both h_0 and h_1 is

$$F_2 = c^2\frac{(4!)^2(n-6)!2^n}{n!2^n} \in O\left(\frac{1}{n^6}\right)$$

and, therefore, the proportion of permutations that have at least one of h_0 or h_1 is

$$2 \times c\frac{4!(n-3)!2^n}{n!2^n} - F_2. \tag{5}$$

We generalize (5) to count the proportion of permutations with at least one of the hurdles $h_0, h_1, \ldots, h_{\lfloor n/4 \rfloor}$; this proportion is at least

$$\left\lfloor \frac{n}{4} \right\rfloor \times c\frac{4!(n-3)!2^n}{n!2^n} - \binom{\lfloor n/4 \rfloor}{2}F_2 \tag{6}$$

which is $\Omega(n^{-2})$ since the second term is $O(n^{-4})$.

Now we turn to the much rarer fortresses.

Theorem 2. *The probability that a random signed permutation on n elements includes a fortress is* $\Theta(n^{-15})$.

Lemma 5 (Upper bound). *The probability that a random signed permutation on n elements includes a fortress is* $O(n^{-15})$.

Proof. We bound the probability that at least three superhurdles occur in a random permutation by bounding the probability that three non-overlapping bad components of length seven exist. We divide the analysis into three cases depending on the number l of elements spanned by a bad component.

1. For one of the three FCIs we have $n - 14 \le l \le n - 11$.
2. For one of the three FCIs we have $17 \le l \le n - 15$.
3. For all FCIs we have $7 \le l < 17$.

As we did in Lemma 2 (equation 1), we can bound the probability that we get an FCI of length l starting at a particular position by

$$Pr(F_l = 1) \le \frac{1}{2(n-1)} \binom{n-2}{l-2}^{-1}. \tag{7}$$

In the first case the probability that the FCI is a superhurdle is $O(n^{-11} \cdot 2^{-n})$ if the FCI is not fragmented and $O(n^{-15})$ if it is (using the same technique as for the proof of Lemma 3). In the second case the probability is at most

$$n \sum_{l=17}^{n-15} F_l = n \sum_{k=15}^{n-17} \frac{1}{2(n-1)} \binom{n-2}{k}^{-1}$$

which, by the same reasoning used for equation 2 to derive $O(n^{-2})$, is $O(n^{-15})$. Thus the first two cases both give us an upper bound of $O(n^{-15})$.

Fact 3 tells us that any pair of FCIs can overlap only on their endpoints. Thus, if we first consider the probability of finding a smallest FCI, we know that no other FCI will have an endpoint inside it. So the probability of having a second FCI, conditioned on having a smaller first one, is dependent only on the size of the first. The same reasoning extends to the probability of having a third conditioned on having two smaller FCIs. Since each of the three FCIs spans less than seventeen elements, the probability of each FCI appearing is at most $n \sum_{l=7}^{17} F_k = O(n^{-5})$, and the probability of there being at least three of them is $O(n^{-15})$.

We now turn to the lower bound. Consider the probability of the existence, in a random permutation, of a permutation with exactly three superhurdles spanning seven elements each. A lower bound on this probability is a lower bound on the probability of existence of a fortress in a random permutation.

Lemma 6 (Lower bound). *The probability that a random signed permutation on n elements includes a fortress is $\Omega(n^{-15})$.*

Proof. Denote by $F_{3,7}(n)$ the number of permutations on n elements with exactly 3 superhurdles spanning 7 elements each. To create such a permutation, choose a permutation of length $n-18$ (with zero adjacencies and without hurdles), select three elements, and extend each of these three elements to a superhurdle, renaming the elements of the permutation as needed. That is, replace element $+i$ by the framed interval of length 7 $f = {}^+(i)^+(i+2)^+(i+4)^+(i+3)^+(i+5)^+(i+1)^+(i+6)$ and rename all the elements with magnitude j to have magnitude $j+6$ (for those with $|j| > |i|$). After extending the three selected elements, we get a permutation on n elements where there are exactly 3 superhurdles each spanning 7 elements.

From Lemma 1 and the results about the rarity of hurdles from the previous section, we have

$$F_{3,7}(n) > \frac{(n-18)! 2^{n-18}}{2} \left(1 - O(n^{-2})\right) \binom{n-18}{3}$$

where $\frac{(n-18)!2^{n-18}}{2}(1-O(n^{-2}))$ is a lower bound for the number of permutations of length $n-18$ (with zero adjacencies and without hurdles) and $\binom{n-18}{3}$ is the number of ways to choose the elements for extension. Therefore we have

$$\frac{F_{3,7}(n)}{n!2^n} > \frac{(n-18)!2^{n-18}}{2}\left(1-O(n^{-2})\right)\binom{n-18}{3}\frac{1}{n!2^n}$$

$$\in \Omega(n^{-15}) \tag{8}$$

4   On the Proportion of Unsafe Cycle-Splitting Inversions

Denote the two vertices representing a permutation element π_i in the breakpoint graph by π_i^- and π_i^+ (π° can denote either). Think of embedding the breakpoint graph on a circle as follows: we place all $2n$ vertices on the circle so that:

1. π_i^+ and π_i^- are adjacent on the circle,
2. π_i^- is clockwise-adjacent to π_i^+ if and only if π_i is positive, and
3. a π_i° is adjacent to a π_{i+1}° if and only if π_i and π_{i+1} are adjacent in π.

For two vertices $v_1 = \pi_i^\circ$ and $v_2 = \pi_j^\circ$ ($i \neq j$) that are adjacent on the circle, add the edge (v_1, v_2)—a reality edge (also called a black edge); also add edges (π_i^+, π_{i+1}^-) for all i and (π_n^+, π_1^-)—the desire edges (also called gray edges). The breakpoint graph is just as described in [6], but its embedding clarifies the notion of orientation of edges, which plays a crucial role in our study of unsafe inversions.

In the breakpoint graph two reality edges on the same cycle are *convergent* if a traversal of their cycle visits each edge in the same direction in the circular embedding; otherwise they are *divergent*. Any inversion that acts on a pair of divergent reality edges splits the cycle to which the edges belong; conversely, no inversion that acts on a pair of convergent reality edges splits their common cycle. (An inversion that acts upon a pair of reality edges in two different cycles simply merges the two cycles.)

An inversion can be denoted by the set of elements in the permutation that it rearrange; for instance, we can write $r = \{\pi_i, \pi_{i+1}, \ldots, \pi_j\}$. The permutation obtained by applying a inversion r on a permutation π is denoted by $r\pi$. Thus, using the same r, we have $r\pi = (\pi_1 \ldots \pi_{i-1} {}^-\pi_j \ldots {}^-\pi_i \pi_{j+1} \ldots \pi_n)$. We call a pair (π, r) *unsafe* if π does not contain a hurdle but $r\pi$ does. A pair (π, r) is *oriented* if $r\pi$ contains more adjacencies than π does. A pair (π, r) is *cycle-splitting* if $r\pi$ contains more cycles than π does. (When π is implied from the context, we call r unsafe, oriented, or cycle-splitting, respectively, without referring to π.) Note that every oriented inversion is a cycle-splitting inversion. A inversion r on a permutation π is a *sorting* inversion if $d(r\pi) = d(\pi) - 1$.

Let π be a random permutation without hurdles and r a randomly chosen oriented inversion on π. Bergeron *et al.* [3] proved that the probability that the pair (π, r) is unsafe is $O(n^{-2})$. However, not every sorting inversion for a permutation without hurdles is necessarily an oriented inversion; on the other hand, it is necessarily a cycle-splitting inversion. The result in [3] thus applies only to a small fraction of all sorting inversions. We now proceed to study *all* inversions that can increase the cycle count. We show that, under Sankoff's randomness hypothesis (stated below), the proportion of these inversions that are unsafe is $O(n^{-2})$.

In [9], Sankoff builds graphs by effectively fixing desire edges, one to each vertex, and then randomly connecting each vertex to exactly one reality edge. We choose to equivalently view this process as randomly linking each vertex, with reality edges already fixed, by exactly one desire edge. It should be noted that the orientation of a reality edge in the breakpoint graph is not independent of the orientation of the other reality edges, but for this random generation process where they are independent, we may generate a graph that does not correspond to a permutation. Sankoff [9] proposed a *Randomness Hypothesis* in this regard; it states that the probabilistic structure of the breakpoint graph is asymptotically independent of whether or not the generated graph is consistent with a permutation. In the randomly constructed graphs, every reality edge induces a direction independently and each direction has a probability of $\frac{1}{2}$, so the expected number of reality edges with one orientation equals that with the other orientation; our own experiments support the randomness hypothesis in this respect, as illustrated in Figure 1, which shows the number of edges inducing a clockwise orientation

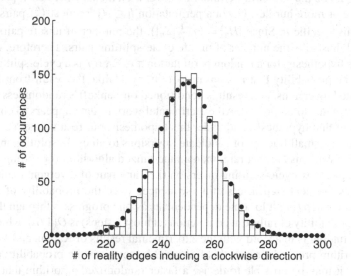

Fig. 1. The number of edges inducing a clockwise direction in cycles of length 500, taken from random permutations. Black dots are the expected values from the binomial distribution while white bars are experimental values.

on a cycle of length 500 from 2000 random permutations of length 750. Observations (the vertical bars) match a binomial distribution (the black dots). This match is important inasmuch as it is simpler to analyze a random breakpoint graph than a random signed permutation.

The number of cycle-splitting inversions in a permutation π equals the number of pairs of divergent same-cycle reality edges in the breakpoint graph for π. Consider a cycle containing L reality edges and let k of them share the same orientation; the number of pairs of divergent reality edges in this cycle is then $k(L-k)$. Thus, under the randomness hypothesis, the expected number of pairs of divergent reality edges for a cycle containing L reality edges is given by

$$\sum_{k=0}^{L} \binom{L}{k} \left(\frac{1}{2}\right)^{L} k(L-k) = \frac{1}{4}L(L-1).$$

The maximum number of pairs of divergent reality edges for a cycle with L reality edges is $\frac{1}{4}L^2$. Thus at least half the number of cycles with L reality edges have at least $\frac{1}{4}L^2 - \frac{1}{2}L$ pairs of divergent reality edges (for $L > 2$).

Using the randomness hypothesis, Sankoff *et al.* [9] have shown that in a random breakpoint graph (with $2n$ vertices) the expected number of reality edges in the largest cycle is $\frac{2}{3}n$. Since the maximum number of reality edges in the largest cycle is n, at least half the random breakpoint graphs have a cycle with at least $\frac{1}{3}n$ reality edges. So, for all random breakpoint graphs, at least $\frac{1}{4}$ of them have at least $\frac{1}{36}n^2 - \frac{1}{6}n$ pairs of divergent reality edges. Hence, under the randomness hypothesis, the number of pairs (π, r), where r is a cycle-splitting inversion in π, is $\Theta(n^2)|\Sigma_n|$.

Let $H_n \in \Sigma_n$ be the subset of permutations over n elements where each permutation contains one or more hurdles. Given a permutation $h \in H_n$, at most $\binom{n}{2}$ pairs of (π, r) can yield this specific h. Since $|H_n| = \Theta(\frac{1}{n^2}|\Sigma_n|)$, the number of unsafe pairs (π, r) is $O(|\Sigma_n|)$ and thus so is the number of unsafe cycle-splitting pairs. Therefore, under the randomness hypothesis, for a random permutation $\pi \in \Sigma_n$, if r is a cycle-splitting inversion on π, the probability that r is unsafe is $O(n^{-2})$. Unlike the result from Bergeron about oriented inversions, this result is conditioned on Sankoff's randomness hypothesis, which remains to be proved. All experimental work to date appears to confirm the correctness of that hypothesis; and under this hypothesis, our result generalizes that of Bergeron from a small fraction of candidate inversions to all cycle-splitting inversions.

If unsafe inversions are that rare, then a randomized algorithm for sorting by inversions could pick any cycle-splitting inversion (i.e., any pair of divergent reality edges) and use it as the next step in a sorting sequence; since the probability of failure is $\Theta(n^{-2})$ at each step (modulo some dependencies as one progresses through the steps), the overall probability of failure at completion (at most n steps) is $O(1/n)$, which is very small. This finding is in accord with the experimental results of Kaplan and Verbin [8], whose algorithm proceeds in just this fashion. Moreover, as the probability of failure is so small, it may be possible to devise a faster randomized algorithm that does not maintain an exact record of all reality edges and cycles (the major time expense in the current algorithms); such an algorithm would suffer from additional errors (e.g., using a pair of edges that is not divergent), but would remain usable as long as the probability of error at each step remained $O(1/n)$ and bounded by a fixed constant overall.

5 Conclusions

We have both extended and simplified results of Bergeron and Caprara on the expected structure of signed permutations and their behavior under inversions. These extensions demonstrate the mathematical power of the framed common interval framework developed by Bergeron and the potential uses of the randomness hypothesis proposed by Sankoff to bind the asymptotic properties of valid and randomized breakpoint graphs. Our results confirm the evasive nature of hurdles (and, even more strongly, of fortresses);

indeed, these structures are both so rare and, more importantly, so hard to create accidentally that, as our title suggests, they can be safely ignored. (Of course, if a permutation does have a hurdle, that hurdle must be handled if we are to sort the permutation; but handling hurdles takes only linear time—the cost comes when attempting to avoid creating a new one, i.e., when testing cycle-splitting inversions for safeness.) Moreover, the possibility of not testing candidate inversions for safeness suggests that further information could be discarded for the sake of speed without harming the convergence properties of a randomized algorithm, thereby potentially opening a new path for faster sorting by inversions.

References

1. Bader, D.A., Moret, B.M.E., Yan, M.: A linear-time algorithm for computing inversion distance between signed permutations with an experimental study. J. Comput. Biol. 8(5), 483–491 (2001); A preliminary version appeared in WADS 2001, pp. 365–376
2. Bergeron, A.: A very elementary presentation of the Hannenhalli–Pevzner theory. Discrete Applied Mathematics 146(2), 134–145 (2005)
3. Bergeron, A., Chauve, C., Hartman, T., Saint-Onge, K.: On the properties of sequences of reversals that sort a signed permutation. In: JOBIM, 99–108 (June 2002)
4. Bergeron, A., Stoye, J.: On the similarity of sets of permutations and its applications to genome comparison. In: Warnow, T.J., Zhu, B. (eds.) COCOON 2003. LNCS, vol. 2697, pp. 68–79. Springer, Heidelberg (2003)
5. Caprara, A.: On the tightness of the alternating-cycle lower bound for sorting by reversals. J. Combin. Optimization 3, 149–182 (1999)
6. Hannenhalli, S., Pevzner, P.A.: Transforming cabbage into turnip (polynomial algorithm for sorting signed permutations by reversals). In: Proc. 27th Ann. ACM Symp. Theory of Comput. (STOC 1995), pp. 178–189. ACM Press, New York (1995)
7. Hannenhalli, S., Pevzner, P.A.: Transforming mice into men (polynomial algorithm for genomic distance problems). In: Proc. 36th Ann. IEEE Symp. Foundations of Comput. Sci. (FOCS 1995), pp. 581–592. IEEE Computer Society Press, Piscataway (1995)
8. Kaplan, H., Verbin, E.: Efficient data structures and a new randomized approach for sorting signed permutations by reversals. In: Baeza-Yates, R., Chávez, E., Crochemore, M. (eds.) CPM 2003. LNCS, vol. 2676, pp. 170–185. Springer, Heidelberg (2003)
9. Sankoff, D., Haque, L.: The distribution of genomic distance between random genomes. J. Comput. Biol. 13(5), 1005–1012 (2006)
10. Sturtevant, A.H., Beadle, G.W.: The relation of inversions in the x-chromosome of drosophila melanogaster to crossing over and disjunction. Genetics 21, 554–604 (1936)
11. Sturtevant, A.H., Dobzhansky, Th.: Inversions in the third chromosome of wild races of drosophila pseudoobscura and their use in the study of the history of the species. Proc. Nat'l Acad. Sci., USA 22, 448–450 (1936)
12. Tannier, E., Sagot, M.: Sorting by reversals in subquadratic time. In: Sahinalp, S.C., Muthukrishnan, S.M., Dogrusoz, U. (eds.) CPM 2004. LNCS, vol. 3109, pp. 1–13. Springer, Heidelberg (2004)

Internal Validation of Ancestral Gene Order Reconstruction in Angiosperm Phylogeny

David Sankoff[1], Chunfang Zheng[1], P. Kerr Wall[2], Claude dePamphilis[2], Jim Leebens-Mack[3], and Victor A. Albert[4]

[1] Dept. of Mathematics & Statistics and Dept. of Biology, University of Ottawa, Ottawa, ON, Canada K1N 6N5
{sankoff,czhen033}@uottawa.ca
[2] Biology Department, Penn State University, University Park, PA 16802, USA
{pkerrwall,cwd3}@psu.edu
[3] Department of Plant Biology, University of Georgia, Athens, GA 30602, USA
jleebensmack@plantbio.uga.edu
[4] Department of Biological Sciences, SUNY Buffalo, Buffalo, NY 14260, USA
vaalbert@buffalo.edu

Abstract. Whole genome doubling (WGD), a frequent occurrence during the evolution of the angiopserms, complicates ancestral gene order reconstruction due to the multiplicity of solutions to the genome halving process. Using the genome of a related species (the outgroup) to guide the halving of a WGD descendant attenuates this problem. We investigate a battery of techniques for further improvement, including an unbiased version of the guided genome halving algorithm, reference to two related genomes instead of only one to guide the reconstruction, use of draft genome sequences in contig form only, incorporation of incomplete sets of homology correspondences among the genomes and addition of large numbers of "singleton" correspondences. We make use of genomic distance, breakpoint reuse rate, dispersion of sets of alternate solutions and other means to evaluate these techniques, while reconstructing the pre-WGD ancestor of *Populus trichocarpa* as well as an early rosid ancestor.

1 Introduction

The reconstruction of the gene order in ancestral genomes requires that we make a number of choices, among the data on which to base the reconstruction, in the algorithm to use and in how to evaluate the result. In this paper we illustrate an approach to making these choices in the reconstruction of the ancestor of the poplar *Populus trichocarpa* genome. This species has undergone whole genome duplication [3,11,14] followed by extensive chromosomal rearrangement, and is one of four angiosperm genomes, along with those of *Carica papaya* (papaya), *Vitis vinifera* (grapevine) and *Arabidopsis thaliana*, that have been sequenced to date, shown in Figure 1.

C.E. Nelson and S. Vialette (Eds.): RECOMB-CG 2008, LNBI 5267, pp. 252–264, 2008.
© Springer-Verlag Berlin Heidelberg 2008

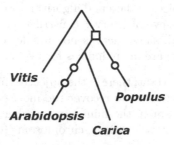

Fig. 1. Phylogenetic relationships among angiosperms with sequenced genomes. The circles indicate likely whole genome doubling events. The circle in the *Populus* lineage, representing the locus of the WGD event at the origin of the willow-poplar family, and the square, representing the ancestor of the rosid dicotyledons, indicate the target ancestors we reconstruct in this paper.

We have been developing methods to incorporate descendants of whole genome doubling into phylogenies of species that have been unaffected by the doubling event. The basic tool in analyzing descendants of whole genome doubling is the halving algorithm [4]. To overcome the propensity of the genome halving procedure to produce numerous, widely disparate solutions, we "guide" the execution of this procedure with information from genomes from related species [18,10,17,19,20], which we call outgroups. This, *ipso facto*, integrates the whole genome doubling descendant into the phylogeny of the related species.

Issues pertaining to data include

Homology sets. Can we use defective sets of homologs, i.e., those which have only one copy in the duplicated genome or are missing the ortholog completely in the guide genome?

Singletons. Should we purge singletons from the data, i.e., sets of homologous markers that have no homologous adjacent markers in common in the either the duplicated genome or the outgroup?

Contigs. Can we use guide genomes that are not fully assembled, but are available only as sets of hundreds or thousands of contigs?

Another choice to be made during reconstruction has to do with the guided halving algorithm itself. The original genome halving problem, with no reference to outgroup genomes, can be solved in time linear in the number of markers [4]. We can introduce information from an outgroup in order to guide this solution, without compromising the optimality of the result and without serious increase in computing time [17,20]. We call this *constrained* guided halving. The true, *unconstrained*, guided halving problem, however, where the solution ancestor need not be a solution of the original halving problem, is likely to be NP-hard [12]. In the heuristics necessary for these two approaches, there is a trade-off between the speed and quality of constrained halving versus the unbiased and possibly better solution obtainable by unconstrained halving.

Once we make our choices of data and algorithm, we may ask how to evaluate the results. As with most evolutionary reconstructions, this evaluation is necessarily completely internal, since there is no outside reference to check against, except simulations. There are many indices for evaluating a reconstruction.

Distance. Most important, there is the objective function; here our genomic distance definition attempts to recover the most economical explanation of the observed data, namely the minimum number of rearrangement events (reversals, reciprocal translocations, chromosome fusions/fissions, transpositions) required.

Reuse rate. Each rearrangement operation can create at most two breakpoints in the gene-by-gene alignment of two genome and its ancestor. When rearranged genomes are algorithmically reconstructed, however, some breakpoints may be reused. If d is the number of rearrangements and b the number of breakpoints, the reuse [6] variable $r = 2d/b$ can take on values in $1 \leq r \leq 2$. Completely randomized genomes will have r close to 2, so that if an empirical comparison has $r \sim 2$, we cannot ascribe much significance to the details of the reconstruction [9]. This is particularly likely to occur for genomes that are only very distantly related.

Dispersion. The motivation for guided halving is to resolve the ambiguities inherent in the large number of solutions. One way to quantify the remaining non-uniqueness is to calculate the distances among a sample of solutions.

In this paper we will refer repeatedly to a main tabulation of results, Table 1, in which we discover the unexpected rapid evolution of the *Carica* gene order in comparison with that of *Vitis*. In Section 2, we report on the origin and processing of our gene-order data and the construction of the full and defective homology sets. Then, in Section 3, we discuss the formulation of genomic distances and the halving problems, and sketch a new algorithm for unconstrained guided halving. In Section 4 we evaluate the utility of singletons and of defective homology sets. Then, in Section 5 we assess the two guided halving algorithms on real and simulated data. Section 6 proposes a way to use unassembled genome sequence in contig form as input to the reconstruction algorithm, an approach that could potentially have wide use in gene order phylogeny. In Section 7 we demonstrate the phylogenetic validity of reconstructing the *Populus* ancestor using either *Vitis* or *Carica*, or both, as outgroups. Note that we have not included *Arabidopsis* in our analyses; as will be explained in Section 8, this was dictated by a paucity of data in the appropriate configurations.

2   The Populus, Vitis and Carica Data

Annotations for the *Populus, Vitis* and *Carica* genomes were obtained from databases maintained by the U.S. Department of Energy's Joint Genome Institute [14], the French National Sequencing Center, Genoscope [5], and the University of Hawaii [8], respectively. An all-by-all BLASTP search was run on a data set including all *Populus* and *Vitis* protein coding genes, and orthoMCL

[7] was used to construct 2104 full and 4040 defective gene sets, in the first case, denoted PPV, containing two poplar paralogs (genome P) and one grape ortholog (genome V), and in the second case, denoted PV or PP, missing a copy from either P or V. This was repeated with *Populus* and *Carica*, genomes P and C, respectively, to obtain 2590 full (PPC) and 4632 defective (PC or PP) sets. The location on chromosomes (or contigs in the case of *Carica*) and orientation of these paralogs and orthologs was used to construct our database of gene orders for these genomes. Contigs containing only a single gene were discarded from the *Carica* data.

3   Genome Distance, Breakpoint Graph, Guided Halving

Genome comparison algorithms generally involve manipulations of the bicoloured breakpoint graph [1,13] of two genomes, called the black and the gray genomes, on the same set of n genes, where two vertices are defined representing the two ends of each gene, and an edge of one colour joins two vertices if the corresponding gene ends are adjacent in the appropriate genome. Omitting the details pertaining to the genes at the ends of chromosomes, the genomic distance d, i.e., the minimum number of rearrangements necessary to transform one genome into the other, satisfies $d = n - c$, where c is the number of alternating colour cycles making up the breakpoint graph [16].

Then the genome halving problem [4] asks, given a genome T with two copies of each gene, distributed in any manner among the chromosomes, to find the "ancestral" genome, written $A \oplus A$, consisting of two identical halves, i.e., two identical sets of chromosomes with one copy of each gene in each half, such that the rearrangement distance $d(T, A \oplus A)$ between T and $A \oplus A$ is minimal. Note that part of this problem is to find an optimal labeling as "1" or "2" of the two genes in a pair of copies, so that all n copies labeled "1" are in one half of $A \oplus A$ and all those labeled "2" are in the other half. The genome A represents the ancestral genome at the moment immediately preceding the WGD event giving rise to $A \oplus A$.

The guided genome halving problem [18] asks, given T as well as another genome R containing only one copy of each of the n genes, find A so that $d(T, A \oplus A) + d(A, R)$ is minimal. The solution A need not be a solution to the original halving problem.

In previous studies [18,10,17], we found that the solution of the guided halving problem is often a solution of the original halving problem as well, or within a few rearrangements of such a solution. This has led us to define a *constrained* version of the guided halving problem, namely to find A so that $A \oplus A$ is a solution to the original halving problem and $d(T, A \oplus A) + d(A, R)$ is minimal. This has the advantage that a good proportion of the computation, namely the halving aspect, is guaranteed to be rapid and exact, although the overall algorithm, which is essentially a search among all optimal A, remains heuristic. Without sketching out the details of the lengthy algorithm, the addition of gray edges representing genome A to the breakpoint graph, as in Figure 2, must favour

Table 1. Guided halving solutions with and without singletons, constrained and unconstrained heuristics, *Vitis* or *Carica* as outgroup, and all combinations of full and defective homolgy sets. A: pre-doubling ancestor of *Populus*, $A \oplus A$: doubled ancestor, PPV, PPC: full gene sets, PP: defective, missing grape or papaya ortholog, PV,PC: defective, missing one poplar paralog. d: genomic distance, b, number of breakpoints, $r = 2d/b$: the reuse statistic.

data sets	genes in A, with singletons	$d(A, Vitis)$ d	b	r	$d(A \oplus A, Populus)$ d	b	r	total d
Solutions constrained to also be solutions of genome halving									
PPV	2104	638	751	1.70	454	690	1.32	1092
PPV,PP	2940	649	757	1.71	737	1090	1.35	1386
PPV,PV	5308	1180	1331	1.77	1083	1457	1.49	2263
PPV,PP, PV	6144	1208	1363	1.77	1337	1812	1.48	2545
Solutions unconstrained									
PPV	2104	593	734	1.62	512	733	1.40	1105
PPV, PP	2940	616	752	1.64	778	1119	1.39	1394
PPV,PV	5308	1121	1307	1.72	1147	1486	1.54	2268
PPV,PP,PV	6144	1129	1328	1.70	1437	1871	1.54	2566

data sets	genes in A, with singletons	$d(A, Carica)$ d	b	r	$d(A \oplus A, Populus)$ d	b	r	total d
Solutions constrained to also be solutions of genome halving									
PPC	2590	896	1152	1.56	565	823	1.37	1461
PPC, PP	3478	905	1158	1.56	884	1282	1.38	1789
PPC,PC	6334	1892	2314	1.64	1262	1700	1.48	3154
PPC,PP,PC	7222	1925	2341	1.64	1541	2065	1.49	3466
Solutions unconstrained									
PPC	2590	864	1125	1.54	628	870	1.44	1492
PPC, PP	3478	873	1125	1.55	951	1318	1.44	1824
PPC,PC	6334	1859	2277	1.63	1321	1742	1.52	3180
PPC,PP,PC	7222	1877	2313	1.62	1617	2126	1.52	3494

data sets	genes in A, without singletons	$d(A, Vitis)$ d	b	r	$d(A \oplus A, Populus)$ d	b	r	total d
Solutions constrained to also be solutions of genome halving									
PPV	2020	560	661	1.69	346	541	1.28	906
PPV,PP	2729	594	690	1.72	453	714	1.27	1047
PPV,PV	4203	573	686	1.67	751	1031	1.46	1324
PPV,PP, PV	4710	675	797	1.69	856	1211	1.41	1531
Solutions unconstrained									
PPV	2020	545	652	1.67	375	564	1.33	920
PPV, PP	2729	567	681	1.67	493	745	1.32	1060
PPV,PV	4203	544	674	1.61	782	1034	1.51	1326
PPV,PP,PV	4710	631	785	1.61	916	1250	1.47	1547

data sets	genes in A, without singletons	$d(A, Carica)$ d	b	r	$d(A \oplus A, Populus)$ d	b	r	total d
Solutions constrained to also be solutions of genome halving									
PPC	2464	772	1014	1.52	412	607	1.36	1184
PPC, PP	3226	812	1058	1.53	536	809	1.33	1348
PPC,PC	4651	779	1054	1.48	774	1050	1.47	1554
PPC,PP,PC	5234	898	1206	1.49	892	1253	1.42	1790
Solutions unconstrained									
PPC	2464	758	1001	1.51	454	639	1.42	1212
PPC, PP	3226	796	1046	1.52	584	839	1.39	1380
PPC,PC	4651	764	1041	1.47	804	1090	1.48	1568
PPC,PP,PC	5234	861	1178	1.46	952	1303	1.46	1813

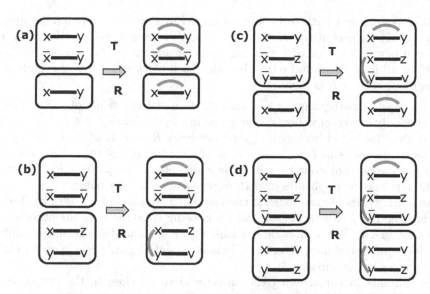

Fig. 2. Choice of gray edge to add at each stage of the reconstruction of A and $A \oplus A$. Each black edge in the diagram represents either an adjacency in T or R or an alternating colour path with a black edge at each end point. If vertex w is copy "1" in T then $\bar{w}$ is copy "2", and vice versa. (a) Configuration requiring the creation of three cycles, two in the breakpoint graph of T and $A \oplus A$, and one in the breakpoint graph of A and R. (b) Configuration requiring the creation of two cycles in the breakpoint graph of T and $A \oplus A$, necessary for $A \oplus A$ to be a solution of the genome halving problem. (c) Alternative configuration if solution of guided halving $A \oplus A$ is not also required to be a solution of the halving problem. (d) Look-ahead when there are no configurations (a), (b) or (c). Here the addition of three gray edges creates a configuration (c).

configuration (b) over (c), even though there are as many cycles created by (c) as by (b). This is a consequence of the original halving theory in Ref. [4]. Otherwise $A \oplus A$ may not be a halving solution. This, however, may bias the reconstruction of A towards T and away from R. For the present work, we implemented a new version of the algorithm, as sketched in Section 3.1, treating configurations (b) and (c) equally in constructing A. The choice among two or more configurations of form (b) or (c) is based on a look-ahead calculation of what effect this choice will have on the remaining inventory of configurations of form (b) and (c). The new algorithm requires much more computation, but its objective function is better justified.

3.1   The New Algorithm

First we define paths, which represent intermediate stages in the construction of the breakpoint graph comparing T and $A \oplus A$ and the breakpoint graph comparing A and R. Then we define pathgroups, which focus on the three current paths leading from three "homologous" vertices in the graph, namely two copies in T and one in R. Note that each vertex represents one of the two ends of a gene.

Paths. We define a path to be any connected fragment of a breakpoint graph, namely any connected fragment of a cycle. We represent each path by an unordered pair $(u, v) = (v, u)$ consisting of its current endpoints, though we keep track of all its vertices and edges. Initially, each black edge in T is a path, and each black edge in R is a path.

Pathgroups. A pathgroup, as in Figure 2, is an ordered triple of paths, two in the partially constructed breakpoint graph involving T and $A \oplus A$ and one in the partially constructed breakpoint graph involving R and A, where one endpoint of one of the paths in T is the duplicate of one endpoint of the other path in T and both are orthologous to one of the endpoints of the path in R. The other endpoints may be duplicates or orthologs to each other, or not.

In adding pairs of gray edges to connect duplicate pairs of terms in the breakpoint graph of T versus $A \oplus A$, (which is being constructed), our approach is basically greedy, but with a careful look-ahead. We can distinguish four different levels of desirability, or priority, among potential gray edges, i.e., potential adjacencies in the ancestor.

Recall that in constructing the ancestor A to be close to the outgroup R, such that $A \oplus A$ is simultaneously close to T, we must create as many cycles as possible in the breakpoint graphs between A and R and in the breakpoint graph of $A \oplus A$ versus T. At each step we add three gray edges.

- Priority 1. Adding the three gray edges would create two cycles in the breakpoint graph defined by T and $A \oplus A$, by closing two paths, and one cycle in the breakpoint graph comparison of A with the outgroup, as in Figure 2a.
- Priority 2. Adding three gray edges would create two cycles, one for T and one for the outgroup, or two for T and none for the outgroup, as in Figure 2b and c.
- Priority 3. Adding the gray edges would create only one cycle, either in the T versus $A \oplus A$ comparison, or in the R versus A comparison. In addition, it would create a higher priority pathgroup, as in as in Figure 2d.
- Priority 4. Adding the gray edges would create only one cycle, but would not create any higher priority pathgroup.

The algorithm simply completes the steps suggested by the highest priority pathgroup currently available, choosing among equal priority pathgroups according to a look-ahead to the configuration of priorities resulting from competing moves.

At each step, we must verify that a circular chromosome is not created, otherwise the move is blocked. As with Ref. [4] this check requires a constant time. The algorithm terminates when no more pathgroups can be completed. Any remaining pathgroups define additional chromosomes in the ancestor A.

4   On the Utility of Singletons and Defective Homology Sets

From the last column of Table 1, it is clear that of the four factors, inclusion/exclusion of singletons, inclusion/exclusion of defective homology sets, outgroup species and heuristic, the largest effects on total genomic distance are due to the choice of homology sets and inclusion of singletons, while the heuristic used has a much smaller effect. We will return to the differences between the algorithms in Section 5, and to the choice of outgroup in Section 7, but we can observe here that the inclusion of the homology sets defective by virtue of one missing *Populus* copy increases the genomic distances disproportionately and also reduces the quality of the inference, as measured by r in all the analyses containing singletons, and all the *Populus-A* $\oplus$ *A* comparisons.

At the same time the inclusion of singletons had a major effect on the distance, especially where the PV or PC homology sets are included. In addition, by comparing all the sub-tables with singletons, in the top half of the table, with the corresponding sub-table without singletons, in the bottom half, the inclusion of singletons degrades the analysis, with few exceptions, as measured by an increase in the two r statistics, the one pertaining to the duplicated genome and the one pertaining to the outgroup.

5   Comparison of the Heuristics

In Table 1, the constrained guided halving algorithm always does better than the unconstrained guided halving heuristic, as measured by the total distance in the last column. At the same time, the unconstrained heuristic had a clear effect in reducing the bias towards *Populus*, in each case decreasing the distance to the outgroup, compared to the constrained heuristic. This decrease was accompanied by a small decrease in r for the outgroup analysis.

In fact the decrease in the bias was far greater than the increase in total cost, meaning that if bias reduction is important, then this heuristic is worthwhile, despite its inability to find a minimizing ancestor and its lengthy execution time.

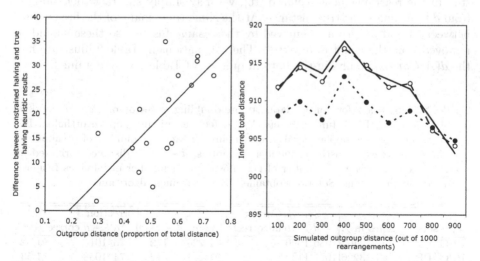

Fig. 3. Performance of the constrained and unconstrained heuristics as a function of the real (left) or simulated (right) distance of the outgroup from A

To further investigate the behaviour of the new algorithm, we simulated evolution by M inversions and translocations (in a 10:1 proportion) from a genome A to produce a genome R and $1000 - M$ rearrangements from genome $A \oplus A$ to produce a genome T. We then applied the constrained and the new algorithms, showing that the new one was superior when $M < 800$, but not for $M \geq 1000$, as seen in Figure 3 (right). Considering the 16 comparisons between the constrained and the new algorithm, the change in the total distance also shows a distinct correlation ($\rho^2 = 0.5$) with the distance from the outgroup and A. We point this out even though the constrained algorithm, as we have seen, seems superior when the distance between R and A is more than 20 % of the total distance. This is plotted in Figure 3 (left).

The difference between the simulations, where the new method is always superior, and the real analysis, where the new method would seem to be superior only when the outgroup is very close to the ancestor, must be ascribed to some way the model used for the simulations does not fit the data. One clue is the relatively high reuse rate in the comparison between the outgroup and A, compared with that between *Populus* and $A \oplus A$.

6   Rearrangements of Partially Assembled Genomes

Our analyses involving *Carica* have incorporated an important correction. The genomic distance between *Carica* and A counts many chromosome fusion events that reduce the number of "chromosomes" in *Carica* from 223 to the 19. These are not a measure of the true rearrangement distance, but only of the current state of the *Carica* data. Since these may be considered to take place as a first step in the rearrangement scenario [16], we may simply subtract their number from d to estimate the true distance. At the same time, many of the breakpoints between A and *Carica* are removed by these same fusions, so these should be removed from the count of b as well. The calculations in Table 2 illustrate how the $d(A, Carica)$ results in the bottom quarter of Table 1 were obtained.

Table 2. Correction for contig data. A: pre-doubling ancestor of *Populus*, $A \oplus A$: doubled ancestor, PPC: full gene sets, PP: defective, missing papaya ortholog, PC: defective, missing one poplar paralog. d: genomic distance, b: number of breakpoints, $r = 2d/b$: the reuse statistic, c: number of contigs, $d - c + 9$: distance corrected for excess of contigs over true number of chromosomes, a: number of 'obvious fusions". Data without singletons. Solutions obtained by constrained algorithm.

| data sets | genes in A | $d(A, Carica)$ | | | | correction | | | |
		d	b uncorrected r		c	$d - c + 9$	a	$b - a$	corrected r
PPC	2464	986	1090	1.81	223	772	76	1014	1.52
PPC, PP	3226	1027	1132	1.81	224	812	74	1058	1.53
PPC,PC	4651	1084	1177	1.84	314	779	123	1054	1.48
PPC,PP,PC	5234	1214	1318	1.84	325	898	112	1206	1.49

Figure 4 (left) shows experimental results on how the increasing fragmentation of a genome into contigs, using a random fragmentation of *Vitis* grenome, decreases the estimated distance between *Vitis* and *A*. This is understandable, since the freedom of the contigs to fuse in any order without this counting as a rearrangment step, inevitably will reduce the distance by chance alone. But the linearity of the result suggests that this decrease is quite predictable, and that the estimates of the distance between *Carica* and *A* are actually underestimates by about 10 %.

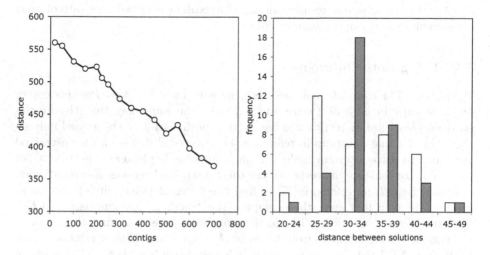

Fig. 4. Left: Effect of increasing fragmentation of *Vitis* into "contigs" on the distance between the reconstructed *A* and *Vitis*. Right: Distributions of distances among solutions for *A* based on *Vitis* data (white bars) and among solutions for *Vitis* fragmented into contigs in different random ways (gray bars).

Figure 4 (right) shows that creating contigs by randomly breaking the *Vitis* genome does not create excessive variability among the solutions, only the same as the dispersion of alternate solutions for the original *Vitis* data, a few percentage points of the distance itself.

7 A Comparison of the Outgroups

Perhaps the most surprising result of this study is that the *Vitis* gene order is decidedly closer to *Populus* and its ancestor *A* than *Carica* is. Both the Tree of Life and the NCBI Taxonomy Browser currently exclude the Vitaceae family from the rosids, though some older taxonomies do not make this distinction.

Before interpreting this result, we should correct two sources of error in the comparison of *Vitis* and *Carica*. The first is that the *Carica* distances are based on a larger gene set; without singletons and defective homology sets PPC is

22 % larger than PPV. As a rule of thumb, we can expect distances to be approximately proportional to the number of genes. This overestimation of the *Carica*-ancestor distance might account for about half the difference in the distances. But the other source of error is due to the contig data, and this results in an *underestimate* of the *Carica*-ancestor distance. From Figure 4, we can estimate that the *Carica* distances are underestimated by about 10 % because of the 223 contigs in the *Carica* data. So this increases the discrepancy between the two outgroups, restoring it almost to what it was before the corrections.

We may conclude that this difference is genuine and substantial. Then assuming that *Populus* and *Carica* have a closer phylogenetic relationship, or even a sister relationship, our results can only be explained by a faster rate of gene order evolution in *Carica* than in *Vitis*.

7.1   Using Both Outgroups

There are 1734 complete homologous gene sets including two *Populus* copies and one copy in each of *Carica* and *Vitis*. In the same way that the unconstrained algorithm in Section 3 is based on a modification of the guided halving algorithm for one outgroup in reference [17], we could define an unconstrained version of the two-outgroup guided halving algorithm implemented in that earlier work. For convenience, however, we use the constrained version of two-outgroup guided halving from reference [17] to find the ancestor (small circle) genome in Figure 5(a) as a first step, then compute the "median" genome based on this ancestor, *Carica* and *Vitis*. The median problem here is to find the genome, the sum of whose distances from ancestor A, *Carica* and *Vitis* is minimal. This problem is NP-hard [12] and solving it is barely feasible with the 1734 genes in our data, requiring some 300 hours of MacBook computing time.

This initial result unfortunately inherits the same defect as the *Carica* data, i.e., it is composed of contigs rather than true chromosomes. In this case, the median genome contains 118 "contig-chromosomes". And in the same way, we may correct it by subtracting the number of contigs in excess of a reasonable number of chromosomes (19 in the median) from the distance in order to obtain a corrected distance. This corresponds to disregarding the fusions counted in the original distance that are essentially carrying out an optimal assembly, modeling an analytical process, not a biological one. This produces the corrected values in Figure 5(b).

Let us compare the distance from *Vitis* and from *Carica* to ancestor A, passing through the median, in Figure 5 (517 and 577, respectively), with the minimum distances[1] in Table 1, and proportionately adjusted for the reduced number of genes ($560 \times \frac{1734}{2020} = 481$ and $772 \times \frac{1734}{2464} = 543$, respectively). Passing through the median modestly augments (by 36 and by 34, respectively) both trajectories. But using the median diminishes the total cost of the phylogeny, i.e., in comparison with a phylogeny where there is no common evolutionary divergence of the outgroups from *Populus* from $481+170+543+170 = 1364$ to $517+577+170 = 1264$.

---

[1] Constrained analyses, no singleton or defective homology sets.

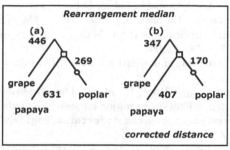
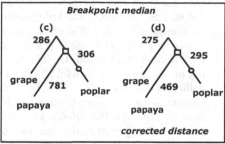

Fig. 5. Branch lengths in angiosperm phylogeny, using two estimates of the median, and applying the contig correction

There is one version of guided halving that is of polynomial complexity [12]. This involves a "general breakpoint model" for multichromosomal genomes, which does not explicitly refer to rearrangements. Running this algorithm, requiring only 15 MacBook minutes, on the three angiosperm genomes results in a median with only 30 contig-chromosomes. Calculating the rearrangement distances from this median to ancestor A, *Carica* and *Vitis* gives the results in Figure 5(c); correcting them for excess contigs gives the results in Figure 5(d).

Figures 5(b) confirm that the papaya genome has evolved more rapidly than the grapevine one. Figure 5(d) shows an even greater distance, although this is not based on the rearrangement median.

8   Conclusions

The main contributions of this paper are:

- The discovery of the rapid rate of gene order evolution in *Carica* compared to *Vitis*,
- A way to use incompletely assembled contigs in genome rearrangement studies,
- A new unbiased algorithm for guided genome halving, and
- The systematic use of reuse rates to show that the inclusion of defective homology sets and singletons are not helpful in ancestral genome reconstruction.

In this work, we have not considered the *Arabidopsis* genome. The main reason is not any algorithmic issue, but the paucity of full homology sets containing four *Arabidopsis* copies as well as copies from one or more outgroups.

References

1. Bafna, V., Pevzner, P.: Genome rearrangements and sorting by reversals. SIAM Journal of Computing 25, 272–289 (1996)
2. Bergeron, A., Mixtacki, J., Stoye, J.: A unifying view of genome rearrangements. In: Bücher, P., Moret, B.M.E. (eds.) WABI 2006. LNCS (LNBI), vol. 4175, pp. 163–173. Springer, Heidelberg (2006)

3. Cui, L., Wall, P.K., Leebens-Mack, J.H., Lindsay, B.G., Soltis, D.E., Doyle, J.J., Soltis, P.S., Carlson, J.E., Arumuganathan, K., Barakat, A., Albert, V.A., Ma, H., de Pamphilis, C.W.: Widespread genome duplications throughout the history of flowering plants. Genome Research 16, 738–749 (2006)

4. El-Mabrouk, N., Sankoff, D.: The reconstruction of doubled genomes. SIAM Journal on Computing 32, 754–792 (2003)

5. Jaillon, O., Aury, J.M., Noel, B., Policriti, A., Clepet, C., et al.: The grapevine genome sequence suggests ancestral hexaploidization in major angiosperm phyla. Nature 449, 463–467 (2007), http://www.genoscope.cns.fr/externe/English/Pro-jets/Projet_ML/data/annotation/

6. Pevzner, P.A., Tesler, G.: Human and mouse genomic sequences reveal extensive breakpoint reuse in mammalian evolution. Proceedings of the National Academy of Sciences USA 100, 7672–7677 (2003)

7. Li, L., Stoeckert Jr., C.J., Roos, D.S.: OrthoMCL: identification of ortholog groups for eukaryotic genomes. Genome Research 13, 2178–2189 (2003)

8. Ming, R., Hou, S., Feng, Y., Yu, Q., Dionne-Laporte, A., et al.: The draft genome of the transgenic tropical fruit tree papaya (Carica papaya Linnaeus). Nature 452, 991–996 (2008), http://asgpb.mhpcc.hawaii.edu

9. Sankoff, D.: The signal in the genomes. PLoS Computational Biology 2, e35 (2006)

10. Sankoff, D., Zheng, C., Zhu, Q.: Polyploids, genome halving and phylogeny. Bioinformatics 23, i433–i439 (2007)

11. Soltis, D., Albert, V.A., Leebens-Mack, J., Bell, C.D., Paterson, A., Zheng, C., Sankoff, D., dePamphilis, C.W., Wall, P.K., Soltis, P.S.: Polyploidy and angiosperm diversification. American Journal of Botany (in press, 2008)

12. Tannier, E., Zheng, C., Sankoff, D.: Multichromosomal median and halving problems under different genomic distances. In: Workshop on Algorithms in Bioinformatics WABI 2008 (in press, 2008)

13. Tesler, G.: Efficient algorithms for multichromosomal genome rearrangements. Journal of Computer and System Sciences 65, 587–609 (2002)

14. Tuskan, G.A., Difazio, S., Jansson, S., Bohlmann, J., Grigoriev, I., et al.: The genome of black cottonwood, Populus trichocarpa (Torr. & Gray). Science 313, 1596–1604 (2006)

15. Velasco, R., Zharkikh, A., Troggio, M., Cartwright, D.A., Cestaro, A., et al.: A high quality draft consensus sequence of the genome of a heterozygous grapevine variety. PLoS ONE 2, e13–e26 (2007)

16. Yancopoulos, S., Attie, O., Friedberg, R.: Efficient sorting of genomic permutations by translocation, inversion and block interchange. Bioinformatics 21, 3340–3346 (2005)

17. Zheng, C., Zhu, Q., Adam, Z., Sankoff, D.: Guided genome halving: hardness, heuristics and the history of the Hemiascomycetes. Bioinformatics 24, i96–i104 (2008)

18. Zheng, C., Zhu, Q., Sankoff, D.: Genome halving with an outgroup. Evolutionary Bioinformatics 2, 319–326 (2006)

19. Zheng, C., Zhu, Q., Sankoff, D.: Descendants of whole genome duplication within gene order phylogeny. Journal of Computational Biology 15 (in press, 2008)

20. Zheng, C., Wall, P.K., Leebens-Mack, J., Albert, V.A., dePamphilis, C.W., Sankoff, D.: The effect of massive gene loss following whole genome duplication on the algorithmic reconstruction of the ancestral Populus diploid. In: Proceedings of the International Conference on Computational Systems Bioinformatics CSB 2008 (in press, 2008)

Author Index

Lecture Notes in Bioinformatics

Vol. 3909: A. Apostolico, C. Guerra, S. Istrail, P.A. Pevzner, M. Waterman (Eds.), Research in Computational Molecular Biology. XVII, 612 pages. 2006.

Vol. 3886: E.G. Bremer, J. Hakenberg, E.-H.(S.) Han, D. Berrar, W. Dubitzky (Eds.), Knowledge Discovery in Life Science Literature. XIV, 147 pages. 2006.

Vol. 3745: J.L. Oliveira, V. Maojo, F. Martín-Sánchez, A.S. Pereira (Eds.), Biological and Medical Data Analysis. XII, 422 pages. 2005.

Vol. 3737: C. Priami, E. Merelli, P. Gonzalez, A. Omicini (Eds.), Transactions on Computational Systems Biology III. VII, 169 pages. 2005.

Vol. 3695: M. R. Berthold, R.C. Glen, K. Diederichs, O. Kohlbacher, I. Fischer (Eds.), Computational Life Sciences. XI, 277 pages. 2005.

Vol. 3692: R. Casadio, G. Myers (Eds.), Algorithms in Bioinformatics. X, 436 pages. 2005.

Vol. 3680: C. Priami, A. Zelikovsky (Eds.), Transactions on Computational Systems Biology II. IX, 153 pages. 2005.

Vol. 3678: A. McLysaght, D.H. Huson (Eds.), Comparative Genomics. VIII, 167 pages. 2005.

Vol. 3615: B. Ludäscher, L. Raschid (Eds.), Data Integration in the Life Sciences. XII, 344 pages. 2005.

Vol. 3594: J.C. Setubal, S. Verjovski-Almeida (Eds.), Advances in Bioinformatics and Computational Biology. XIV, 258 pages. 2005.

Vol. 3500: S. Miyano, J. Mesirov, S. Kasif, S. Istrail, P.A. Pevzner, M. Waterman (Eds.), Research in Computational Molecular Biology. XVII, 632 pages. 2005.

Vol. 3388: J. Lagergren (Ed.), Comparative Genomics. VII, 133 pages. 2005.

Vol. 3380: C. Priami (Ed.), Transactions on Computational Systems Biology I. IX, 111 pages. 2005.

Vol. 3370: A. Konagaya, K. Satou (Eds.), Grid Computing in Life Science. X, 188 pages. 2005.

Vol. 3318: E. Eskin, C. Workman (Eds.), Regulatory Genomics. VII, 115 pages. 2005.

Vol. 3240: I. Jonassen, J. Kim (Eds.), Algorithms in Bioinformatics. IX, 476 pages. 2004.

Vol. 3082: V. Danos, V. Schachter (Eds.), Computational Methods in Systems Biology. IX, 280 pages. 2005.

Vol. 2994: E. Rahm (Ed.), Data Integration in the Life Sciences. X, 221 pages. 2004.

Vol. 2983: S. Istrail, M.S. Waterman, A. Clark (Eds.), Computational Methods for SNPs and Haplotype Inference. IX, 153 pages. 2004.

Vol. 2812: G. Benson, R.D.M. Page (Eds.), Algorithms in Bioinformatics. X, 528 pages. 2003.

Vol. 2666: C. Guerra, S. Istrail (Eds.), Mathematical Methods for Protein Structure Analysis and Design. XI, 157 pages. 2003.